S. 280

Benchmark Papers
in Behavior

Series Editors: Martin W. Schein
West Virginia University
and
Stephen W. Porges
University of Illinois
at Urbana-Champaign

**Benchmark Papers
in Behavior / 15**

A BENCHMARK® Books Series

MAMMALIAN
SEXUAL BEHAVIOR
Foundations for
Contemporary Research

Edited by
DONALD A. DEWSBURY
University of Florida

Hutchinson Ross Publishing Company

Stroudsburg, Pennsylvania

Copyright © 1981 by **Hutchison Ross Publishing Company**
Benchmark Papers in Behavior, Volume 15
Library of Congress Catalog Card Number:
ISBN: 0–87933–396–0

83 82 81   1   2   3   4   5
Manufactured in the United States of America.

LIBRARY OF CONGRESS CATALOGING IN PUBLICATION DATA
Main entry under title:
Mammalian sexual behavior.
  (Benchmark papers in behavior; 15)
  Includes indexes.
  1. Mammals—Behavior—Addresses, essays, lectures. 2. Sexual behavior in animals—Addresses, essays, lectures. I. Dewsbury, Donald, Donald A., 1939–   . II. Series.
QL739.3.M35   599.05′6   81–6231
ISBN 0–87933–396–0   AACR2

Distributed world wide by Academic Press,
a subsidary of Harcourt Brace Jovanovich,
Publishers.

CONTENTS

Contents

Contents

vii

Contents

PART IV: EVOLUTION

PART V: FUNCTIONS

SERIES EDITOR'S FOREWORD

It was not too many years ago that virtually all research publications dealing with animal behavior could be housed within the covers of a very few hard-bound volumes that were easily accessible to the few workers in the field. Times have changed! The present-day students of behavior have all they can do to keep abreast of developments within their own area of special interest, let alone in the field as a whole; and of course we have long since given up attempts to maintain more than a superficial awareness of what is happening "in biology," "in psychology," "in sociology," or in any of the broad fields touching upon or encompassing the behavioral sciences.

It was even fewer years ago that those who taught animal behavior courses could easily choose a suitable textbook from among the very few that were available; all "covered" the field, according to the bias of the author. Students working on a special project used *the* text and *the* journal as reference sources, and for the most part successfully covered their assigned topics. Times have changed! The present-day teacher of animal behavior is confronted with a bewildering array of books to choose among, some purported to be all-encompassing , others confessing to strictly delimited coverage, and still others being simply collections of recent and profound writings.

In response to the problem of the steadily increasing and overwhelming volume of information in the area, the Benchmark Papers in Behavior was launched as a series of single-topic volumes designed to be some things to some people. Each volume contains a collection of what an expert considers to be *the* significant research papers in a given topic area. Each volume, then, serves several purposes. To teachers, a Benchmark volume serves as a supplement to other written materials assigned to students; it permits in-depth consideration of a particular topic while at the same time confronting students (often for the first time) with original research papers of outstanding quality. To researchers, a Benchmark volume serves to save countless hours digging through the various journals to find *the* basic articles in their area of interest; often the journals are not easily available. To students, a Benchmark volume provides a readily accessible set of original papers on the topic, a set that forms the core of the more extensive bibliography that they are likely to compile;

it also permits them to see at first hand what an "expert" thinks is important in the area, and to react accordingly. Finally, to librarians, a Benchmark volume represents a collection of important papers from many diverse sources, thus making readily available materials that might otherwise not be economically possible to obtain or physically possible to keep in stock.

The choice of topics to be covered in this series is no small matter. Each of us could come up with a long list of possible topics and then search for potential volume editors. Alternatively, we could draw up long lists of recognized and prominent scholars and try to persuade them to do a volume on a topic of their choice. For the most part, we have followed a mix of both approaches: match a distinguished researcher with a desired topic, and the results should be outstanding. So it is with the present volume.

Dr. Dewsbury enjoys a well-earned reputation as a leading researcher in comparative animal behavior, with emphasis on mammalian sexual behavior. He is a prodigious writer and a careful thinker. Therefore, we are especially pleased that he has agreed to take on the task of assembling this volume and providing us with his valuable insights on the topic. It was an enormous assignment, since the literature on sexual behavior is extensive. However, few know the literature better, and fewer still have the perspective, as does Dr. Dewsbury, to allow for sorting through the ideas and trends and then reassembling them into a reasonable and cohesive sample of the whole. We are pleased with the outcome and we consider the present volume a valuable addition to the Benchmark Papers in Behavior series.

MARTIN W. SCHEIN
STEPHEN W. PORGES

PREFACE

As contemporary animal behaviorists, it is easy for us to focus on all that is new. We have at our disposal an armament of modern techniques that include videotapes, online computers, autoradiographic analyses, assays for brain monoamines, and electrophoresis. However, these techniques become valuable only when used in answering meaningful questions. Although techniques come and go, research questions often are more enduring. Every generation of students of sexual behavior has been concerned with the sensory bases of behavior, neural correlates, adequate quantification of behavior, the consequences of mating, and so on. The objective of this volume is to bring together a representative sample of literature directed at various enduring research questions in the study of sexual behavior. The goal is to help clarify the development and evolution of these areas of research activity. The volume was assembled in the belief that perspective regarding the origins and development of the fundamental research areas can facilitate adherence to meaningful versus trivial scientific questions.

As the major thrust of the volume is historical, many selections are old by contemporary standards. Thus many of the methods described in these papers are outmoded, and many of the conclusions require revision in the light of more recent data. I have tried to elaborate some of these limitations in the editorial commentaries. The reader would do well, however, to remember that contemporary methods and conclusions are likely to be viewed in a similar light by future scientific generations. In my judgment, these authors made effective use of the available techniques in the pursuit of meaningful questions. Their work had impact on ours. It is in this sense that the materials included in this volume provided the *foundations* for contemporary research. There are, of course, new wrinkles in every research area. However, in reviewing this literature, one cannot help but get the impression that no matter what fundamental questions one considers, Beach, Larsson, Stone, Young, or one of their predecessors had already considered it.

When Martin Schein first asked me to edit this volume, I anticipated little difficulty in selecting papers for inclusion. Soon, however, I came into contact with the practical realities of the necessary 300-page limit on reprinted material. There is simply no way in which to do justice to the literature on mammalian sexual behavior in just 300 pages of reprinted papers. To make the task almost manageable, the volume is limited to mammals, but excludes humans. Because the intent was to explore the "foundations" for research in the 1970s and 1980s, no paper published after 1970 was included. Because C. S. Carter covered the area of hormones and sexual behavior in Volume 1 of this series, material from that area was minimized. Even so, I am certain to have omitted papers that many readers will regard as crucial and to have offended many scientists whose work I respect and admire. Rather than attempt to pick a "Top 40," I have tried to provide a representative sample of a variety of research areas, by researchers, with diverse backgrounds, working with a variety of species. With the severe page limit, one reinforces brevity, diversity, and, where possible, priority. I have tried to find papers that either represent a major advance beyond previous work in an area or that provide excellent, concise examples of the kind of work done by several researchers. I can only apologize to the authors of the really fine work that could not be included. In the editorial commentaries I have tried to acknowledge some of the papers that had to be excluded. In addition, I have tried to place the works selected in historical perspective by summarizing previous work, concurrent research, and especially more recent work. It is my view that much of the relevance of older work stems from an understanding of its relation to what has come since.

One important point does not come through from a single reading of the printed page. For the contemporary researcher there is little stigma attached to research on sexual behavior. We live in an era conditioned by Kinsey, *Playboy*, Masters and Johnson, motion-picture nudity, and birth-control pills. Much of the research in this volume was done in a climate not conditioned by these factors. In many instances the conduct of research on sex required pure courage as much as skill in research design. The Committee for Research in Problems of Sex, National Research Council, played a major role in facilitating this work. It is appropriate to acknowledge the courage and foresight of the pioneers in the study of animal sexual behavior and to express our indebtedness to them.

I thank C. Sue Carter, Benjamin D. Sachs, and Daniel G. Webster for their comments on a draft of this manuscript. Martin Schein originally suggested this volume and was most helpful throughout its production. Frank A. Beach aided both in consultation at the conception of the work and in commenting on an earlier draft of it. Obviously, I accept final responsibility for decisions made during its development. Alice Nippes and Katherine O'Dea did a fine job of typing the manuscript and related correspondence. Finally, I thank my family, Joyce, Bryan, and Laura, for bearing with me throughout the production of the volume.

DONALD A. DEWSBURY

CONTENTS BY AUTHOR

MAMMALIAN SEXUAL BEHAVIOR

INTRODUCTION

There are good reasons for the intense interest scientists have shown in the study of mammalian sexual behavior. First, sexual behavior is of obvious biological importance, as successful reprocution lies very close to the heart of the concept to biological fitness. Second, sexual behavior provides an excellent "preparation" for laboratory study. With proper care one can reliably elicit a stable, yet very complex behavioral pattern with little effort involved in conditioning animals. The effects of a wide array of manipulations on complex, biologically important behavior can then be assessed relatively easily. Third, information on sexual behavior is of obvious potential value for application in practical settings. In many cases, humans wish to increase or decrease the reproductive rates of various nonhuman species. There is in addition a general belief that information on sexuality in nonhumans will be helpful in understanding both normal and abnormal sexual processes in humans. Finally, such research is intrinsically interesting.

Scientists attempting to answer different questions have created a literature on mammalian sexual behavior that is remarkably diverse. The problem for an editor is that of organizing this vast array of information. The method selected for this volume is to use the system proposed by Tinbergen (1963) for the orderly investigation of animal behavior. This organizational schema has proven useful in other such endeavors (Dewsbury, 1978). According to Tinbergen, the first task in the study of behavior is to conduct careful observation in order to provide detailed description. Without a solid foundation of descriptive information, the accuracy of conclusions regarding more complex problems is questionable. However, observation and description are not ends in themselves but rather are means to enable consideration of four classes of questions—questions of the development, control, evolution, and function. Questions of development relate to changes occurring throughout the lifespan of the individual organism, including the role of genetic and experiential factors operating over time. Questions of control relate to the short-term determinants of behavior, including sensory factors, neural and endocrine correlates, motivational considerations, and social influences. Questions of evolutionary history transcend individual organisms and are concerned with changes in behavior across species and over the span of evolutionary time. Finally, questions of function relate to adaptive significance and to the role behavior plays in survival and reproduction. As stressed by Tinbergen, the complete analysis of behavior requries answers to all four classes of questions. This volume is organized so that questions of description, development, control, evolution, and function are considered in turn.

What kinds of behavior should be included under the rubric of "sexual"? In this volume, we have followed the usage of Beach (1947): "As used in this article the term 'sexual behavior' refers exclusively to the overt acts comprising heterosexual copulation, to those contiguous reactions commonly designated as 'courtship' or 'precoital play,' and to a variety of non-copulatory sexual responses such as those involved in auto-erotic and homosexual activities" (p. 240).

According to Beach (1976a) there are four phases in the complete mating interaction. The *sexual-attraction* phase relates to bringing two or more organisms together. *Appetitive behavior*, often termed "courtship," includes variable behavior occurring as a prelude to actual copulating. Courtship often functions to set the occasion for copulation. Appetitive behavior in females is referred to as "proceptive" (Beach, 1976b). Although often underemphasized in earlier work, its importance is now generally recognized. The phase of *consummatory behavior* concerns the actual stereotyped pattern of copulation. *Postconsummatory behavior* includes the immediate behavioral consequences of mating activity.

It is important to remember that sexual behavior takes place within the total social and environmental context of each organism. Although it is often important to study sexual activity under controlled conditions, as in most of the papers in this book, eventually the behavior must be returned to its natural context. Sexual activity may occur seasonally, within various mating systems in different species, and in particular locations. A summary of some of the broad questions regarding the context of sexual behavior has been presented by a group of ethologists (McBride, 1976). For the present purposes it is sufficient to note that sexual behavior often occurs under more complex conditions than those of the laboratory and that questions related to those conditions are important.

The objective in this volume is to bring together papers that will help to elucidate the development and evolution of the various research areas and thus to present a summary of some of the work that provides the *foundations for contemporary research.* Questions of description, development, evolution, and function are considered in turn. For additional information the reader is referred to the editorial commentaries and to the list of supplementary readings provided for each section.

REFERENCES

Beach, F. A., 1947, A review of physiological and psychological studies of sexual behavior in mammals, *Physiol. Rev.* **27**:240–307.

Beach, F. A., 1976a, Cross-species comparisons and the human heritage, *Arch. Sex. Behav.* **5**:469–485.

Beach, F. A., 1976b, Sexual attractivity, proceptivity, and receptivity in female mammals, *Horm. Behav.* **7**:105–138.

Dewsbury, D. A., 1978, *Comparative Animal Behavior*, McGraw-Hill, New York, 1978.

McBride, G., 1976, The study of social organization, *Behaviour* **59**:96–115.

Tinbergen, N., 1963, On aims and methods of ethology, *Z. Tierpsychol.* **20**:410–429.

SUGGESTED READINGS

Beach, F. A., 1961, *Hormones and Behavior*, Harper, New York.

Beach, F. A., ed., 1965, *Sex and Behavior*, Wiley, New York.

Bermant, G., and J. M. Davidson, 1974. *Biological Bases of Sexual Behavior*, Harper, New York.

Beyer, C., ed., 1979, *Endocrine Control of Sexual Behavior*, Raven, New York.

Carter, C. S., ed., 1974, *Hormones and Sexual Behavior*, Dowden, Hutchinson & Ross, Stroudsburg, Penn.

Daly, M., and M. Wilson, 1978, *Sex, Evolution, and Behavior*, Duxbury, North Scituate, Mass.

Diamond, M., ed., 1968, *Perspectives in Reproduction and Sexual Behavior*, Indiana University Press, Bloomington, Ind.

Ford, C. S., and F. A. Beach, 1951, *Patterns of Sexual Behavior*, Harper, New York.

Hutchinson, J. B., ed., 1978. *Biological Determinants of Sexual Behavior*, Wiley, New York.

McGill, T. E., D. A. Dewsbury, and B. D. Sachs, eds., 1978, *Sex and Behavior: Status and Prospectus*, Plenum, New York.

Montagna, W., and W. A. Sadler, eds., 1974, *Reproductive Behavior*, Plenum, New York.

Wickler, W., 1972, *The Sexual Code: The Social Behavior of Animals and Men*, Doubleday, Garden City, N.Y.

Young, W. C., ed., 1961, *Sex and Internal Secretions*, 2 vols., Williams & Wilkins, Baltimore.

Part I

DESCRIPTION

Editor's Comments
on Papers 1 Through 5

HETEROSEXUAL COPULATORY BEHAVIOR

Before one can meaningfully consider questions of development, control, evolution, and function, it is essential that adequate descriptive information be available regarding the behavior under study. Fortunately, there are many fine descriptions in the literature on sexual behavior. These descriptions generally come from two kinds of sources. Some are the work of specialists in the study of sexual behavior and are done expressly to provide a basis for further work in that area. Examples include the studies by Whalen (1963) on cats and Young and Grunt (1951) on guinea pigs. Other descriptions are provided by zoologists interested in the general biology of particular species and who provide descriptive material in the context of broader study. Examples of this approach include the studies of Buechner and Schloeth (1965) on Uganda kob, Eisenberg (1963) on heteromyid rodents, and Pearson (1944) on short tailed shrews.

One probably could trace the history of pretwentieth-century description of copulatory behavior in philosophical and natural historical literature. For example, in his "Historia Animalium," Aristotle stated that hedgehogs copulate belly to belly, presumably because he could not imagine such spiny-skinned animals copulating in any other way. (In fact, males mount from the rear, see Reed, 1946). However, most such casual descriptions had little impact on contemporary research. Therefore, papers selected for inclusion herein have been limited to those published in this century.

With a vast amount of descriptive material available, an editor must decide between including relatively long selections providing detail on many aspects of sexual behavior in a few species and sampling shorter descriptions. I have opted for something of a compromise. Paper 1 is a superb description by Carpenter of the copulatory behavior of rhesus monkeys in the field. The decades of the 1960s and 1970s have seen an explosion of field studies on primates due in large part to the efforts of this pioneering comparative psychologist. In this paper, Carpenter characterizes both estrus and the events surrounding copulatory activity in the context of normal reproductive activity in the field. In a companion paper, Carpenter (1942) went on to describe homosexual, autoerotic, and non-conformist behavior. These two papers provided important exemplars for later description. Interest in rhesus monkeys has continued, and precise descriptions under both field and laboratory conditions are now available (for example, Michael and Saayman, 1967).

Because of their close taxonomic affinity, humans have been especially interested in the sexual behavior of the great apes. The work of Yerkes (Paper 2) was a pioneer effort at attaining a comprehensive understanding of the behavior of chimpanzees in captivity. Miller (1931) had proposed that, in contrast to most mammals, chimpanzees display a pattern in which "the sexual psychology of the female as well as that of the male has been liberated from strict periodical oestrous control" (p. 385). Yerkes demonstrated that this was not the case. These results were relevant to theories of the origins of human social behavior, as it had been proposed that the opportunity for continuous sexual activity provided the "glue" for social structure. The early work on primate sexuality by Yerkes, Kempf (1917), Bingham (Paper 12), O. L. Tinklepaugh, and others was summarized by Miller (1928) and represents an important part of the foundation for contemporary research. As expected, most researchers found more variability in the behavior of primates than in other species. More recent work on sexual behavior in the apes has been conducted by Nadler (1976, 1977) at the Regional Primate Research Center, named for Yerkes, and has verified many of the conclusions drawn in the pioneering work. In addition, research in natural and seminatural settings has helped put ape sexual behavior in its proper perspective (for example, VanLawick-Goodall, 1968; Tutin and McGrew, 1973).

Laboratory rats are easily the most frequently studied animals in work on mammalian sexual behavior. There are three fundamental classes of events in the copulatory behavior of rats. On *mounts*, the male mounts the female from behind but fails to gain vaginal penetration. On *intromissions*, the male gains penetration for approximately one third of a second but dismounts without ejaculating. On *ejaculations*, the male mounts, gains intromission, and ejaculates. Mounts, intromissions, and ejaculations occur in organized *series*, with each series terminated by an ejaculation and separated from other series by a postejaculatory refractory period. The history of the description of this pattern provides an excellent example of progressive refinements as methodology has been improved. In 1922 Stone (Paper 3) provided the first scientific description of copulatory behavior in rats. It is interesting to note that in reviewing relevant literature, Stone could find only studies of wood frogs, crayfish, doves, and a few primate species for citation. The temporal relationships of rat copulatory behavior were outlined by Stone and Ferguson (1940).

The work of Beach and Jordan (Paper 4) provided a further refinement of previous methods of quantification. Both Beach and Jordan and Knut Larsson (1956) used a set of measures including the mount latency, intromission latency, ejaculation latency, intromission frequency, mount frequency, interintromission interval (or copulatory speed), and postejaculatory interval, that were to become standard in many studies of the sixties and seventies. Adoption of this standardized set of measures marked a major shift in emphasis from studies of the presence or absence of sexual activity as a result of various experimental manipulations to a more precise analysis of the quantitative functional relationships between independent and dependent variables.

Interest in describing rat copulatory behavior has continued unabated. A number of methods have been used to determine the actual durations of copulations. Stone and Ferguson (1940) used cinematographic techniques. Peirce and Nuttall (1961) wired the male and female in series into an electrical circuit so that the moist tissue contact characteristic of copulations completed the circuit and provided the measure of copulation duration. Dewsbury (1967) noted that in typical tests of copulatory behavior, rats spend less than 1 percent of their time in actual copulation and quantified the behavior occurring in the remaining time. Recent descriptions have been so refined as to be expressed in terms of milliseconds and have resulted from utilization of frame by frame analyses of motion-picture film (Diakow, 1975; Pfaff and Lewis, 1974) and x-ray cinematography (Pfaff et al., 1978). Noting that much important precopulatory sexual behavior is omitted in the small test cages used in most studies, McClintock and Adler (1978) provided a description of both wild and domesticated Norway rats in larger, more complex test enclosures. This progressive refinement of description from 1922 to 1978 has rendered it possible to detect very subtle effects of various treatments with considerable precision.

Paper 5 by Bermant, Clegg, and Beamer represents an application of the precise methodology developed with rats to another species, in this case sheep. As in rats, copulatory behavior in sheep can be seen to generate a set of highly reliable quantitative measures susceptible to further experimentation.

As valuable as descriptions of copulatory activity can be as a foundation for further research, they can be of even more value when organized into some coherent schema. Several authors have engaged in attempts to summarize and integrate comparative descriptive information (for example, Beach, 1949; Ford and Beach, 1951). Young (1941) wrote an important review of patterns of sexual behavior in mammalian females. Reed (1946) provided an excellent summary of research on the copulatory behavior of small mammals, bringing together material from a wide array of sources. Building on Reed's work, Dewsbury (1972) attempted to structure the descriptive information on mammalian copulatory patterns. Dewsbury noted that without a taxonomic system, we have no means with which to classify and categorize patterns of copulatory behavior, and thus no basis for systematic comparisons. Dewsbury proposed that copulatory patterns can be classified according to four criteria: (1) whether or not there is a lock or mechanical tie between penis and vagina, (2) whether or not intravaginal thrusting occurs, (3) whether multiple intromissions are prerequisite to ejaculation, and (4) whether or not multiple ejaculations occur in a single episode. This system produces 2^4 or 16 possible patterns of copulatory behavior. Dewsbury then reviewed available information on different species of mammals in relation to this classificatory system. The reader interested in descriptions of a broader range of mammalian species is referred to that paper.

9

It should be noted throughout several of the readings in this section that the active role of females in the initiation of sexual activity often appears. The topic of female sexuality has received increased interest in recent years, probably reflecting changing values in the culture. One issue that has not yet been full resolved is that of the phylogenetic generality of female orgasm. Surprisingly little research has been conducted on the problem (Fox and Fox, 1971; Sherfey, 1966). The "clutching reaction" displayed by female rhesus and stumptail macaques may be an indicant of orgasm in these species (Chevalier-Skolnikoff, 1974; Zumpe and Michael, 1968). More definitive information on the occurrence of orgasm in nonhuman species requires evidence from psychophysiological and electromyographic measures. With contemporary technology in telemetry and such physiological methods, definitive evidence on this question should soon be available.

Together, these five papers and the other work cited herein provide a sample of the kind of comparative descriptive information available on mammalian sexual behavior. Additional information can be found in papers dealing with the behavioral patterns of different species throughout this volume.

REFERENCES

Beach, F. A., 1949, A cross-species survey of mammalian sexual behavior, in *Psychological Development in Health and Disease*, P. H. Hoch and J. Zubin, eds., Grune and Stratton, New York, pp. 52–78.

Buechner, H. K., and R. Schloeth, 1965, Ceremonial mating behavior in Uganda kob (*Adenota kob thomasi* Neuman), *Z. Tierpsychol.* **22:** 209–225.

Carpenter, C. R., 1942, Sexual behavior of free-ranging rhesus monkeys (*Macaca mulatta*). II. Periodicity of estrus. homosexual, autoerotic and non-conformist behavior, *J. Comp. Psychol.* **33:**143–162.

Chevalier-Skolnikoff, S., 1974, Male-female, female-female, and male-male sexual behavior in the stumptail monkey, with special attention to the female orgasm, *Arch. Sex. Behav.* **3:**95–116.

Dewsbury, D. A., 1967, A quantitative description of the behavior of rats during copulation, *Behaviour* **29:**154–178.

Dewsbury, D. A., 1972, Patterns of copulatory behavior in male mammals, *Quart. Rev. Biol.* **47:**1–33.

Diakow, C., 1975, Motion picture analysis of rat mating behavior, *J. Comp. Physiol. Psychol.* **88:**704–712.

Eisenberg, J. F., 1963, The behavior of heteromyid rodents, *Univ. Calif. Pub. Zoology* **69:**1–100.

Ford, C. S., and F. A. Beach, 1951, *Patterns of Sexual Behavior*, Harper, New York.

Fox, C. A., and B. A. Fox, 1971, a comparative study of coital physiology with special reference to the sexual climax, *J. Reprod. Fert.* **24**:319–336.

Kempf, E. J., 1917, The social and sexual behavior of infra-human primates with some comparable facts in human behavior, *Psychoanal. Rev.* **4**:127–154.

Larsson, K., 1956, Conditioning and sexual behavior in the male albino rat, *Acta Psychol. Gothoburg.* **1**:1–269.

McClintock, M. K., and N. T. Adler, 1978, The role of the female during copulation in wild and domestic Norway rats (*Rattus norvegicus*), *Behaviour* **67**:67–96.

Michael, R. P., and G. S. Saayman, 1967, Individual differences in the sexual behavior of male rhesus monkeys (*Macaca mulatta*) under laboratory conditions, *Animal Behav.* **15**:460–466.

Miller, G. S., 1928, Some elements of sexual behavior in primates and the possible influence on beginning of human social development, *J. Mammalogy* **9**:273–279.

Miller, G. S., 1931, The primate basis of human sexual behavior, *Quart. Rev. Biol.* **6**:379–410.

Nadler, R. D., 1976, Sexual behavior of captive lowland gorillas, *Arch. Sex. Behav.* **5**:487–502.

Nadler, R. D., 1977, Sexual behavior of captive orangutans, *Arch. Sex. Behav.* **6**:457–475.

Pearson, O. P., 1944, Reproduction in the shrew (*Blarina brevicauda* Say), *Am. J. Anat.* **75**:39–93.

Peirce, J. T., and R. Nuttall, 1961, Duration of sexual contact in the rat, *J. Comp. Physiol. Psychol.* **54**:584–586.

Pfaff, D. W., and C. Lewis, 1974, Film analysis of lordosis in female rats, *Horm. Behav.* **5**:317–335.

Pfaff, D. W., C. Diakow, M. Montgomery, and F. A. Jenkins Jr., 1978, X-ray cinematographic analysis of lordosis in female rats, *J. Comp. Physiol. Psychol.* **92**:937–941.

Reed, C. A., 1946, The copulatory behavior of small mammals, *J. Comp. Psychol.* **39**:185–206.

Sherfey, M. J., 1966, *The Nature and Evolution of Female Sexuality*, Random House, New York.

Stone, C. P., and L. W. Ferguson, 1940, Temporal relationships in the copulatory acts of adult male rats, *J. Comp. Psychol.* **30**:419–433.

Tutin, G. E. G., and W. C. McGrew, 1973, Chimpanzee copulatory behavior, *Folia Primatol.* **19**:237–256.

VanLawick-Goodall, J., 1968, The behavior of free-living chimpanzees in the Gombe Stream Reserve, *Animal Behav. Monog.* **1**:161–311.

Whalen, R. E., 1963, Sexual behavior of cats, *Behaviour* **20**:321–342.

Young, W. C., 1941, Observations and experiments on mating behavior in female mammals, *Quart. Rev. Biol.* **16**:135–156, 311–335.

Young, W. C., and J. A. Grunt, 1951, The pattern and measurement of sexual behavior in the male guinea pig, *J. Comp. Physiol. Psychol.* **44**:492–500.

Zumpe, D., and R. P. Michael, 1968, The clutching reaction and orgasm in the female rhesus monkey (*Macaca mulatta*), *J. Endocrinol.* **40**:117–123.

1

Reprinted from pp. 113–120, 131–142 of *J. Comp. Psychol.* **33**:113–142 (1942)

SEXUAL BEHAVIOR OF FREE RANGING RHESUS MONKEYS (MACACA MULATTA)[1]

I. Specimens, Procedures and Behavioral Characteristics of Estrus

C. R. CARPENTER, Ph.D.

The Pennsylvania State College and The School of Tropical Medicine, San Juan, Puerto Rico

Received May 7, 1941

PURPOSE

An essential part of a systematic comparative study of non-human primate behavior and social relations is that of describing similarities and differences in patterns of reproductive activities in representative genera and species. For those types which have been adequately studied, either in the laboratory or in the field, wide variations have already been observed and recorded. Further data are required on additional species of primates which are either little known or as yet completely unknown behaviorally in order to permit making adequate comparisons and charting the contiguities and divergencies in functional evolution.

It is especially important, at this time, to supplement what is known of the sexual behavior of the Rhesus monkey (*Macaca mulatta*) both because of its phylogenetic status and because of its important role in many branches of contemporary scientific research. Though this primate type has been extensively studied in laboratories, recent observations of it in a free range situation have yielded important supplementary facts. It is the purpose of this article to describe definitively the sexual behavior of Rhesus monkeys living in free ranging groups on a small island off Puerto Rico.

[1] This work was supported by a grant-in-aid from the National Social Science Research Council. The establishment and maintenance of the Santiago Primate Colony was made possible by a non-departmental grant to Columbia University by the John and Mary Markle Foundation. The designated directors of the grant were Dr. P. E. Smith and Dr. George W. Bachman with the collaboration of Dr. Earl T. Engle and the author.

During the period of establishing the Colony and while conducting the field work for this study, the author held an appointment as Assistant Professor in the Department of Anatomy, College of Physicians and Surgeons, Columbia University and he was assigned to the School of Tropical Medicine, San Juan, Puerto Rico.

LOCATION OF THE STUDY

The field work for this report was done on Santiago Island, a small *Cay* of about 37 acres which lies five-eighths of a mile off the Southeast coast of Puerto Rico. The nearest point on the mainland is La Playa de Humacao and it is from there that boats leave to cross the narrow channel to the Island. The typography of Santiago Island is irregular. Its surface is covered by a cocoanut palm grove, scrub trees, grass and weeds, rocks and new plantings of shade, fruit and orna-

SANTIAGO ISLAND
PUERTO RICO

mental trees (see map on this page). The Island offers a situation which, in many respects, is uniquely favorable for a breeding colony of Rhesus monkeys. Here according to original plans, field and laboratory research were to be closely integrated.

SANTIAGO RHESUS COLONY

The original stock of the Santiago Colony was collected by the author in India during September, 1938. Three months later, 409 monkeys were released on Santiago Island. On March 1, 1940 the Rhesus Colony had about 350 monkeys in various stages of development ranging from infancy to senility. There were

24 adult males and four sub-adult males who lived in heterosexual groupings, while 12 sub-adult males lived in unisexual male groupings, thus making a total of 40 males in the Colony.   There were approximately 150 adult females, making a ratio of more than 6 adult females to each adult male.   In addition there were 3 sub-adult females, 48 infants born in India during July and August, 1938, 90 infants born in the Colony between May 19 and December 31, 1939 and 20-odd infants born since January 1, 1940.[2]   Besides these Rhesus monkeys there was a male, a female and an infant of the pig-tail type (*Macaca nemestrinus*) constituting a small family.

The animals organized themselves into 6 heterosexual groups ranging in size from 3 in the case of the pig-tail group to 147 for the largest Rhesus group.   The smallest heterosexual Rhesus group had 13 individuals.   The average Rhesus group contained approximately 70 animals.

This report is based on observations of 40 different Rhesus females of the Colony who passed through 45 periods of sexual receptivity between February 29 and April 27, 1940.

When this study began the Rhesus Colony had just been released from isolation cages, in which they had been held for six weeks for tuberculosis testing. Following their release, the females passed through a kind of mating season, during which more females became receptive than during any other comparable time in the history of the Colony.   Thus, it developed that the period from February 29 to April 27 was unusually favorable for the study of sexual behavior in these primates living under free ranging conditions.

PROCEDURES

Since the monkeys of the Santiago Colony were relatively tame, being semi-domesticated by hand feeding, they could usually be approached without unduly disturbing their normal behavior.   The numbering on the thigh of the monkey with figures large enough to be read at close quarters by the naked eye or at a distance with the aid of binoculars, made it possible to identify *individuals*. Later many could be recognized on sight because of individual differences in

[2] I am indebted to Mr. M. I. Tomilin, primatologist in charge of the Santiago Rhesus Colony, for these figures which were tabulated when the monkeys were trapped for the annual census and were tested for tuberculosis.   The composition of groups will be described in the next paper of this series.

It is a pleasure to express our obligation to Dr. R. M. Mugrage, Director of the Ryder Memorial Hospital in Humacao, Puerto Rico, and to Mrs. Mugrage, who provided Mrs. Carpenter and me with living accommodations, without which it would have been impossible to conduct this and other studies of the Santiago Colony.

I am indebted to Professor Robert M. Yerkes and Professor George B. Wislocki for valuable editorial suggestions.

appearance. This made it possible to keep daily systematic records of the animals being studied. Females could be observed repeatedly throughout their estrous periods and records could be kept of their matings with a single male or with different males.

The observations were begun shortly after 6:30 a.m. each day and continued until 4:00 p.m. with the exception of short periods for breakfast and lunch. During more than two months of field work, observations for only two days were missed. Each morning, females who had previously shown receptive behavior were noted and each group was closely observed for varying periods of time, ranging from about one hour to four hours, in order to detect other females who might have come into estrus in the meantime.[3]

PERTINENT LITERATURE

The author has described the periodicity of receptive sexual behavior for the howler monkey (*Alouatta palliata*) (4) living under natural conditions on Barro Colorado Island, Panama Canal Zone. It was estimated on the basis of systematic observations that the estrous period for females of this species lasted for four or five days, during which the female had a succession of close *consort relations* and copulated (see page 137) with males of her own group.

Attempts were made to check the validity of these observational records on estrous behavior of howlers. By special permission of the management of Barro Colorado Island, three females were shot after they had shown estrous behavior for several days and their uteri and ovaries were removed and fixed for histological study. Dempsey (8) has recently described the histology of the ovaries and uteri. He reports corpora lutea and other histological evidence of recent ovulation. The histology of the uteri was quite similar to that of Rhesus monkeys during the late follicular phase of the menstrual cycle.

Later studies (5) of the red spider monkey (*Ateles geoffroyi*) led to the belief that this New World type, like the howler, had a definite period of sexual activity which resembled estrus in lower mammals. The collection of a rather large series of female reproductive tracts showed a number of vaginal plugs in the vaginas of enlarged reproductive tracts.

Hamlett (11) in 1939 concluded from a study of Cebus monkeys, another New

---

[3] The equipment used in this study consisted of two pairs of binoculars; one seven and the other eight power magnification. Photographic records were made both by still camera (Leica) and by a moving picture camera (Bell and Howell 70 DA, 16 mm.).

A documentary film comprehensively picturing the behavior and social relations of Rhesus monkeys of this Colony was made shortly after this study was completed. The film was sponsored by the American Film Center, 630 Fifth Avenue, New York City and was produced in collaboration with the New York Zoological Society. A short 400 feet color film is available from the author at State College, Pennsylvania.

World type, that they have a very restricted period of sexual behavior. On the basis of the occurrence of vaginal plugs and not by observation of behavior, Hamlett states, "Copulation took place only at or shortly prior to the time of ovulation" (p. 187).

Zuckerman (23) for the baboon, Tinklepaugh (16), Yerkes and Elder (20) for the chimpanzee and Ball and Hartman (3) and Ball (2) for the Rhesus monkey have described varying degrees of sexual periodicity in the female for these Old World types. Ball and Hartman (3) have, as late as 1935, shown a mode of sexual behavior for the Rhesus monkey but have confirmed in part a long standing belief previously supported by Corner (7), Allen (1) and Miller (14) that "Primates differ from lower mammals in accepting copulation throughout the cycle instead of only short periods at the time of ovulation" (p. 2).

It is the opinion of the present author that Rhesus monkeys living under free range conditions have limited periods of sexual activity within the menstrual cycle. These periods terminate and then re-occur after non-fertile cycles. During the limited estrous period there is a marked frequency of copulation and likewise marked changes in associated behavior traits and social relations. This period corresponds in many respects to "rut," "heat" or estrus in "lower" mammals such as the dog. Nevertheless, there are important differences.

It has become customary to think of estrus as being a physiological state, especially closely connected with the endocrines. It is possible that this concept is too narrow and it would seem more logical to use the term *estrus* to refer to the total complex of associated changes which occur near the time of ovulation and lead to possibilities of fertilization. This would include gross anatomical changes, changes in endocrine balances, variations in *thresholds* of stimulation in the receptors, perhaps also in parts of the nervous system, and subsequent changes in behavior. The latter is a necessary link in the fertilization process. Heape (12) in 1900 defined "oestrus" as the "special period of sexual desire of the female" and again it has been defined as that period when *the female actively seeks coitus with a male.* Two facts should be noted in this connection: 1) Estrus is a complex of changes of which behavioral changes are a prominent and essential part, and 2) Estrus is not mere passive receptivity but involves aggressive sexual behavior in the female. The term estrus in this report is used in a broad sense to include anatomical, physiological and behavioral changes. It is illogical to consider behavioral changes as mere indicators of estrus; behavior traits are essential parts of estrus.

In addition to defining the periodicity of sexual behavior in the Rhesus monkey, descriptions will be given of the normal characteristics of sexual activity and social relations in this primate. An attempt will be made to formulate for the reader a *norm* of sexual activities for *Macaca mulatta*, first by presenting a normal case study.

TYPICAL CASE OF ESTRUS IN A FREE RANGING RHESUS FEMALE

Female 126 was a member of Group II which consisted of approximately 147 animals. There were 10 adult males who varied widely in dominance, prestige or social status. Female 126 was marked by a notch in the lobe of her right ear and she was distinctively colored so that once her number had been learned, she could be certainly identified. She had an infant which in all probability had been born during the Summer of 1939, making it about 9 or 10 months old.

This female was first observed on the morning of April 1, when she was bluffingly attacked by male 153-2,[4] an old male second in the order of dominance in Group II. While f 126 was being attacked by male 153-2, she was joined by several other females in making a counter attack on him. F 126 became closely and persistingly associated in a typical *consort relation* with male 153-2. She was hyperactive, explorative and aggressive. She made frequent attacks on females who approached her. She had marked corrugated sexual swelling of a deep raspberry red, while her face was a lighter red. On the following day, the second of April, she continued to associate very closely with male 153-2 in a true consort relationship. She groomed him diligently and followed him wherever he moved. There was coagulated ejaculate in the labial folds of the female. On the next day, f 126 continued the consort relation with male 153-2. She associated closely with him and expressed intense sexual excitement during copulation by the *spasmodic arm reflexes* (quick, jerky movements of the arm) and by *sham attacks* on nearby females and young. The consorts, f 126 and male 153-2 and the infant of f 126 were sometimes separated from other animals of Group II. She was not observed on the fourth but again on the fifth, she was seen to be closely associated with male 153-2. While the pair was being observed male 163-1, the most dominant male in Group II, made a vicious biting attack on f 126 and left her wounded and bleeding freely. Male 153-2 was apparently indifferent to the attack on the female and showed no antagonism to male 163-1, but after the attack was over, he continued to follow her. At this time, f 126 had a vaginal overflow ("vaginal plug") indicating that a number of copulations had taken place.

The most dominant male of the group, male 163-1, now replaced male 153-2. Twice male 163-1 made strong attacks on f 126. She seemed to avoid him while seeking to approach male 164-4, a mild, subordinate male of the same group. On the sixth, male 163-1 was closely associated with f 126. He guarded her

---

[4] 153 indicates the number which was tattooed on the animal's thigh. The order of intra-group dominance is given in the order 1, 2, 3, etc., 1 being the highest rank. This dominance order ranking was based on numerous indicators of aggression and group status which will be reported in a forthcoming paper dealing with that subject. Male 153-2 was second in dominance order in his group at the time of the observation.

against other males and females and together with her infant, they formed a compact sub-grouping. The consort relationship between f 126 and male 163-1 continued during the seventh and eighth. When f 126 made approaches to male 164-4, male 163-1 would repeatedly make *strong driving attacks on her, but not on male 164-4.* Male 163-1 carried out several complete copulations with f 126 on the ninth. The consort relationship persisted through the tenth and eleventh, after which f 126 was again observed mixing with other females and young. Male 163-1, having left f 126, established a consort relation with another receptive female. During this period of eleven days, the sexual skin swelling in f 126 had decreased in turgidity and paled in color, while during the last few days her face became a brighter red, which persisted for some days after she was no longer associated with males.

TRAITS, CHARACTERISTICS AND INDICATORS OF ESTRUS

In this study the following observed characteristics were used as a means of gathering descriptive data regarding some aspects of the estrous period:

1. *Sexual skin swelling and color* are well known characteristic responses of the *young* Rhesus female to estrogenic hormone balance.

2. *Increased general activity* has been shown by Richter (15) and others to be associated with estrus in rodents. There are as yet no laboratory studies which correlate the amount of general activity and sexual behavior in primates, but it seems logical to expect increased general activity to be associated with increased sexual motivation.

3. *Affinitive approaches and gestures* have been found in previous studies (4, 3) to characterize the estrous period in primates. A sexually motivated Rhesus female is not only receptive for the male but she actively approaches him and at the same time she may exhibit rhythmic lip movements, given with elevated eyebrows, projected muzzle and a typical facial expression.

4. *Presenting* consists of the female approaching a male (or of being approached) and orienting her hind quarters to him as a part of the complex "invitational" behavior (3). Presenting is the initial stage of posturing for copulation. Closely associated are slight changes in posture and locomotion.

5. *The act of copulation* in the Rhesus monkey consists of a series of *mountings* with intromission, ranging in number from 3 to more than 100, until the series reaches a climax when the male is stimulated to ejaculate. Mountings, even with erections and intromission, are differentiated from a true *copulation series* which culminates in ejaculation.

6. *Coagulated seminal fluid* may sometimes be observed in the labial folds of the female after a number of copulations. This vaginal overflow is part of the "vaginal plug".

7. *The sporadic arm reflex* was shown by preliminary observations to be associated with intense sexual excitement in the Rhesus female. The movement occurs in the dominant hand and arm and consists of a rapid flexion and extension of the hand and arm. It is to be seen most frequently in a strongly aroused female during a copulation series.

8. *Increased agressiveness and wounds* are obviously correlated with estrus in females. Males characteristically attack females who are becoming sexually active and who are trying to form consort relations. Females who are shifting from their normal, non-estrous group statuses and females shifting from one male to another often attack more subordinate associates and are frequently the objects of agression. Furthermore, females who are becoming non-receptive and are returning to their normal non-estrous group status likewise show reciprocal aggression.

9. *Grooming*, which consists of one animal cleaning the fur of another or of itself, is at all times a prominent part of Rhesus behavior, but it is markedly increased during estrus as shown by preliminary observations.

10. *The consort relation* first observed in howlers.was found to characterize the estrous female's relation to the male in Rhesus monkeys. The relationship can be easily identified by an experienced observer. It consists of a male and female forming an association with a high degree of reciprocally interactive behavior and "rapport".

11. *Minor signs of estrus* may be listed as follows: a. Scuffed hair at the base of the female's tail as a result of repeated mountings by the male. b. Scuffed and calloused popliteal region of the hind legs where a male grasps the legs of the female with his feet when mounting. c. A strained walk and evidence of general fatigue. These signs may show up clearly toward the end of a long estrus.

These clearly definable and observable traits, characteristics or signs occur in various combinations and are either a part of estrus or indicate estrous behavior in the female Rhesus monkey. These criteria have been used in this study to ascertain the typical sexual behavior and periodic dispersion of this activity within the menstrual cycle. It is important to note that while the sporadic arm reflex, the complete and repeated copulation series and the true consort relationship are specifically associated as a part of estrus, the other traits, when taken singly, may or may not be specific for the estrous period. The reliability of these traits and indicators increases when they are found in combination.

[*Editor's Note*: Material has been omitted at this point.]

Sexual swelling was usually accompanied by redness of a raspberry shade during the initial phase of estrus and this paled gradually toward the terminal phase of the period. Shortly after breaking off the consort relation with the final male, the sex skin of a female which had shown swelling and redness would usually bleach out and acquire a whitish or pale pink appearance. These observations confirm in general the descriptions of a number of workers who have used this reaction as an indicator of estrogen effects (24, 9).

About one out of ten estrous females developed an intense redness of the face. This facial coloring appeared in marked degree following the reddening and swelling of the sex skin but it lasted after detumescence and paling of the sex skin, thus showing a different threshold to the estrogens than does the sexual skin reaction.

2. *Hyperactivity* (exp.): A female in the early phase of estrus may first become conspicuous by her restless, irritable and explorative behavior. She leaves her normal sub-group within the large group and actively approaches males. One or more of the males may follow her as she makes excursions away from the focus of the main grouping. If her consort is the most dominant male, his following her will cause the entire group to travel much farther than normally. The increase in attacks made on her, her aggressiveness toward other animals, plus various aspects of sexual behavior add greatly to the amount of activity exhibited. Therefore, not only is the receptive female herself more active but also her hyperactivity stimulates other animals within the group, especially the males, to greatly increased activity.

The fact that the female changes from her normal non-estrus group status and takes a position in closer relation to a male or males, produces in the group a state of flux which is shown by increased general activity, restlessness and increased number of intra-group conflicts.

Characteristically, estrous females show toward the end of the period a diminution of general activity to a level at which the female is sluggish, unresponsive and sleepy. This hypoactive phase of estrus is especially marked in those females which show protracted estrus and engage frequently in copulations with a number of males. The condition of fatigue leading to unresponsiveness, at times frustrates the male and provokes attacks by him on the female.

3. *Affinitive approaches and gestures:* The estrous Rhesus female in the initial phase of receptivity actively approaches males of her group until one or more of them responds adequately. These repeated approaches usually meet initially with *resistance* on the part of the male and he may repeatedly attack her. The approaches, the direction of attention, the stance of the female, and certain sterotyped gestures such as the elevated eyebrows and rhythmic lip

20

movements, function as signs to the male of the receptive attitude or physiological state of the female. Perhaps odor and more specific visual cues are also important. Normally most non-estrous females withdraw when a dominant male approaches them. Positively oriented behavior is differential stimulation which functions as a kind of *invitation* (3, 4) for the male, serving as a cue to the female's receptivity for copulation.

4. *Presenting:* Presenting normally follows the approach and affinitive gestures, if they are given. A male in a state of sexual readiness will respond to this situation by mounting the female, unless he is inhibited by some distraction or his attention is otherwise engaged.

Presenting is not a behavior pattern specific to the female estrous cycle. Juveniles present to other juveniles during play; females present to females, males to males, and juveniles to males. Presenting under some of these conditions may be classified as a kind of greeting response. At other times, presenting is a response to generalized, non-sexual excitement, as when animals are too closely approached by an observer or when an intra-group fight occurs. Presenting may also occur as an act of submission to a more dominant animal. Consequently this behavior is not exclusively characteristic of estrus, but it is normally only one of the activity patterns of an estrous female.

5. *Copulation* (cop.): True copulation consists of a succession of mountings and these are carried out as follows: The male places his hands on the hips or mid-spinal region of the female while supporting most of his weight on her hind legs which he grasps with his feet in the popliteal region. The male's entire weight is borne by the female. In this position the male usually makes from 2 to 8 piston thrusts with a single intromission, except when ejaculation occurs, but the number of thrusts tends to increase as the mountings are repeated. Most mountings do not lead to ejaculation and after these, the male dismounts and the behavior is repeated after an interval in the same series which ranges from less than 1 minute to more than 20 minutes. The length of the interval between mountings depends both upon the strength of the aroused sexual drives in the participating male and female and upon the virility characteristic of the particular male.

Mounting of a female by a male, even with intromission, may be a response to a true sex drive or it may occur as a greeting response, as an expression of an affinitive relationship or as a substitute act to foil an attack ("displacement" or an "instrumental act"). Mounting is not specific for estrus nor, indeed, for the male-female relation because males mount males, and females mount females. True copulation in the Rhesus monkey consists of a *series* of mountings with intromission and several piston thrusts prior to ejaculation. A copulation series may have from 3 mountings in highly motivated males to over

100 mountings in males nearing sexual satiation. As the threshold to stimulation of ejaculation is heightened, the number of mountings prior to ejaculation increases.

Those mountings which lead to ejaculation are clearly indicated: The number of thrusts is greater and the series of thrusts ends with the male showing general tenseness or mild rigor, accompanied by a baring of the teeth and usually by a squeal. The female twists and turns, reaches back and touches or holds the male with one hand and often turns her muzzle as if she were trying to see the male. Characteristically after dismounting, the male collects the ejaculate which adheres to his penis and eats it. The female meanwhile usually grooms the male.

The following example of a normal copulation series will give the reader a rather clear conception of how the complex series of copulatory behavior is carried out. The time intervals between the mountings are noted along with the complete time which elapsed from the beginning to the end of the copulation series. Although more than the average number of mountings occurred in this series, which took place with this particular female after the middle of an estrous period, it is not an extreme example.

COPULATION SERIES BETWEEN OM-1 AND f 73

Time: 9:26 a.m. First mounting (1) 2 thrusts, yawning by the male; interval 1 min., (2) 1 thrust, more yawning by the male; interval 1 min., (3) 3 thrusts, yawning, self grooming by the male; interval 2 min., (4) 2 thrusts, female crouching and extremely submissive; interval 30 sec., (5) 4 thrusts; interval 30 sec., (6) 2 thrusts; interval 1 min., (7) 3 thrusts; interval 1 min., (8) 2 thrusts; interval 1 min., (9) 2 thrusts, rather quick movement of hands and head by female; interval 3 min., (10) 3 thrusts, female grooming male; interval 1 min., (11) 3 thrusts; interval 30 sec., (12) 3 thrusts, female grooming male; interval 1 min., (13) 3 thrusts, female grooming male; interval 30 sec., (14) 4 thrusts, squeal by female; interval 1½ min., (15) 5 thrusts, female shows quick, jerky movement of right arm and sidewise turning of head; interval 30 sec., (16) 6 thrusts, female purses her lips and turns her face to the male, slight attack in the form of a slap by male; interval 1 min., (17) 5 thrusts, squeal by the female, male yawns; interval 1 min., (18) 4 thrusts, male slaps the female upon dismounting, at almost all intervals now female shows jerky arm reflex; interval 30 sec., (19) 3 thrusts; interval 2 min., (20) 5 thrusts; interval 30 sec., (21) 4 thrusts, after each mounting with intromission the male squeals; interval 30 sec., (22) 3 thrusts, male squeals; interval 30 sec., (23) 3 thrusts, male squeals; interval 30 sec., (24) 3 thrusts; interval 2 min., (25) 3 thrusts, bluffing, jerky movements of female or her bluff toward the observer or other nearby animals stimulate the male to begin mounting; interval 1 min., (26) 3 thrusts; interval 1 min., (27) 5 thrusts, slight attack made by male on female; interval 2 min., (28) 5 thrusts, another mild attack, intromission squeal by male; interval 1 min., (29) 5 thrusts; interval 30 sec., (30) 7 thrusts, squeal; interval 1½ min., (31) 5 thrusts; interval 30 sec., (32) 4 thrusts; interval 30 sec., (33) 7 thrusts; interval 30 sec., (34) 7 thrusts; interval 20 sec., (35) 5 thrusts; interval 1 min., (36) 5 thrusts; interval 30 sec., (37) 22 thrusts, rigidity in male with ejaculation and squeal. The female turns her head to the right and upward as if trying to put her mouth to that of the male. Immediately on dismounting, the female begins to groom the male. Time 10:06. Total time elapsed 40 minutes.

At the onset of estrus in free ranging Rhesus females, the sexual motivation rises rapidly in intensity. The increment in the strength of the drive, as indicated by the frequency of copulations and eagerness with which the female approaches the male, is very marked during the early part of the period, while the decrement toward the end of the period takes place much more gradually, trailing off slowly into non-receptivity and avoidance of the male. The peak of the period of receptivity is estimated to occur during the second fourth, or the mid-portion of the estrous period, judging from estimated frequencies of copulations and the expressed preferences of the most dominant males.

A summary of table 1 shows that the length of the estrous period, as observed, ranged from 4 to 15 days with an average duration of 9.2 days and an average deviation of 2.90 days. The *median* period lasted 8 days and the mode of the distribution fell at 7 days.

The frequency of copulation is a function of the interacting sex drives of both the female and the male consorts. The copulation frequency is highest following the establishment of the first persisting consort relationship and each time the female shifts from one male to another, there occurs a mode in the frequency of copulations. This increase in frequency occurs even though the second male has become relatively satiated by his previous female consort. As observed by Hamilton (10), the stimulus value of a new female is greater than that of a previous female associate and as a result of the change there is an increase in the sexual expression of the male. It is equally clear from this study that the sex hunger of an estrous Rhesus female during the estrous period greatly surpasses that of any single male during the same period. This is shown by the fact that on the average each female had 2.93 males, average deviation 2.04, during the estrous period. Furthermore, throughout the greater part of the estrous period, females eagerly seek to initiate copulation with a male. Females like 29, 37, 47, 115, 145, and 147 satisfied as many as three of the most dominant males during their individual estrous periods. As pointed out by Yerkes (19) for the chimpanzee, the significance of this difference in periodic sexual capacities is an important factor in determining various qualities of socio-sexual behavior.

The season of mating observed during February, March and April, 1940 for the Santiago Colony is confirmed by the lack of primary sexual behavior during the following months, especially May and June and also by a mode of births during August, September and October. However, it is equally clear to the author that the breeding season observed in the Santiago Colony thus far since its establishment, relates to the management of the Colony, e.g., the time of capture, periods of confinement in cages, release, and change of climate and cannot be taken as an entirely valid indication that Rhesus monkeys have a corresponding breeding season in their natural Indian habitat, although informed animal dealers in Calcutta (Mr. and Mrs. A. W. Chater) report a season of births each June, July and August.

6. *Coagulated seminal fluid or vaginal overflow* (v.p.): Most females may be observed to have "vaginal plugs," or more properly as observed, vaginal overflows, at times of frequent copulations during the estrous period. In this study 34 instances of vaginal overflows were observed one or more times in 45 estrous periods. These observations indicate, among other things, the enormous excess of spermatic fluids delivered to females beyond the amount necessary for fertilization. It is also suggested that in the economic management of a Rhesus colony, controlled matings should replace the excessive copulations which occur under uncontrollable free range conditions. The number of true copulation series during an estrous period is estimated to range from 25 to 50 or 60, and the same male may engage a female in three or four series on the same day. Under properly controlled matings a short series of copulations near the time of ovulation should lead to fertilization.

It is an interesting fact that whereas males remove and eat seminal fluids which cling to their own genitalia, males never showed any interest in the overflow of fluids coagulated in the labial folds of the females.

The excess seminal fluids of the Rhesus monkey emit a characteristic odor which is perceptible to humans and which may be a source of additional stimuli attracting males to receptive females and causing the males to follow or to compete for them.

7. *The spasmodic arm reflex* (a.r.): Females with intense sexual excitement consistently show the reflex which involves rapid flexion and extension of the dominant hand and arm as shown in the example of a copulation series just described. Characteristically this occurs when a copulation series is in progress and the female is strongly stimulated. Then, when the arm reflex is exhibited, it stimulates the male to mount the female. It would seem that in some instances this reflex has come to have a *stimulus value*, i.e., it is used by the female to incite the male to mount. In this context it may be termed an "instrumental act." The reflex seems to be highly specific to the estrous cycle and apparently this relationship has been observed for the first time in this study.

8. *Aggresiveness and wounds* (agg., att., and w.): Table 1, page 122, shows that in the 45 estrous periods, 22 females were observed to be attacked with varying degrees of vigor one or more times. Sixteen estrous females were wounded, some severely. Fourteen estrous females were seen to make re-

peated attacks, usually on other females but sometimes on subordinate males. Although the non-estrous Rhesus females make attacks and are attacked, it is the conviction of the author that this behavior is markedly increased during estrus.

The estrous female, furthermore, is in the vortex of stressful social relationships in her group. During her receptive period, her social status in the group shifts and she becomes a sexual incentive for the group's males. She actively approaches males and must overcome their *usual resistance* to close association, hence she becomes an object of attacks by them. Even other females attack her as a result of her shifted social status.

Several examples may properly emphasize this point: Females 49, 105, 126, 144, "f.n." and 109 were all severely wounded during their estrous periods. Female 105 lost parts of both ears, was cut severely on the arm and received a network of wounds over her face and muzzle. Female 144 had a leg wound which compelled her to walk on three legs for several days. Female 145 was deeply cut on the thighs. Female "f.n." had a badly bruised nose while female 109 had a long, deep gash and her infant was wounded so severely that it died. Female 126 had a motor nerve cut in her left hip which caused her to be a permanent cripple. The number of cases could be extended to show that wounds are highly coincidental with estrus.

A final, partial explanation of the increase in fighting is that aggression may constitute a sadistic component of the normal mating behavior of Rhesus monkeys. A necessary condition to serial copulation and the consort relation seemingly is that the female be driven to a state of submission, of "awe" and of complete "rapport" in relation to the male. In Rhesus monkeys *the male is at all times completely dominant over the female even during the estrous period.* (Likewise, females playing the dominant role in homosexual behavior attack the subordinate female.) There does not seem to be a reversal of dominance in the Rhesus monkey during estrus as reported by Yerkes (19) for the chimpanzee. There is a change which seems to involve an increased *tolerance* of the male for the estrous female. Dominant, aggressive males will allow estrous females to feed with them from small (2 ft. square) food trays, but this did not occur with non-estrous females. This *tolerance* is expressed in the close spacial association characteristic of consorts; a necessary condition to which is the receptiveness of the female.

These observations show that: 1) Females are more subject to attacks during estrus than during the non-estrous phase of the menstrual cycle. 2) Females in estrus are far more aggressive than non-estrous females. The increased aggressiveness of the female results in her ascending in the female dominance scale, i.e., becoming more dominant and often, *with the reinforcement of her male consort,* she becomes temporarily the most dominant of all the females in her group and, in addition, she may occupy the most preferred place of safety in the most dominant male's sphere of protection.

9. *Grooming* (gr.): Receptive females are very frequently observed participating in reciprocal social grooming either with their male consorts, with other females or even with young animals. Often it seems that females gain the tolerance of their male consorts both by copulating with them and by grooming them; thus grooming may be used to divert the direction of any to-be-avoided behavior. It is characteristic for the receptive female to groom her male persistently, especially immediately after copulation but also during intervals between the mountings. Frequently, too, receptive females invite grooming from any nearby animals by assuming suitable relaxed postures. It is not unusual to see several females and young juveniles grooming a receptive female showing marked sexual swelling. The assumption, that the estrogens produce physiological changes in the skin that increases the need for grooming seems possible and provides suggestions for experimentation. It is probable also that the high incidence of wounds in estrous females, since wounds normally stimulate associated individuals to dress them, would accentuate grooming behavior both in the groomer and the groomee.

Grooming, like copulation, functions to increase the degree of "rapport" between the male and female consorts, to strengthen the social bonds and thus to prevent attacks by the male. In this case the behavior is used as an "instrumental act." Active grooming may be used by the female as a means of avoiding attacks by the male or to gain a more favorable relationship with him.

10. *The consort relation* (c.r.): A prominent characteristic of mating behavior of Rhesus monkeys is that *females during estrus associate closely with a succession of males.* These associations typically include a male and a receptive female and the relationships usually are mutually exclusive of other animals except during transitional phases, e.g., when the female is shifting from one male to another or when a male is breaking off the association with one female and establishing an association with another one. In the latter case the just-previous female may, for a short time, continue the association in a secondary position (see table 1). This possibility seems to depend upon the quasi-conditioned relationship between the two females involved and on the degree to which the first female has been sexually satiated and previously "conditioned" to the male. Usually females are rather antagonistic and competitive with each other during estrus. With the reinforcement of the male, the preferred female is abnormally aggressive toward other females. However, in one case f 124 and f 179, the competing females engaged in homosexual behavior. Thus, at times females in secondary positions gain the preferred female consort's tolerance by grooming her or by submitting to being mounted by her.

Table 1, page 122, lists the associations between males and females during estrus, shows the shifts which were observed and indicates the females that were in secondary positions. For example, female 29 was first associated with male 150-3 (a young adult male, third in the male dominance gradient). She was then taken over by male 174-2, and following a short association with him, f 29 was possessed by male 160-1 for the remainder of her estrous period. The last four columns summarize important facts listed in detail in table 1. Column 2 shows the number of *different* males who were associated with females during single estrous periods. The range in number of different males is from 1 to 9, the median is 2 and the mean 2.93 (A.D. 2.04). Most females, about 69 per cent, had either 1, 2 or 3 males during a single estrous period. The number of shifts which occurred is greater than the number of different males, showing that a single male may be associated more than once with a female during a single estrous period. The number of shifts ranged from 0 to 14. Column 3 shows the estrous females making the various numbers of shifts and these shifts may be taken as a *measure of promiscuity* in Rhesus monkeys. It is noteworthy that 20 per cent made no shifts but continued the association with one male. These constant associations occur under three conditions: 1) When there is only one male in the group. 2) When there is one receptive female in the group or when at the same time the supremely dominant male inhibits competition from subordinate males. 3) When the supremely dominant male and subordinate males each simultaneously have preferred females which completely occupy them. Somewhat fewer, 17.8 per cent shifted only once, while 11 per cent shifted twice or had three males during a single estrous period.

The pattern of mateship of the howler monkey (4), the gibbon[5] and other studies (5, 23), along with the present investigation show the marked species differences which exist among primates in mateship patterns. In Rhesus monkeys *a temporary mateship with usually one female at a time and this only during the female's estrus*, characterizes the species and is shown, furthermore, by the fact that females were observed 248 times with only one male on a single day, while more than one male was observed with a female on the same day only 49 times. Rarely did a single male possess more than one female at the same time. For this species, therefore, the position on the monagamous-promiscuous scale can be quantitatively indicated.

Zuckerman (23) has observed the social status of the baboon female during her period of maximum swelling.

"The 'chief wife' in the harem is the female that shows greatest swelling, her prior position being manifested by the fact that she keeps closer to the overlord than do the other females of the harem. At such times she seldom moves

---

[5] Carpenter, C. R., 1940, A field study in Siam of the behavior and social relations of the gibbon (*Hylobates lar*), Comp. Psychol. Monog., **16**, no. 5, 1–212.

away from him, although when her sexual skin is quiescent they may be separated by as much as twenty feet. The 'chief wife' grooms him more than do his other females, and it is upon her that he concentrates his sexual activities. She is also allowed to take from him a limited supply of food, although he would usually attack her if she were to do the same when her sexual skin is quiescent" (p. 226).

It would seem that *within the harem* of an autocratic baboon, shifts in social status among females occur in a manner somewhat comparable to shifts which occur in the large, heterosexually structured Rhesus society but the main differences are: 1) Whereas in the baboon, the estrous female is restricted to one dominant male, in the Rhesus monkey, she may be possessed successively by several males exercising varying dominance behavior. 2) Whereas the autocratic baboon male possesses a harem, the supremely dominant Rhesus male usually has temporarily only one preferred female consort but sometimes there may be another in a secondary position, and 3) Whereas the supremely dominant baboon exclusively possesses all possible estrous females, in Rhesus groups males of varying degrees of dominance may possess the estrous females successively.

Following estrus, the female remains in the original group but she is less closely associated with males, the consort relationship ends and she becomes a part of a female and young sub-grouping or cluster. She returns to approximately the same social status which she occupied before estrus, although this status in relation to the males with whom she was associated during estrus may have been somewhat changed. The non-estrous female may be tolerated at closer distances by some males than by others. Some of the females, usually the more dominant ones, may form a kind of central cluster around the most dominant male but the relationship in such a sub-grouping is easily differentiated from the male-female consort relationship formed by an estrous female with a responsive male.

Of all the characteristics of estrus the most reliable and indicative is the consort relationship with the associated reciprocal behavior patterns, especially repeated copulation series.

11. *Minor signs of estrus:* Scuffed hair at the tail base, calloused popliteal regions and strained walk are all indicators of estrus which show up after numerous copulations and toward the end of the estrous cycle. They are sometimes accompanied by evidences of general exhaustion and greatly reduced activity and responsiveness. During the final stage of the estrous period, the male consort may place the female in a position for mounting by the use of his hands. Manual control replaces signaling as a means of communication. During this phase, too, the number of mountings before ejaculation may greatly

increase toward the upper observed limits of 100 or more times and these mountings tax the strength of some females, since each time she must bear the entire weight of the male. Thus, during the final stages of estrus, the negative may outweigh the positive incentives and further copulations are avoided. The period terminates as the internal sex drive simultaneously wanes.

In summary, it should be noted that estrous behavior includes all the behavioral characteristics and indicators which have been described, but in addition there are anatomical and physiological changes which could not be observed in the field. The sporadic arm reflex, the vaginal overflow, the true copulation series and the persisting consort relationship are the most reliable observable characteristics of estrus in the free ranging Rhesus female.

SUMMARY

The more general results of this study will be given in a following paper. Meanwhile certain generalizations summarize important details of the findings thus far and characterize important aspects of the normal sexual behavior of Rhesus monkeys living under free range conditions.

1. Sexual swelling and coloration along with facial coloring occur in approximately 50 per cent and 10 per cent respectively of the estrous females observed during this study. These changes vary greatly with individuals and with age.

2. Females in estrus are hyperactive during most but not all of the period of receptivity. A phase of hypoactivity and fatigue occurs during the last of the cycle and for a short recovery period immediately following.

3. The motivation for sexual activity is communicated by the female to available intra-group males (and occasionally to males of other groups) by means of gross posturing, orientations and stereotyped invitational gestures.

4. Presenting by females anticipatory to copulation is not specific to the estrous period but it does occur as a component of their primary sexual behavior.

5. Normal copulatory and associated behavior patterns have been described in detail. True copulation is mainly restricted to the female's estrous period. The duration of this period varies around a mean of 9 days, about one-third of the menstrual cycle.

6. Sporadic mountings even with intromission to which females sometimes submit may occur as a means of foiling an attack or as a means of gaining some end ("displacement" or an "instrumental act") or as a submissive response. Grooming may also function in a similar manner.

7. The Rhesus monkey shows a characteristic consorting mateship, consisting of a temporary close association between a male and a female. Such consort relationships are limited to the estrous period. Although typically the male-

female consort relation excludes other females, occasional y females with lessened sex drive occupy a secondary position to the preferred consort of the male.

8. Female Rhesus monkeys during an estrous period, shift from one male to another under certain conditions of grouping structure, and on the average a female enters successively into consort relationships with about 3 males during a single estrus. Nevertheless some females (9 instances in 45 estrous periods) are possessed by only one male during the entire estrous period. This constant type of mateship only occurs under certain stated conditions.

9. The process by which a male possesses a female ("courtship") involves vigorous, aggressive attacks by the male. During these attacks many of the females are wounded and all are driven into a state of submissive "rapport" and responsiveness to the male. These attacks occur most frequently during the initial part of the estrous period, when the female is shifted from one male to another and occasionally when the female becomes unresponsive near the end of her estrous period.

10. A definite, though perhaps an artificial, mating season prevailed in the Santiago Colony during this field study, following which the amount of primary sexual activity was greatly reduced.

11. Estrous females show a marked increment in their aggressiveness during most of the receptive period, as compared with non-estrous parts of the reproductive cycle and with non-receptive females.

12. An increment in self and social grooming occurs during estrus. It is suggested that this increase in grooming may bé due to the action of sex hormones on the skin and to wounds which are inflicted on many free ranging estrous females. A female may groom a male as a means of overcoming or deflecting his aggression.

13. Males seemingly resist the initial approaches of females in the early phase of estrus but subsequent copulations and grooming result in a *tolerance* by the male of the estrous female. This relationship contrasts sharply with the aggressive and intolerant reactions of dominant males to non-estrous females who are not permitted to eat with them from the same food trays.

14. A sporadic arm reflex occurs in the excited, sexually receptive Rhesus female. This reflex is judged to be both an expression of heightened physiological sex tension and a behavior pattern which stimulates the male to mount the female and thus continue the copulation series.

15. Throughout this study it has been found that the sexual hunger of the female and her capacities for copulation during her estrous period greatly exceed that of any single male during an equal time. A single estrous female may satiate, entirely or in part, several sexually vigorous males; although when a period of such intense sexual activity terminates, the female may show extreme fatigue bordering on exhaustion.

LITERATURE CITED

(1) ALLEN, EDGAR 1928 Sex characteristics in monkeys. Proc. Soc. Exper. Biol. and Med., **25**, 325–327.

(2) BALL, JOSEPHINE 1937 Sexual responsiveness and temporally related physiological events during pregnancy in the rhesus monkey. Anat. Rec., **67**, no. 4, 507–512.

(3) BALL, J., AND HARTMAN, C. G. 1935 Sexual excitability as related to the menstrual cycle in the monkey. Amer. Journ. Obst. and Gynec., **29**, 117.

(4) CARPENTER, C. R. 1934 A field study of the behavior and social relations of howling monkeys (Alouatta palliata). Comp. Psychol. Monog., **10**, 1–168.

(5) CARPENTER, C. R. 1935 Behavior of the red spider monkey (Ateles geoffroyi) in Panama. Journ. Mammal., **16**, 171–180.

(6) CARPENTER, C. R. 1940 Rhesus monkeys (Macaca mulatta) for American laboratories. Science, **92**, no. 2387, 284–286.

(7) CORNER, G. W. 1923 Ovulation and menstruation in Macacus Rhesus. Contr. to Embryol., Carnegie Inst. of Wash. Pub. 332, 73–101.

(8) DEMPSEY, E. W. 1939 The reproductive cycle in new world monkeys. Amer. Journ. Anat., **64**, no. 3, 381–405.

(9) ELDER, J. H., AND YERKES, R. M. 1936 The sexual cycle of the chimpanzee. Anat. Rec., **67**, no. 1, 119–143.

(10) HAMILTON, G. V. 1914 A study of sexual tendencies in monkeys and baboons. Journ. Anim. Beh., **4**, 295–318.

(11) HAMLETT, G. W. D. 1939 Reproduction in American monkeys. I. Estrous cycle, ovulation and menstruation in Cebus. Anat. Rec., **73**, no. 2, 171–187.

(12) HEAPE, W. 1900 The "sexual season" of mammals and the relation of "pro-oestrum" to menstruation. Quart. Journ. Micros. Sci., London, **44**, 1–70.

(13) MILLER, G. S. 1928 Some elements of sexual behavior in primates and the possible influence on beginning of human social development. Journ. Mammal., **9**, 273–279.

(14) MILLER, G. S. 1931 The primate basis of human sexual behavior. Quart. Rev. Biol., **6**, no. 4, 379–410.

(15) RICHTER, C. P. 1927 Animal behavior and internal drives. Quart. Rev. Biol., **2**, 307–343.

(16) TINKLEPAUGH, O. L. 1933 Sex cycles, other cyclic phenomena in a chimpanzee during adolescence, maturity and pregnancy. Journ. Morph., **54**, 521–547.

(17) YERKES, R. M. 1933 Genetic aspects of grooming, a socially important primate behavior pattern. Journ. Soc. Psychol., **4**, 3–25.

(18) YERKES, R. M. 1939 Sexual behavior in the chimpanzee. Human Biol., **11**, no. 1, 78–111.

(19) YERKES, R. M. 1940 Social behavior of chimpanzees: Dominance between mates, in relation to sexual status. Journ. Comp. Psychol., **30**, no. 1, 147–186.

(20) YERKES, R. M., AND ELDER, J. H. 1936 The sexual and reproductive cycles of chimpanzee. Proc. Nat. Acad. Sci., Wash., **22**, no. 5, 276–283.

(21) YERKES, R. M., AND ELDER, J. H. 1936 Oestrus, receptivity, and mating in chimpanzee. Comp. Psychol. Monog., **13**, no. 5, 1–39.

(22) YOUNG, WILLIAM C. 1942 Observations and experiments on mating behavior in female mammals. Quar. Rev. of Biol. (in press).

(23) ZUCKERMAN, S. 1932 Social life of monkeys and apes. London, Kegan Paul, pp. 357.

(24) ZUCKERMAN, S., AND VAN WAGENEN, G. 1935 The sensitivity of the new-born monkey to oestrin. Journ. Anat., **69**, part 4, 497–500.

2

Reprinted from pp. 78, 109–111 of *Human Biol.* **11**:78–111 (1939)

SEXUAL BEHAVIOR IN THE CHIMPANZEE[1]

BY ROBERT M. YERKES

Yale Laboratories of Primate Biology

I. INTEREST AND OBJECTIVES

IN THE forefront of human needs is the wise control of sexual behavior and of the quantity and quality of population. For the satisfaction of these necessities our knowledge and understanding of the nature, conditions, and relations of sexual and reproductive phenomena are wholly inadequate. Fortunately research in this field of biology commands universal interest, and its results are promptly seized upon for critical scrutiny, evaluation, and application. The continuing study of chimpanzee behavior, from which the materials of this report are derived, is motivated primarily by interest in human problems and by the conviction that in its sexual and reproductive life this anthropoid ape is sufficiently manlike, amidst experimentally advantageous differences, to render it peculiarly useful as substitute for man in many types of inquiry. For whereas because of social taboos, self-consciousness, modesty, shame, and deception, it is extremely difficult or impossible to study the sexual behavior of man experimentally and objectively, the opposite is true of the chimpanzee and of certain other primates. Through intensive study of the reproductive life of this ape, research leads and methods may be discovered, insights may be achieved, and modes of modification and control developed which will find fruitful application in human social biology. This is the faith which underlies our relatively indirect and comparative attack on problems of sexual behavior. It is not to improve the chimpanzee, nor even to satisfy our curiosity about its patterns of behavior and their mechanisms that we study the organism persistently. Instead, our concern is with the extension of knowledge of life and the improvement of its quality. As experimental material, the ape is a means, not an end.

---

[1] The assistance of the Committee for Research in Problems of Sex, National Research Council, is gratefully acknowledged.

[*Editor's Note*: Material has been omitted at this point.]

VII. SUMMARY

1. With a previously published account of oestrus in the chimpanzee as background, this paper presents further results of inquiry into the nature and conditions of sexual responsiveness and mating behavior. Methodologically it differs from our previous study in that a rating scale was systematically used to measure the responsiveness and mutual acceptability of consorts. Observations were made in a total of 233 experimental matings, for which six females and three males served as subjects. A vasectomized male was available to minimize the risk of undesired pregnancies. So far as feasible, observations were made throughout sexual cycles at intervals of one to two days.

2. Female receptivity exhibits periodicity, as does genital swelling, and concurrently with it. In general, its fluctuations during the five-week sexual cycle range from zero, during the menstrual and post-menstrual cycle phases, to a maximum strength at a point, presumably hormonally determined, between the middle and end of maximal swelling. With the onset of detumescence it falls to zero and continues in abeyance for the remainder of the cycle.

3. A period of "heat," or, in our terminology, maximal receptivity, is characteristic of the chimpanzee cycle. Its duration has not been satisfactorily determined, since the methods employed do not eliminate the influence of the male, and consequently measure a social response which is determined by the physiological status of the two consorts. It is indicated that receptivity waxes gradually, achieves a maximum, and presently disappears suddenly. The duration of "heat" probably can

be reliably determined only by direct measurement of the isolated female. Suitable methods are being sought.

4. Completion of the mating pattern by copulation does not necessarily imply female receptivity even in low degree, since the male consort may dominate and command the female and she may respond defensively, protectively, or accommodatingly *in the experimental mating situation,* whatever her sexual status, desire, or preference.

5. A fully receptive female may on occasion repulse the advances of a responsive male, or the reverse may occur. In either case the consort which was non-coöperative may mate promptly with another individual. Such selectiveness usually seems to be due to physical incompatibility, individual preference, or unfavorable affective relations. The ability of females to control a male in the mating situation differs extremely.

6. Highly indicative of the nature of receptivity are these items of behavior: (a) The initial movement of the consorts when the barrier between their cages is removed, for F usually rushes into M's cage if highly receptive, while he awaits her presentation; (b) the fact and manner of presentation by F to M, for if fully receptive she runs to him and prostrates herself before him; (c) the duration of sexual union and the behavior of F during the act; (d) efforts of F to remain with M after the completion of the mating experiment.

7. Female sexual preference and perfection of coöperation in mating appear to vary directly with such male traits as sexual vigor, robustness, penis length, and the nature of the copulatory act, and inversely with male dominance and aggressiveness. The correlations evidently are worthy of careful statistical determination and verification.

8. Typically, sexual receptivity is absent or low during menstruation, and copulation rarely occurs, and then by male domination. During pregnancy it ordinarily is absent except in periods of genital swelling. When a pregnant female exhibits marked swelling she tends to excite the male sexually and copulation may occur. In some instances it appears that receptivity is of considerable strength, while in others mating obviously is due to the expectancy and responsiveness of the male as induced by the visual sign, swelling. Under such circumstances, and lacking opportunity to escape, the female may not safely risk struggle with the male. Instead she responds protectively and accommodatingly, irrespective at times of sexual desire.

34

9. Our observations suggest the hypothesis that "heat," as periodicity of female receptivity, is characteristic of the primate as of many of the mammals; that it varies in frequency from type to type; that the inadequacy or unreliability of observations and interpretations probably is responsible for erroneous statements and prevalent misunderstandings concerning the pertinent facts of human life. At present our observational techniques are inadequate. Obviously methods should be developed which will render possible measurement of sexual receptivity and responsiveness objectively in the isolated individual. Meantime, the rating scale, as used in this inquiry, has taken us a step nearer our goal.

REFERENCES

BALL, JOSEPHINE. 1937. A test for measuring sexual excitability in the female rat. *Comp. Psychol. Monog.*, vol. 14, no. 1. Pp. 37.

——————. 1937a. Sexual responsiveness and temporally related physiological events during pregnancy in the rhesus monkey. *Anat. Rec.*, 67:507-512.

——————, and HARTMAN, CARL G. 1935. Sexual excitability as related to the menstrual cycle in the monkey. *Am. J. Obstet. Gynecol.*, 29:117-119.

ELDER, J. H. 1938. The time of ovulation in chimpanzees. *Yale J. Biol. Med.*, 10:347-364.

——————, and YERKES, ROBERT M. 1936. The sexual cycle of the chimpanzee. *Anat. Rec.*, 67:119-143.

HEMMINGSEN, A. 1933. Studies on the oestrus-producing hormone (oestrin). *Skand. Arch. f. Physiol.*, 65:97-250.

Moss, F. A. 1924. Study of animal drives. *J. Exper. Psychol.*, 7:165-185.

STONE, CALVIN P. 1932. "Sexual drive." In Sex and Internal Secretions. Edited by Edgar Allen. Williams and Wilkins, *Baltimore.* Pp. 828-879.

——————, and STURMAN-HUBLE, M. 1927. Food vs. sex as incentives for male rats on the maze-learning problem. *Am. J. Psychol.*, 38:403-408.

TINKLEPAUGH, O. L. 1933. Sex cycles and other cyclic phenomena in a chimpanzee during adolescence, maturity, and pregnancy. *J. Morphol.*, 54:521-547.

WARDEN, C. J. 1931. Animal motivation: experimental studies on the albino rat. Columbia Univ. Press, *New York.* Pp. xii + 502.

WARNER, L. H. 1927. A study of sex behavior in the white rat by means of the obstruction method. *Comp. Psychol. Monog.*, vol. 4, no. 5. Pp. 68.

YERKES, ROBERT M. 1933. Genetic aspects of grooming, a socially important primate behavior pattern. *J. Social Psychol.*, 4:3-25.

——————, and ELDER, JAMES H. 1936. Oestrus, receptivity, and mating in chimpanzee. *Comp. Psychol. Monog.*, vol. 13, no. 5. Pp. 39.

——————, and ——————. 1937. Concerning reproduction in the chimpanzee. *Yale J. Biol. Med.*, 10:41-48.

ZUCKERMAN, S. 1930. The menstrual cycle of the primates. Part I. General nature and homology. *Proc. Zool. Soc. Lond.*, 691-754.

3

Reprinted from pp. 95–105, 145–153 of *J. Comp. Psychol.* 2:95–153 (1922)

THE CONGENITAL SEXUAL BEHAVIOR OF THE YOUNG MALE ALBINO RAT

CALVIN P. STONE[1]

University of Minnesota

INTRODUCTION

1. Purpose of investigation

The general use of experimental technique in the study of congenital behavior of lower and higher animals reflects a gradually changing trend of psychological interest in this subject. Interest is shifting from problems centered in the determination of significant relationships between conscious and organic processes to those defining units of response and analyzing the stimuli by which they are activated. As yet, however, in the field of animal psychology, investigations of congenital behavior have not been sufficiently intensive to give a complete account of either the overt patterns of activity or the activating stimuli. They have furthered an understanding of limited aspects of particular responses, given valuable suggestions along lines of methodology, and copiously illustrated the erroneous nature of many of the older conceptions concerning native behavior; but, in general, the scope of these accounts is too limited to place our knowledge of the subject as a whole on a fundamental working basis. For this reason there is need, at the present time, of comprehensive studies of fundamental types of native behavior. Intensive studies will yield data that may be used as guides in the reorganization and expansion of current views of these activities.

For the advancement of the science of animal behavior, there is need of intensive genetic studies of complex patterns of native behavior to ascertain the life histories of both the constituent elements and the patterns as wholes. With respect to the

[1] I wish hereby to acknowledge my indebtedness to Dr. K. S. Lashley at whose suggestion and under whose direction this experiment was conducted.

nature of activating stimuli our knowledge is very incomplete. For the most part investigators have dealt with the external stimuli alone. They have overlooked the internal factors. Now, to really understand the activation of native patterns of response, it is necessary to take the latter into account as well; for the external stimuli merely discharge the elaborate and complex neuromuscular and glandular mechanisms pre-organized by the action of potent internal factors operating within the organism. A preliminary survey of the internal factors operating would seem to indicate that at least two distinct types can be recognized. There is a group which underlies the development and integration of the neuro-muscular and glandular mechanisms, and another that serves to keep the mechanisms in a state of readiness for response to particular patterns of external stimulation.

The field of congenital sexual activity offers a favorable opportunity for an intensive and comprehensive study of a type of behavior both typical and universal. It is a type which may be satisfactorily studied in the ordinary laboratory animals; and it should yield data of both theoretical and practical importance for many general phases of congenital activity. On this account congenital sexual behavior was subjected to experimental study in this investigation. The study is concerned primarily with the copulatory act of the young male albino rat.

On the whole the rat is a very satisfactory animal for observation in a study of sexual behavior. It thrives well in confinement, becomes sexually mature in a relatively short time (sixty to ninety days), and displays its sexual behavior, under appropriate stimulation, throughout the entire year. The copulatory act of the animal is one that may be studied from many aspects. Its appearance in post-natal life is late enough to make possible a genetic study of its development; and it persists throughout a span of the rat's life sufficiently long to enable one to observe the alterations with age, experience, and modified conditions of the general metabolism.

In this paper an attempt will be made to set forth salient data concerning: (1) The constituent elements of the copulatory pattern of activity and the serial and functional relationships they manifest, and (2) the kinds of external stimuli adequate to initiate the copulatory act and the manner in which their influences are mediated through the receptors. In a subsequent paper data bearing upon internal factors underlying sexual functions will be presented.

2. Previous studies of sexual behavior

It will be of value to consider briefly the experimental studies of sexual behavior of the rat and those of other animals, as well, which have a direct bearing on the present study. This literature will be, therefore, the subject of immediate consideration.

Banta ('14) gives an excellent account of the mating of the wood frog. Any moving object which came within the radius of several feet attracted the male and caused pursuit. Masculine aggression was increased or lessened by the response of the object clasped. The vigorous struggles of a male when clasped by another male were followed by relaxation of the clasp and dismounting of the latter. Frequently several males were seen competing for a single female. If one succeeded in getting into a favorable position on the back of the female and fixed his forelimbs firmly about her he usually succeeded, by vigorous kicks of the hind legs, in forcing other competitors to release their holds. Sexual excitement reached its highest pitch at 52°F. and was almost wholly wanting at 45°F.

The conjugation of the crayfish was studied by Andrews ('14). He made a very detailed analysis of the pattern of copulatory acts. Seventeen distinct elements were discovered. Apparently the first elements of the pattern were set off by purely local stimuli afforded by other crayfish in the immediate presence of the male. Through the amputation of bodily structures the author tried to ascertain to what extent the later elements of the act would appear without complete expression of the preceding elements. It was found that the later elements were carried on with as much precision as the animal's imperfect mechanical equipment, due to the loss of essential bodily structures, permitted. Recognition of sex arose through trial at copulation. The submissive response of the female, as contrasted with the vigorous response of the male, together with the inability to perform the conjugal act on the latter caused the agressor to relax his clasp and to dismount.

The observations of Craig ('14) on the sexual behavior of the dove reared in isolation bring out many interesting points concerning the adequate stimuli for the initiation of the copulatory act. Male doves, prior to sexual experience with female doves, attempted to perform the sexual act upon the hand of the observer. Copulatory attempts were made also upon such inanimate objects as the perch of the cage and the shoe of the observer. When placed with the receptive female very imperfect coördination in mounting attempts were made prior to the successful execution of the first complete act. One male mounted obliquely across the body of the female and failed totally to effect the union. Following the first series of successful consummations of the sexual act a male showed marked alteration in his general behavior, characterized chiefly by increased agressiveness. An irregularity in the pattern of the copulatory act which persisted for many months was noticed in the case of another male. The preliminary billing that usually precedes copulation was entirely omitted by this male.

Kirkam ('13) studied the breeding habits of the rat. His findings indicate that the male, though present at the time of parturition and for several days subsequently, does not kill the offspring. One pair of young rats mated before either was sixty days of age. This seems to be a very unusual instance of sexual precocity. Observations on the copulatory action of brown rats were reported by Miller ('11). He noted that the females in heat did not allow the young males to copulate. Size and ability to dominate were the determining factors. Frequently adult females were seen chasing about the cage small males which had attempted to copulate. When placed with an adult male the same females permitted copulation, offering little or no resistance. He concluded (from evidence that seems quite insufficient) that bodily odors served as a basis for sexual recognition.

A very elaborate study of the sexual behavior of monkeys and baboons was made by Hamilton ('14). The animals were given far more freedom in their place of confinement than is customary in such studies. Thus reactions closely approximating those found in the natural habitat were observed. Some of the outstanding features of this study are: the great amount of sexual intercourse performed by the males, homosexual tendencies of both males and females, and the use of homosexual behavior on the part of the weaker individuals to divert the attacks of larger males in their attempts at copulation. Under normal conditions of life, the young male, the author believes, does not form habits of masturbation.

Motani ('16) describes the dominating behavior of a male chimpanzee directed toward its cohabitant. When intercourse was desired the male hit the floor of the cage with his hand, whereupon the female approached and assumed the copulatory position before him. Copulation during pregnancy was frequent.

Studies dealing with the reproductive glands as the chief organs exerting a regulatory influence over sexual phenomena are now quite numerous. Although the data of these reports are to some degree relevant to those to be reported in this paper, they can be more advantageously considered in connection with the subsequent study which deals with the internal factors underlying sexual behavior. Such literature will, therefore, be reviewed at a later time.

3. Technique—General

The rats were confined in a uniformly heated room. An extra allowance of whole milk was given to the mothers throughout the period of lactation in order to assure favorable conditions for normal development in the suckling young. Approximately half of the litters were weaned when twenty-one days of age; the other half, at the age of twenty-five days. Although the latter gained weight more rapidly, during the subsequent five or ten days, than those weaned at twenty-one days, this difference in rate of gain was only temporary and so small that it was entirely concealed by the individual variations in rate of growth appearing prior to the age of sexual maturity. Males were separated from females at weaning time. All were segregated for observation in wire-mesh cages which rested on one-half inch of wood shavings contained in galvanized-iron trays. When not in observation cages, all animals except those used in experiments involving isolation, were permitted to mingle freely with others of the same sex and of approximately the same age. Individuals kept in isolation were confined in cages of the dimensions 12 by 8 by 8 inches.

Feeding hours were regular and fell daily between 6.00 and 8.00 p.m. The diet, uniform throughout the period of study, consisted of white bread and whole milk served together in small tin containers during six days of the week. On the remaining day a grain mixture (McCallum diet), made into a mash by the addition of water, was substituted for the bread and milk. Once a week fresh cabbage leaves were added to the above diet. Weight records were taken at five-day intervals or oftener in experiments requiring more detailed records of weight.

The animals whose sexual behavior was observed from day to day were removed from the housing cages to individual observation cages (12 by 8 by 8 inches) from four to six hours prior to the time of observation. The purpose of isolating the animals was to allow time for each to become thoroughly cage-adapted, thereby reducing random cage exploration to a minimum during the period of observation.

In general, the method of studying the sexual behavior of the young animals was systematic observation of normal and operated animals in confinement. Special methods and technique employed in particular phases of the work will be described in detail in connection with the report of experimental data in the following sections of the paper.

THE COPULATORY ACT

The constituent elements of the copulatory act are well coordinated at the time of the initial copulation. No more fundamental problem confronts the student of animal behavior than that of discovering the process by which these constituent elements are integrated into a complex unitary pattern. By the elucidation of this process an important step will have been taken in the analysis of the changes through which elementary activities are being combined and dissolved continually in habit formation throughout the lifetime of living organisms.

A preliminary basis for the study of the integrative process may be established by resolving the copulatory pattern into its chief constituent activities and then ascertaining whether these elements appear in isolated or partially isolated function during the prepuberal life of the young animal or whether they appear for the first time as parts of the complex copulatory pattern. This has been done by observing the prepuberal activity of the young male under natural and experimentally controlled situations. The presentation of these data may be simplified somewhat by considering at the outset the copulatory act of the sexually mature and experienced male. This consideration will furnish an adequate basis for the examination of the prepuberal activities and the initial copulatory act of the young male.

1. Copulatory act as seen in the mature rat

One cannot give a detailed description of the male's sexual activities without taking into account the pattern of feminine activity by which it is initiated and toward which it is directed. That is to say, the conjugal act as a whole involves both a masculine and a feminine pattern of activity, the elements of which have a relationship of functional interdependence, and are themselves, acts of a high degree of complexity. The elements of each pattern appear in serial order and are so synchronized that elements of one pattern act in harmony with those of the other. And thus through mutual furtherance they run their courses in the wonderfully complete and well coördinated manner seen in a cross-sectional view of the reproductive process.

When a female is placed into the cage of a *potent* male that has been thoroughly cage-adapted, the latter begins in a few seconds to examine her ano-vaginal zone. Smelling and licking occur irrespective of oestral cycle, and the amount varies with a given individual from day to day. The non-receptive female, thus placed does a variable amount of cage exploration. While doing this, she is followed by the male with his activities directed almost wholly to her. If he is too vigorous in vaginal sniffing, nibbles too deeply in search for parasites, or impedes her progress by playful assaults, she sometimes resists with a back-kick of the hind foot, or by moving into a corner of the cage where she is less accessible to his assaults. The receptive female usually makes a less extensive round of exploration. When approached by the male in the manner described above, she assumes a tense attitude that involves the general musculature of the body and limbs, Frequently this tension is accompanied by slight vibratory movements of the head and shoulders sufficiently general in its distribution to give to the whole body a shivering effect. The shiver may last from one to two seconds. After a moment of smelling or licking, on the part of the male, the receptive female runs forward in the cage a short distance or bruskly shifts her position so that the root of her tail is directly in front of the male. This running movement resembles the quick irregular gallop of the guinea-pig. If the male is sexually aroused by this movement and pursues her across the cage, as is usually the case after she has moved one or more times, she runs only a short distance and halts upon being overtaken by the male and caught in the copulatory clasp. She halts instantaneously when firmly clasped.

Although the responses just described are typical, there are many variations from this type. Some females, persistently, and nearly all observed at the onset of the oestral period, re-act to the vaginal smelling and licking by running wildly about the cage, kicking with the hind foot, and, not infrequently, biting at the pursuer.

In many cases the male begins to copulate within the first minute after the female is placed in the cage. In some instances, however, a variable number of pre-copulatory activities of short duration precedes the act. These activities, for the most part irrelevant in nature, occur in promiscuous order. (They never suggest a type of courting behavior.) The acts appearing most frequently are: nibbling at the head or body of the female,

43

repeated licking and smelling in the ano-vaginal zone, scratching his own body, and momentarily sniffing at the wires of the cage. Rarely, at such times, he stops to lick the penis.   The copulatory act, when it does appear, comes with such definiteness and orderly sequence of elements that it stands out in marked contrast and is never confused with this background of promiscuous activities.   It is an organized unit of behavior that runs its course with great smoothness when the first element of the series is set into action.

In the act of mounting, the male clasps the sides of the female with his fore-paws encircling her body just posterior to the short ribs.   The palmar surfaces of the fore-paws are directed medialward, in a dorso-lateral, lateral, or ventro-lateral position on the female's sides, varying with the size of the female and the girth of the male's fore-limbs.   Simultaneously with the clasp, or following so closely that no time interval is discernible, rapid vibratory palpation of the female's sides occurs.   To the palpatation, the female responds by depressing the back deeply in the lumbar region.   This movement simultaneously elevates the coccygeal region with the hip-joint serving as a fixed pivot of rotation.   The elevation of the coccygeal region raises the external orifice of the vagina from the floor of the cage to a height more accessible to the male, and at the same time, throws the axis of the vagina into a plane conforming with the direction of piston movements of the penis when vaginal entrance is attempted.

Following the palpation of the female's sides, which serves as an adequate stimulus for the depression of her back with consequent elevation of the coccygeal region, the male makes a series of rapid movements of the pelvic region which serve to direct the erect penis into and out of the vagina in piston-like manner.   With the consummation of the orgasm the palpation and piston movements cease and the male throws himself from the position assumed in mounting by a vigorous backward lunge that clears his body from that of the female by a distance of from three to five or more inches.   With the backward lunge he comes to a sitting position with the weight of the body supported on the root of the tail and toes of the hind feet. At the same time the mouth is brought down to the penis which is still in a semi-erect condition. Vigorous licking of the organ follows. After a few seconds licking the organ becomes flacid and is retracted into its sheath and the male comes to a position of rest on all fours where he remains quietly for a short time or engages in minor activities such as nibbling at parasites, scratching the body, or sniffing warily at the female. Smelling and licking in the vaginal zone is seldom repeated after the first two or three copulatory acts.

By way of summarization of the foregoing description the overt pattern of the copulatory act may be broken up into the following elementary acts or groups of activities:

1. Pursuit and mounting.

2. Palpation of the female's sides.

3. On the part of the female, depression of the lumbar region of the back with consequent elevation of the sacro-coccygeal region.

4. Pelvic movements of the male by which the penis is thrust toward the vagina in piston-like manner.

5. Cessation of palpation of female's sides and the backward lunge.

6. Licking of the penis.

The duration of the complete act, from the time of the initial clasp to the backward lunge is usually only a fraction of a second. The time is so brief that one cannot measure it accurately with a stop-watch. In each instance of its appearance, the pattern is strikingly uniform, yet close observation reveals some minor irregularities. A few of these are deserving of mention.

Piston movements by which vaginal entrance is effected do not invariably follow palpation. Sometimes, apparently, this is due to failure in eliciting coccygeal elevation on the part of the female, but again it seems to be due to failure of erection. Evidence of the latter is seen most frequently during a long series of copulations. The number of such instances increases with the number of copulations.

Withdrawal, as a rule, is made immediately upon the consummation of the orgasm, but in the case of males copulating for a somewhat prolonged period of time, there occurs now and then an instance of failure or retarded withdrawal after cessation of piston movements. Prolongations of from one to five seconds have been observed. In these cases the withdrawal is effected by the male alone, with the elimination of the backward lunge, or by the female's jumping forward and slipping from the relaxed clasp of the male. One female on the occasion of tardy withdrawal invariably squealed sharply and turned savagely on the male, striking at his head with her fore-paws or viciously biting at his face and shoulders.

[*Editor's Note*: Material has been omitted at this point.]

GENERAL REMARKS

The scope of this study is limited to an account of the copulatory pattern and its activation. Data bearing on the internal factors which condition the response will be presented in a subsequent paper. In its present stage of completion this investigation leaves many unsolved problems for future consideration, but, nevertheless, it demonstrates clearly the fruitfulness of the experimental method of approach. Certain data considered briefly in the foregoing pages are of sufficient importance because of their bearing on the general subject of congenital behavior to justify greater emphasis than heretofore given.

1. Three fundamental aspects of congenital behavior

From the standpoint of objective psychology, the chief lines of interest in unmodified congenital behavior center in three fundamental aspects of its nature and operation. These lines of interest call for studies of particular patterns of congenital behavior for the purpose of analyzing: (1) The constituent elements and their temporal, serial, and functional relationships; (2) The kinds of stimuli adequate to evoke the unit of response under observation and the manner in which they activate the effectors through the medium of the receptors and the nervous system; and (3) The internal factors operating within the organism to (a) provide the special structures, if any, involved in the response, (b) to integrate these and other structures into mechanisms for coördinated action, and (c) to sensitize the receptor-effector mechanism for response of a definite pattern to stimuli of a special character.

Having analyzed each of these aspects, the experimenter will then be prepared to give not only a detailed account of the initial appearance of the pattern of behavior studied, but also will have acquired the factual background needed to facilitate an examination of the modification it undergoes during a life of exercise in an ever changing environment. Thus can be brought together a valuable fund of scientific information which will be of service to students of psychology and closely related sciences who are concerned with the application of all available facts in the field of native behavior.

2. The copulatory act an hereditary mechanism

The results of this investigation substantiate the generally accepted view that the copulatory response is the action of an hereditary mechanism. No evidence was found supporting the view expressed in a recent paper (Kuo '21), in which, if my interpretation of the article is correct, the author denies the hereditary organization of such mechanisms. He regards them as products of experience. A careful consideration of the experimental data on this subject indicates quite clearly, it seems, that this latter view is not in harmony with the existing facts.

3. Precision in the initial copulatory act

The copulatory act differs quite strikingly from many of the early post-natal responses with respect to the precision of functioning of its elementary components. Greater precision is to be expected, however, because of the more advanced stage of physiological development reached by the animal prior to the first copulatory act. The slight irregularities described in a foregoing section of the paper may be variously explained. In some instances they are probably the result of physiological immaturity of some of the structures involved; in others, due to incomplete integration of the elementary activities of the copulatory unit or to incomplete dissolution of these elements from other distinctly foreign units with which they have functioned previously; or, again, they may arise from the failure of neural arcs of the copulatory act to gain priority over other arcs aroused by stimuli acting simultaneously with the stimulus adequate to evoke the copulatory act.[6] The vigor of the copulatory mechanism in the young male, as measured by the number of complete copulatory attempts made in a given time interval, closely approximates that of the experienced adult. This, however, cannot be taken as a measure of relative precision in the delivery of the spermatazoa and their fluid media to their proper receptacle in the reproductive tract of the female. Data which indicate the rate and range of improvement in this function are not available.

4. Native and modified native elements comprise the copulatory act

Inspection of the constituent activities of the copulatory pattern reveals a point[7] that has oftimes been described in the lit-

[6] Sherrington, C. S., The Integrative Action of the Nervous System, p. 231.
[7] See especially Morgan, C. Lloyd: Animal Behavior, 1901.

erature on delayed responses. The act is a compound of both native and modified native or acquired responses. The latter have been exercised in connection with other units of response and habitual performances from the time of their initial appearance to the time of their incorporation in the copulatory pattern. It is important to note, however that these elementary acts exercised in prepuberal life were not substitutes for or abbreviated and playful types of the copulatory act. The evidence does not show that their early exercise was, in some mysterious manner, pointed toward acts of greater precision in the reproductive function. Such activities as pursuit of other rats, playful clasping, wrestling, vaginal smelling, and nibbling have no demonstrable connection with the copulatory act prior to their incorporation in this act at or near the age of physiological sexual maturity. The function of penis licking is still a matter of conjecture. Though it may sometimes be shown to be a form of masturbation or in other respects connected with the process of tumescence or detumescence, there is no evidence at the present time which links it with the activation of the copulatory act. Palpation of the female's sides, piston movements of the pelvis, and the backward lunge characteristic of the copulatory act are activities which have never been observed prior to the first real copulatory attempts.

The evidence from this study harmonises, on the whole, with the view that the elements of the copulatory act are integrated into a functional unit of response at a rather definite period of the rat's life. Prior to this integration they do not function in isolation or in smaller units of action which are the anlage of the copulatory act. This point is of special interest because of its bearing on the sexual theories of infancy promulgated chiefly by the Freudian School. Because of the relation such facts have to this interesting subject, it is desirable to extend the present osbservations to other animals and to the human infant as well to determine to what degree these findings are common. In this way data will be made accessible to test the correctness and appraise the value of these theories.

5. The adequate activating stimuli

Analysis of the activation of the copulatory act is not sufficiently advanced, as yet, to enable one to define accurately the stimuli adequate to evoke the copulatory response on the first occasion of its appearance. It may be said, however, on the basis of available data, that the adequate stimuli are probably both mechanical and thermal in nature, and that they are almost certainly mediated through the cutaneous receptors and the receptors of deep sensibility. In some way the movements of the female ordinarily supply the stimuli which under the proper conditions initiate the act. The effectiveness of these stimuli is assured, as will be shown later, only after certain internal factors have done their work toward preparing the receptor effector mechanism for response to this particular pattern of stimulation.

6. Fundamental need of more experimental data on congenital behavior

Heretofore no special comment as to the bearing this study has on the current discussions of "instinct" has been given. The use of this term is unnecessary in the exposition of experimental data; furthermore, its use is undesirable in this connection because of the difficulty in so defining the term as to disentangle it from its many obsolete and unscientific attachments.

It must be obvious to all who have reviewed recent discussions of the term "instinct" that the chief cause of our inability to elucidate the field of congenital activity cannot be attributed primarily, or even in large measure, to the general *abuse* of the term, "instinct," or any of its equivalents. Neither is it due primarily to faulty interpretations of the past or deprivation of plausible interpretations just recently brought forward. The cause is far more fundamental. The difficulties cannot be obviated by discussion alone. They must be overcome through experimentation. The paramount need at the present time is a broader factual acquaintance with the detailed nature and operation of congenital behavior mechanisms. It is a need which must be met by carefully planned and executed studies of native behavior designed to give facts making possible a complete account of patterns of congenital behavior, their activation, their internal conditioning factors, and finally their modifications through exercise. Results of this kind should be a reliable guide in determining what classificatory terms must be discarded, re-defined, shorn of their unscientific encumberances, or enriched by further interpretation.

CONCLUSIONS

The data herein presented seem to justify the following conclusions:

1. On the whole the evidence brought forward indicates that the chief elements of the primary reproductive act appear within a short period of time at the age of puberty. Activities such as smelling the vagina, pursuit and playful clasping seen in wrestling have no definite connection with the copulatory act prior to the age of sexual maturity. At that time the function of smell in the initiation of the act is wholly negligible. Licking the penis appears in prepubertal life at about the forty-fifth day and probably serves the same function thereafter when acting in isolation as when it is incorporated in the copulatory pattern. Palpation of the female's sides, pelvic movements, and the backward lunge are elements of the act which have no exercise in prepubertal life.

2. The *overt* pattern of the sexual act consists of the following elements: (*a*) Pursuit and mountings; (*b*) palpation of female's sides. (*c*) depression of lumber region of back, on the part of the female, with the consequent elevation of the sacro-coccygeal region; (*d*) pelvic movements of the male by which the penis is thrust in piston-like action toward the vagina; (*e*) cessation of palpation of female's sides and the backward lunge; (*f*) licking the penis.

3. Differences between the initial copulatory act and the act as modified by age and experience are found. Minor differences, due to purely physical limitations of size and strength, are found in the location of the fore-paws on the female's sides, vigor of

51

palpation of female's sides, and reduction of force of backward lunge. More important differences are: Increase of aggressiveness with sexual experience; incomplete patterns of the act involving omission of the essential elements such as palpation, pelvic movements and backward lunge; failure to lick the penis after copulation; and mounting in a manner that will not permit the completion of the full copulatory act, i.e., mounting at head.

4. The age at which the initial copulatory act appears is related to general physical development. The earliest age at which the act was initiated in this study was sixty-four days. It does not appear prior to somatic maturity. On the other hand, somatic maturity is not a reliable indication of copulatory ability.

5. The number of copulations following the initial experience compares favorably with that of adult males. The usual number during thirty minutes subsequent to the first act is from twenty-five to thirty-five. The greatest number observed in this study was 53 during the first twenty minutes.

6. The movements of a receptive female appear to supply the adequate stimuli for the initiation of the copulatory act. Movements of adult rats that simulated those of a female in heat brought out sexual acts, complete in pattern from the male's standpoint, in the case of two males that had not had sexual experience. The quick movements of a guinea pig served to arouse great sexual excitement in a young male. Excitement was accompanied by mounting attempts, but no complete acts.

7. Sensory controls reveal the following:

a. Visual stimuli are not essential to the initiation of the copulatory act. No visible effects in their absence are seen in the nature of the pattern, the pursuit tendency in mounting, the age of first copulation, or the rate at which copulations may take place.

b. Olfactory stimuli are not necessary for the arousal of sexual excitement or the initiation of the copulatory act. Their absence neither retards the age at which the act first appears, or modifies the sexual drive as manifested by eagerness to copulate and ability to copulate for a prolonged period of time.

c. Elimination of visual, olfactory, gustatory sense organs and the vibrissac prior to sexual experience produces no changes in the sexual act, either in the pattern, the age at which it appears, or the responsiveness to stimulation of a female in heat. Substitute action of the visual, olfactory, or gustatory senses is, therefore, improbable.

d. Partial elimination of auditory sensitivity in blind and anosmic rats after sexual experience makes no change in the ease or arousing the pairing impulse. Experiments of this evidence that audition is an essential receptor for the initiation of the copulatory act in the rat.

e. The lack of evidence for any important function of each of the other primary senses in the initiation of the copulatory act points to cutaneous and receptors of deep sensibility as the primary organs that mediates the exteroceptive stimuli by which the act is initiated.

REFERENCES

ANDREWS, E. A.: Conjugation in the crayfish. Jour. Exp. Zool., 1914, 9, 235–265.

BANTA, ARTHUR M.: Sex recognition and the mating of the wood frog. Biol. Bull., 1914, 26, 171–184.

VON BECHTREW, W.: Die Funktionen der Nervencentra. 1908, 3, 1.

CRAIG, WALLACE: Behavior of young birds breaking out of the egg. Jour. of Animal Behav., 1913, 2, 296.

CRAIG, WALLACE: Male doves reared in isolation. Jour. of Animal Behav., 1914, 4, 121–133.

DARWIN, CHARLES: Descent of Man. Popular Edition. London, 1901.

ELLIS, HAVELOCK: Studies in the Psychology of Sex, 3, 4.

GROOS, KARL: The Play of Animals. 1898.

HAMILTON, G. V.: A study of sexual tendencies in monkeys and baboons. Jour. of Animal Behav., 1914, 4, 295–318.

HEWER, EVELYN E.: The effect of thymus feeding on the activity of the reproductive organs in the rat. Jour. of Physiol., 47, 479–490.

KIRKHAM, WM. B.: The breeding habits, maturation of eggs and ovulation of the albino rat. Amer. Jour. of Anat., 1913, 15, 291–317.

KUO, ZING YANG: Giving up instincts in psychology. Jour. of Philosophy, 1921, 18, no. 24, 645.

McDOUGAL, WM.: An Introduction to Social Psychology. Sixth Ed. 1912.

MILLER, NEWTON: Reproduction in the brown rat. Amer. Natural., 1911, 45, 623–635.

MONTANE, LOUIS: A Cuban Chimpanzee. Jour. of Animal Behav., 1916, 6, 330–333.

SHERRINGTON, C. S.: The Integrative Action of the Nervous System, 1906.

SMALL, W. S.: Notes on the psychic development of the white rat. Amer. Jour. of Psychol., 1899, 2.

STEINACH, E.: Untersuchungen zur Vergleichenden Physiologie der Männlichen Geschlechtsorgans. Pflüger's Archiv., 1894, 56.

STEINACH, E.: Geschlectstrieb und echt sekundäre Geschlechtsmerkmale als Folge der inneresekretorischen Funktion der Keimdrüsen. Zentralbl. f. Physiolog., 1910, Bd. 24.

STEINACH, E.: Willkürliche Umandlung von Säugetiermännchen in Tiere mit ausgeprägt weiblichen Geschlechtscharakteren der Keim drüsen. Archiv. f. d. ges. Physiol., 1912, Bd. 144.

STEINACH, E.: Feminierung von Männchen und Maskulierung von Weibchen. Zentralbl. f. Physiol., 1913, 27, 717-723.

4

Copyright © 1956 by the Society for Experimental Biology and Medicine

Reprinted from *Quart. J. Exp. Psychol.* **8**:121–133 (1956)

SEXUAL EXHAUSTION AND RECOVERY IN THE MALE RAT*

BY

FRANK A. BEACH and LISBETH JORDAN
From Yale University

Synopsis

Twelve male rats were left with receptive females and allowed to copulate and ejaculate until they reached a criterion of "sexual exhaustion." They were then retested after 1, 3, 6 and 15 days of sexual inactivity. Following these observations males were tested once each day or once every other day and allowed to achieve a single ejaculation.

In the course of a period of unlimited access to the receptive female males usually need approximately 10 intromissions to produce the initial ejaculation, but successive ejaculations are produced by fewer and fewer intromissions. The time to recover from the effects of an ejaculation increases progressively as exhaustion is approached.

Very few animals copulate when tested 24 hours after sexual exhaustion. Considerably more recovery is evident in tests conducted after a 3-day rest, but it is not complete and rats are not capable of achieving as many ejaculations as they tend to achieve after longer periods of inactivity. As measured by ejaculation-frequency, the curve of sexual recovery is negatively accelerated and probably reaches asymptote after 7 to 10 days of rest.

Various other measures in addition to ejaculation-frequency support this conclusion.

Males allowed to ejaculate once each day or every other day are somewhat less responsive than fully rested animals, but do not show any progressive loss in sexual excitability or capacity.

A working hypothesis is proposed to explain most of the findings. It postulates the existence of an *Arousal Mechanism* which is distinct from a *Copulatory Mechanism*. The ways in which these hypothetical mechanisms are affected by sexual performance and sexual rest are discussed.

Introduction

Theoretical Background

Sex is commonly listed with hunger and thirst as one of the "primary drives" (Miller & Dollard, 1941), but it has been subjected to only limited systematic analysis. Psychologists who employ this particular classification presumably do so because they believe that when it *is* investigated "sex drive" will be found to behave in more or less the same manner as hunger and thirst. There are several reasons to question such an assumption. In any event, the term "drive" has acquired so many different meanings and developed so many conflicting connotations that it is of questionable value as a scientific concept. We use it in this introduction in conformation with current fashion, but we do not regard the present investigation as a study of sex "drive." It constitutes an examination of certain factors influencing sexual arousal and copulatory performance.

In the first place, hunger and thirst are related to conditions of tissue need. Sex is not. The notion that tension and pressure in the primary or accessory reproductive apparatus is a source of sex drive is challengable. Male rats deprived of testes, prostate gland and seminal vesicles copulate normally when treated with androgen (Beach, 1947). Females of the same species in which ovaries, tubes, uterus and vagina are congenitally absent execute the mating pattern under the influence of exogenous ovarian hormones (Beach, 1945).

* This study was supported in part by a research grant (M-943) from the National Institutes of Mental Health, Public Health Service. We are indebted to Allan Goldstein for assistance in the statistical analysis of results.

In the second place, sex differs from hunger and thirst in terms of its biological or evolutionary significance. The latter two can appropriately be designated as drives essential to the preservation of the individual. Sex is logically classified as a drive essential to the preservation of the species. In this respect sex is more closely related to maternal drive. Failure to react appropriately to hunger or thirst can have fatal consequences for the organism. Absence of sex drive, or failure to react to it when it is present does not impair the well-being of the individual, although if such a condition were widespread the existence of the species would be imperilled.

Inasmuch as these two types of drive have quite different biological functions, they may well have quite different evolutionary histories. At any rate there is no justification for the *a priori* assumption that sex drive obeys the same laws and has the same characteristics as hunger and thirst. It does not involve a state of physiological need or metabolic deficit. Its "satiation," or better, its exhaustion, is a catabolic rather than an anabolic process, resting upon the expenditure of energy with no resultant "return" to the individual. Satiation of hunger or thirst restores the organism to an optimal condition. The end-point of sexual exhaustion is fatigue.

A period of deprivation following satiation of the hunger or thirst drives is marked by gradually increasing tissue needs which, if not relieved, eventually result in death. The sequel to sexual exhaustion is a period of recovery, and during a protracted term of deprivation the strength of sexual motivation or capacity first rises and then probably stabilizes at some point near the maximal level. There is some reason to believe that if deprivation is continued for a long enough time sexual responsiveness tends to decrease. but this is speculative.

Earlier Studies and Purpose of the Present Experiment

Previous experiments. As indicated above, systematic studies of sex drive in normal animals have not been numerous. In particular the nature of sexual exhaustion and subsequent recovery of sexual capacity have received very little attention. Casual observations and naturalistic descriptions make it clear that pronounced species differences exist. But even for the most frequently studied animal—the domestic rat, the evidence is incomplete and inconclusive.

The most widely cited experiment of the male rat is that of Warner (1927), who used the Columbia Obstruction Box to measure drive strength. Warner's conclusions were as follows:

> The male sex drive appeared to be at its lowest point immediately after a two-hour period during which mating had taken place freely. Recovery was rapid during the first six hours and almost as rapid during the succeeding six. By 24 hours the tendency of males to cross the grid to the female had reached its high point. During the following 6 days there was apparently little change, but what change there was in the direction of a reduction in the strength of this tendency (p. 159).

More recent investigations lead us to conclude that Warner was not measuring sex drive at all, or at least that observation of *copulatory performance* would have shown that recovery is much slower than his conclusions suggest. For example, Schwartz (1955) trained male rats to press a bar in order to obtain access to a receptive female. Rate of pressing on a variable interval schedule was taken as one measure of sexual responsiveness. An independent measure was the copulatory behaviour. Males were sexually exhausted in the apparatus and retested 1, 3 or 6 days later. Consummatory measures showed that the curve of recovery is negatively accelerated and approaches asymptote approximately 6 days after satiation. Rate of bar-pressing proved an insensitive measure of sexual recovery, but was signicantly correlated with changes taking place in the consummatory response during the course of exhaustion.

Purpose. The purpose of the present investigation was twofold: (1) to measure the course of sexual exhaustion in fully-rested male rats, and (2) to measure the degree of sexual recovery at different intervals after exhaustion. The study was conceived as a normative one which would provide empirical information essential to the planning of subsequent experiments.

SUBJECTS, APPARATUS AND PROCEDURE

Selection and Maintenance of Subjects

Experimental subjects were 12 male rats approximately 150 days old at the beginning of experimentation. Both hooded and albino types were included. All males had been born and reared in our laboratory. Sexually active males were selected on the basis of their mating performance in three or four preliminary tests with receptive females. All animals copulated in the preliminary tests, and 10 males ejaculated. Males were housed individually and were maintained in the experimental room in which the light-dark cycle was controlled by an electric clock. Drinking water and Purina chow were available *ad libitum*, and escarole was added twice weekly.

Apparatus

The apparatus included special observation cages and a recording device. The cages were semicircular, having a straight, glass front, sheet-metal sides and back, and a plexiglass top. The cages were 20 inches wide at the front, 12 inches deep, and 10 inches high. The top was hinged to permit introduction and removal of the male. The rear half of the top was surmounted by a cylindrical release can in which a receptive female was placed just before the test. The bottom of the can was hinged and could be dropped by raising a counterweight.

Three observation cages were fastened to a rack in such a position that they could be watched simultaneously by an observer seated approximately 6 feet away. At the observer's side were three counterweights controlling the trap-floors of the release cans.

The observer also operated three microswitches manually, each switch activating a separate marker which recorded on waxed paper pulled by a constant speed kymograph. One inch of paper passed under the markers every 12·6 seconds. The observer recorded different types of sexual responses made by the subjects by closing the microswitch for varying lengths of time. With this arrangement it was possible to observe and record three mating tests simultaneously. Later, when the kymograph records were analysed, the frequency and timing of response could be determined.

Testing Procedure

Incentive females were brought into heat by injections of estrogen and progesterone and their receptivity was pretested by noting their reactions to a non-experimental male. The dosages and timing of injections were those used in earlier studies (Beach & Holz-Tucker, 1949). Only fully receptive females were used in the experimental tests. No attempt was made to pair each male with the same female in different tests.

All testing was done during the dark phase of the 24-hour cycle. Before the start of a test a receptive female was put in the release can and the experimental male was placed in the observation cage and allowed 5 to 15 minutes for cage adaptation. The longer period was used in early tests, but as the experiment progressed shorter periods of adaptation proved sufficient. At the end of the adaptation period the kymograph motor was started and the incentive female was dropped into the cage by lifting the counterweight.

Records made during the test included the number of intromissions, incomplete attempts at intromission, and ejaculations. The time at which each event occurred was automatically indicated by virtue of the fact that the recording paper moved at a constant speed.

All tests were continued until the criterion of sexual exhaustion was met. A male was judged to be exhausted when he allowed 30 minutes to elapse without mounting the incentive female. Males that failed to mount at least once during the first 30 minutes of a test were removed from the cage and scored as negative for that day.

57

Scheduling of Tests

Measuring exhaustion and recovery. The original design was based upon Schwartz's finding that 6 days of rest yielded full recovery from sexual exhaustion in male rats. Accordingly, 3 tests were scheduled to occur 6 days after complete exhaustion. One of these tests was followed by a retest after a 24-hour interval. Another was followed by a test conducted after a rest of 3 days. Tests immediately preceding the 1-day and 3-day tests were in turn preceded by 6 days of rest.

Results obtained in the course of experimentation led us to believe that sexual recovery was not complete in our animals after a 6-day rest. Therefore when the randomized schedule was completed 11 rats were given a rest of 15 days and then retested. Since a few animals had not ejaculated in the test preceding the 15-day rest, the recovery period actually varied from 15 to 21 days in different individuals. Six rats were tested again 1 day later. If no copulatory behaviour occurred they were allowed another 2 days of rest and tested once more a total of 3 days after the 15-day test. The remaining animals were tested 3 days after the 15-day test.

Measuring level of performance at shorter intervals. Beginning from 9 to 16 days after the final exhaustion test 5 males were tested for sexual activity once a day for 13 days. These tests lasted 15 minutes from the first mount or 15 minutes from the introduction of the female if no mounting occurred. Most males of our strain will achieve at least one ejaculation within 15 minutes from the start of copulation if they copulate at all. Six other males plus 3 additional experienced copulators were given 11 or 12 similar tests at 48-hour intervals, i.e. every other day.

RESULTS

The Course of Exhaustion in Fully Rested Males

Because of evidence to be presented below we concluded that males which had not engaged in sexual activity for 15 days could be regarded as fully rested and capable of maximal copulatory performance. Therefore behaviour in the 15-or-more-day-tests was analysed to study the course of sexual exhaustion.

Eleven males were tested after a rest of 15–21 days. In this test 10 animals ejaculated 5 or more times. One rat failed to copulate in the first 30 minutes of the 15-day test and was also negative at 16 days. On the eighteenth day he ejaculated 7 times. The reasons for such failures are obscure, but it is not uncommon for a vigorous male to skip a test occasionally.

Gross measures of sexual capacity. In fully rested males the time needed to reach sexual satiation varied from 61·5 to 141·5 minutes. The mean duration of the test from the introduction of the receptive female to the occurrence of the male's final ejaculation was 89·2 minutes (Median = 82·8). Subtracting from this total the time elapsing before the male first began to mate (*mount latency*), and also the periods of sexual inactivity following all but the terminal ejaculation (*refractory periods*), we find that the average number of minutes spent in actual copulatory activity was 36·7 (Median = 23·1). The range was from 17·4 to 89·7 minutes.

In terms of ejaculatory performance the criterion of sexual exhaustion was reached after a mean number of 6·9 ejaculations (Median = 7·0). Individual variation was considerable and the total number of ejaculations ranged from 5 to 10 for different males. In order to achieve this number of ejaculations the subjects effected intromission an average of 41·7 times (Median = 39·0) and the range was 24 to 67.

In the case of 8 of the 11 males the final ejaculation terminated all sexual activity. Only 5 per cent. of all intromissions and 18 per cent. of all incomplete mounts occurred after the terminal ejaculation. These scores were contributed by 3 rats which resumed copulating but did not ejaculate again and finally met the criterion by failing to mount the female for a period of 30 minutes. These results gave us confidence in our selection of the criterion for sexual exhaustion. Apparently if a previously active male ceases copulating for 30 minutes he would be unlikely to attain another ejaculation in that test even if left with the female for a longer period of time. One or more ejaculations occurred in 57 of all the exhaustion tests, and in 53 (90 per cent.) of these tests all mounting activity was terminated by the final ejaculation.

The average delay between the introduction of the receptive female and the male's first mounting response (*mount latency*) was 74·0 seconds (Median = 8·0). The mean was heavily influenced by negative tests for which an arbitrary mount latency of 1,800 seconds (30 minutes) was assigned. Since there were only 2 negative tests this practice did not affect the median. When negative tests were excluded, the range in individual mount latencies was found to be 2 to 580 seconds.

Mount latency applies to the first mount whether it be a completed copulation with intromission or an abortive copulatory attempt. *Intromission latency* refers to the time between introduction of the incentive female and the occurrence of the first complete copulation. In fully rested males this measure had an average value of 91·3 seconds when negative tests were included (Median = 12·0). Counting only positive tests the range of intromission latencies was 2 to 750 seconds.

Behavioural changes as exhaustion approaches. To analyse the nature of sexual exhaustion we studied the changes in behaviour preceding each successive ejaculation up to the point at which the criterion was met. As noted above, only the records of males rested for 15 or more days were used.

Some of the results are summarized in Table I, which represents the situation up to and including the sixth ejaculation. So few animals ejaculated more than 6 times that group scores are not represenative past this point. The average number of intromissions preceding the first ejaculation was 10·64 and for succeeding ejaculations this value decreased with the result that the sixth ejaculation was preceded by an average of only 4·10 intromissions.

TABLE I

CHANGES IN PERFORMANCE ASSOCIATED WITH SUCCESSIVE EJACULATIONS

Serial number of ejaculation	Intromissions to produce ejaculation	Seconds to achieve ejaculation	Seconds to recover and resume mating
1	10·64	450	324
2	6·00	216	395
3	5·73	198	468
4	5.09	150	495
5	5·09	210	597
6	4·10	132	818

Time scores shown in Table I are based upon group medians. *Ejaculation latency*, or the time from the first mount to the occurrence of the ejaculation was 450 seconds in the case of the first, and 216 seconds in the case of the second ejaculation. There was some tendency for further decrease with successive ejaculations. There was, at the same time, a clear-cut increase in the refractory periods that followed successive ejaculations. Lengthening of the refractory period is a premonitory sign of sexual exhaustion. For example, in the one male which ejaculated 10 times, successive refractory periods were as follows: 330, 389, 432, 432, 244, 364, 465, 881 and 1,082 seconds. The tenth ejaculation was, of course, followed by an unmeasured refractory period which exceeded 30 minutes.

The Course of Recovery from Sexual Exhaustion

It will be recalled that males were tested at 1, 3, 6 and 15 days after sexual exhaustion. Sometimes a male failed to copulate and was retested at some interval not included in this series. Our analysis of the following behavioural data has been confined to results obtained on 70 tests conducted at the stated intervals. Seventeen other tests which occurred at 7, 8, 9, 10, 12, 18 or 21 days have not been used in the statistical treatment although results from them are represented in some of the figures.

Changes in the test as a whole. The percentage of tests in which each type of sexual response occurred at least once after various recovery periods is shown in Table II. Relatively little recovery appears to have taken place during the first 24 hours after exhaustion and the curve is still rising after 3 days of rest. From these data one cannot determine whether full recovery has been achieved at the end of 6 days because the number of 15-day tests is too small to yield a stable score. Other findings to be presented below show that 6 days of rest does not produce complete sexual recovery.

TABLE II

PERCENTAGE OF TESTS IN WHICH EACH TYPE OF RESPONSE OCCURRED AT
DIFFERENT POINTS ON THE CURVE OF SEXUAL RECOVERY

	Days since sexual exhaustion			
	1	3	6	15
Total number of tests	16	23	25	6
Per cent. with no mating activity	69	26	4	17
Per cent. with attempted copulation only ..	25	0	8	0
Per cent. with ejaculation	6	74	88	83

Ejaculation was achieved by only one male in the test conducted 24 hours after exhaustion, and this animal ejaculated but once. For those rats that reached ejaculation in 3-day tests, the median number of ejaculations to exhaustion was 3·0 (R = 1–5). At 6 days this value was 5·0 (R = 4–6), and after a 15-day rest the median number of ejaculations preceding exhaustion was 6·0 (R = 5–8). Figure 1 illustrates the number of ejaculations occurring in all tests including those that took place at intervals other than the 4 standard ones. A curve drawn through the median values for days 1, 3, 6 and 15 would show that sexual recovery is a negatively accelerated function and is probably complete at some point between the tenth and fifteenth days of rest.

TABLE III

MEAN NUMBER OF COMPLETE AND INCOMPLETE COPULATORY RESPONSES
NECESSARY TO PRODUCE EJACULATION

Days since sexual exhaustion	Ejaculations	Copulations per ejaculation	Incomplete copulations per ejaculation	Combined
3	58	8·19	9·10	17·29
6	109	6·20	7·14	13·34
15	33	6·27	5·97	12·24

FIGURE 1

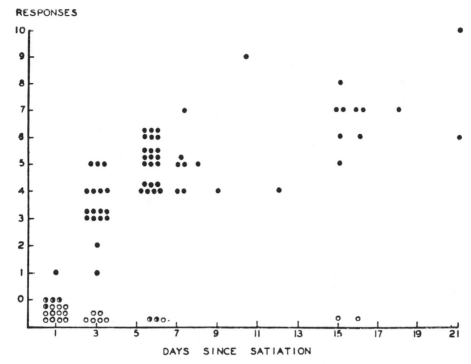

Number of mounting responses or ejaculations to produce exhaustion in tests occurring after varying periods of rest. (Hollow circles represent no copulatory activity. Half-filled circles represent copulation but no ejaculation. Solid dots show the number of ejaculations occurring before the criterion of exhaustion was reached)

In general it appeared more difficult for partially rested males to achieve ejaculation than for other males which were more nearly recovered from sexual exhaustion. This difference is brought out in Table III. Animals that had rested only 3 days had to execute more copulatory responses to attain ejaculation than did males with 6 or 15 days of rest. Differences between the 6-day and 15-day tests were not pronounced, a finding which corroborates the conclusion that the curve of sexual recovery is negatively accelerated and that recovery is almost though not entirely complete at 6 days.

TABLE IV

AMOUNT OF ACTIVITY AND LENGTH OF TIME NECESSARY TO PRODUCE SEXUAL
EXHAUSTION AFTER DIFFERENT PERIODS OF SEXUAL REST

From start of test to final ejaculation	Days since preceding exhaustion		
	3	6	15
Median Number of Minutes	58	60	87
Median Number of Intromissions..	22	29	39
Median Number of Incomplete Copulations ..	22	17	30

Since the number of ejaculations tended to vary with the number of days of recovery it is to be expected that the total amount of sexual activity necessary to produce exhaustion would vary in like manner. This proved to be the case as is shown in Table IV. The number of intromissions and for the most part, of incomplete copulations preceding exhaustion increased in males tested after 3, 6 and 15 days of sexual inactivity. The length of time during which males remained sexually reactive increased similarly.

The median *intromission latency* (i.e. delay before the first completed copulation) in 1-day tests was 3,600 seconds. Comparable scores were 623 seconds for 3-day tests, 135 seconds for 6-day tests and 18 seconds for tests preceded by 15 days of sexual rest. This measure appears to be quite sensitive to sexual exhaustion and recovery. If we compare median *mount latencies* in tests conducted after different periods of rest the results are as shown in Table V. It will be recalled that mount latency refers to the first copulatory response regardless of whether intromission is achieved. Mount latencies like intromission latencies are clearly affected by the degree of recovery from a session of unlimited sexual activity.

TABLE V

EFFECTS OF VARYING PERIODS OF SEXUAL INACTIVITY UPON DELAY PRECEDING
THE INITIATION OF SEXUAL CONTACT

	Days since sexual exhaustion				
	1	3	6	15	Grand median
Median Mount Latency in Seconds ..	1,800	429	30	33	116·5
Per cent. of Tests Above Grand Median	81	65	32	33	—
Per cent. of Tests Below Grand Median	19	35	68	67	—

Changes occurring as the test progressed. The course of sexual exhaustion after different periods of rest is reflected in Figure 2, which shows the time from the beginning of the test at which successive ejaculations occurred. The shape of the curves obtained after 3, 6 or 15 days of rest is similar. The positive acceleration reveals that successive ejaculations were spaced farther and farther apart as the exhaustion point was approached. The overall rate of decline is seen to be a function of the degree of recovery at the start of the test.

The function represented in Figure 2 is actually a combination of two measures: plug latencies and refractory periods. These are plotted separately in Figures 3 and 4. Here it is shown that refractory periods following successive ejaculations increased progressively for all groups and the rate of increase was slowest in males rested for 15 days and most rapid in rats tested after only 3 days of rest.

The curves for ejaculation latencies tell a different story. Time to achieve the first ejaculation was inversely related to duration of the rest period preceding the test. However, for all groups the remaining ejaculations were achieved in shorter times and no intragroup differences are evident. This was due to two factors. The number of complete and incomplete copulations preceding the first ejaculation was higher than that preceding subsequent ejaculations for all 3 groups, and the speed with which one mounting response followed another was increased after the initial ejaculation had been achieved.

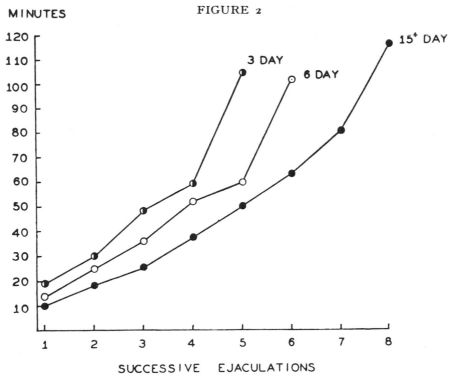

FIGURE 2

Time from beginning of test at which successive ejaculations occurred

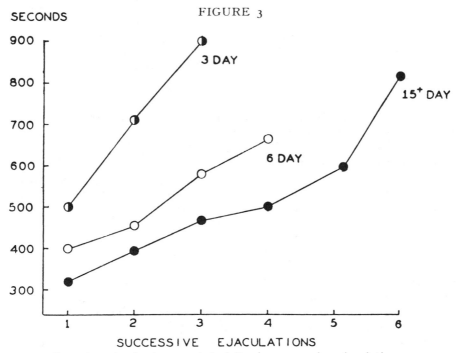

FIGURE 3

Duration of refractory periods following successive ejaculations

Short interval tests. It will be recalled that 9 rats were given 15-minute tests at intervals of every other day, and 5 males were tested once every 24 hours. Ejaculation occurred in 78 per cent. of the 107 tests conducted at 48-hour intervals and in 77 per cent. of the 65 tests given once a day. Ejaculation was present in only 83 per cent. of the tests made after a 15-day rest.

FIGURE 4

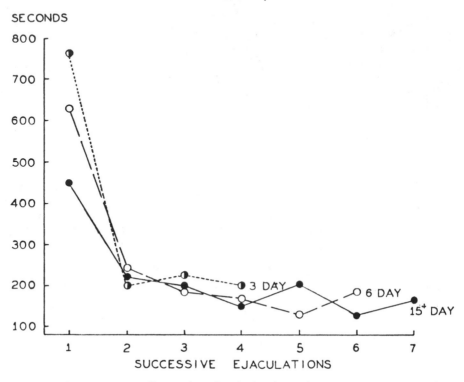

Successive ejaculation latencies

Median mount latency was 10 seconds on the 48-hour tests and 14 seconds on the 24-hour tests. In the 15-day tests the mount latency was 8 seconds. Intromission latency had a median value of 34 seconds for the 1-day, 17 seconds for the 2-day, and 12 seconds for the 15-day tests. The average number of intromissions to achieve one ejaculation was 10·4 in tests given daily, 11·4 for tests occurring on alternate days, and 10·6 in 15-day tests. Median ejaculation latency was 343 seconds in daily tests, 410 seconds in tests given every other day, and 450 seconds in tests preceded by a 15-day rest. Continuous graphs of sexual performance on successive short-interval tests reveal no consistent trends of behavioural change. The records of rats tested every day are somewhat more variable than those of males tested on alternate days, but there are no indications of progressive or cumulative sexual fatigue.

Considered in combination these findings present a reasonably clear picture. Male rats allowed to ejaculate once a day maintain sufficient sexual responsiveness to continue at a constant level of mating activity for considerable periods of time. Nevertheless the motivational level is probably below the maximum. Animals tested every other day have slightly greater sexual capacity but are not as responsive as fully rested males.

DISCUSSION

In an earlier publication the present writer postulated the existence of two physiological mechanisms involved in masculine sexual behaviour (Beach, 1955). The arousal mechanism (*AM*) mediates the awakening and increase of sexual excitement which eventuates in copulation. When the copulatory threshold is reached mounting and intromission occur and these responses are mediated by the copulatory-ejaculatory mechanism (*CM*). Coition produces a qualitative change in the sensory input resulting from genital stimulation. Ejaculation is the end product of a series of appropriately spaced intromissions.

Following an ejaculation the *AM* becomes temporarily unresponsive. As the *AM* regains its reactivity the male begins again to react to sexual stimulation; arousal occurs and the *CM* is once more thrown into action.

Effects of Sexual Exhaustion and Rest upon the AM

Sexual exhaustion. The two indicators of excitability of the *AM* are the intromission latency and the post-ejaculatory refractory periods. Changes in the latter measure as shown in Table I reflect the gradual change occurring in the *AM* during a period of unlimited coital behaviour. The duration of successive refractory periods is progressively increased. This is interpreted to mean that a series of ejaculations has a cumulatively inhibiting effect upon the responsiveness of the *AM*.

The length of time needed for recovery of excitability was increased by each successive ejaculation until finally this interval exceeded 30 minutes and the test was terminated. It is the ejaculation rather than total intromissions which depresses the *AM* because this effect is independent of the number of intromissions needed to produce orgasm.

Sexual rest. Failure to copulate on any given occasion may be due to a variety of reasons and excitability of the *AM* is certainly influenced by factors other than preceding ejaculations. Nevertheless the high proportion of negative tests occurring 24 hours after sexual exhaustion and the still appreciable number after a 3-day rest undoubtedly reflect in part submaximal recovery of the *AM* in some individuals.

More clear-cut evidence is provided by the median mount latencies recorded in positive tests occurring after 1, 3, 6 or 15 days of sexual rest. The respective scores in seconds were 1,800, 429, 30 and 33. An even more sensitive index to the state of the *AM* is the intromission latency. Median values for this measure were 3,600, 623, 135 and 18 seconds. It seems clear that recovery of excitability in the *AM* following sexual exhaustion is a negatively accelerated function and that a maximal state of responsiveness is probably not attained in less than one week.

Measurements of the rapidity with which males recover from the effects of successive ejaculations also reflect the effects of sexual rest upon the *AM*. The median time to resume copulating after the first ejaculation was 504 seconds in tests occurring after a 3-day rest, 397 seconds in 6-day tests and 324 seconds in tests following a rest of 15 or more days. The rate at which the duration of subsequent refractory periods increased was greatest in 3-day tests and slowest in 15-or-more-day tests as can be seen in Figure 3.

Excitability of the *AM* appears to be held slightly below maximum if males are allowed a single ejaculation every 48 hours, and the degree of depression is a bit greater if ejaculations are spaced 24 hours apart. Males that ejaculated once every 24 hours had a median intromission latency of 34 seconds. For those that experienced one ejaculation every other day this figure was 17 seconds, and males with a 15-day rest achieved intromission in 12 seconds.

Effects of Sexual Exhaustion and Rest upon the CM

Sexual exhaustion. Changes in the functional state of the *CM* are reflected in the amount of genital stimulation necessary to produce ejaculation and in the ejaculation latency. Present findings suggest that the first factor is reduced as a result of successive ejaculations. This "sensitizing" effect on the *CM* is most noticeable following the initial ejaculation. As shown in Table I, more than 10 intromissions preceded the first ejaculation in fully rested males, whereas the second orgasm occurred after only 6 such responses. Further reduction in intromissions necessary to produce succeeding ejaculations is indicated by the tabular data.

As measured in this fashion the excitability of the *CM* seems to remain high until the male ceases to make any sexual responses due to extreme loss of reactivity in the *AM*. One male ejaculated 10 times before reaching the criterion of sexual exhaustion. Each of the last two ejaculations was produced by 2 intromissions.

Table I also reveals that the length of time needed to reach ejaculation tends to decrease as the test progresses. This is, of course, due in part to the reduction in intromission frequency, but other factors to be discussed below may be involved.

Sexual rest. The amount of genital stimulation needed to produce the first ejaculation in a test varies according to the male's sexual condition. The total complete and incomplete copulations preceding the initial ejaculation was 23·2 in 3-day tests, 19·1 in 6-day tests and 17·4 in tests following a rest of 15 days.

The time needed to ejaculate for the first time varied inversely with the length of sexual rest. Median first plug latencies were 450 seconds in 15-day tests, 631 in 6-day tests and 756 seconds in tests conducted after 3 days of rest.

It is apparent that the excitability of the *CM* is low after an exhaustion test and recovers gradually as a result of sexual inactivity. At the same time there is a clear-cut tendency for successive ejaculation latencies to decrease in all tests regardless of time since the preceding exhaustion.

One ejaculation every 24 or 48 hours seems to have no detectable effect upon the *CM*. At any rate the number of intromissions necessary to produce one ejaculation was 10·4 in 24-hour tests and 11·4 in 48-hour tests as compared with 10·5 in 15-or-more-day tests.

A measure which as yet we have been unable to interpret in any meaningful fashion might be called *copulatory speed*. It is derived by dividing the number of intromissions to ejaculate into the ejaculatory latency. In effect this indicates the rapidity with which one copulation follows another. In the course of an exhaustion test the curve of copulatory speed first increases and then decreases. In tests of fully rested males the speed was 51·5 seconds to the first ejaculation, 34·9 seconds to the third and 52·7 seconds to the seventh ejaculation. In achieving their final ejaculation males tend, as noted above, to execute relatively few copulations but these are widely spaced in time. It is because of progressive changes in copulatory speeds that the copulatory latencies listed in Table I were said to reflect something more than a mere reduction in intromission frequency.

Estradiol benzoate (Progynon B) and progesterone (Proluton) used in this study were generously supplied by Dr. Edward Henderson, of Schering Corporation, Bloomfield, New Jersey.

REFERENCES

BEACH, F. A. (1945). Hormonal induction of mating responses in a rat with congenital absence of gonadal tissue. *Anat. Rec.*, **92**, 289–292.

BEACH, F. A. (1947). A review of physiological and psychological studies of sexual behaviour in mammals. *Physiol. Rev.*, **27**, 240–307.

BEACH, F. A., and HOLZ-TUCKER, A. M. (1949). Effects of different concentrations of androgen upon sexual behaviour in castrated male rats. *J. comp. physiol. Psychol.*, **42**, 433–453.

BEACH, F. A. (1955). Neural and Chemical Regulation of Behaviour. Wisconsin Symposium on Interdisciplinary research. In press.

MILLER, N. E., and DOLLARD, J. (1941). *Social Learning and Imitation*. New Haven.

SCHWARTZ, M. (1955). Instrumental and consummatory measures of sexual capacity in the male rat. *J. comp. physiol. Psychol.* In press.

WARNER, L. H. (1927). A study of sex behaviour in the white rat by means of the obstruction method. *Comp. psychol. Monog.*, **4** (22).

5

Reprinted from *Animal Behav.* **17**:700–705 (1969)

COPULATORY BEHAVIOUR OF THE RAM, *OVIS ARIES.* I: A NORMATIVE STUDY

By GORDON BERMANT, M. T. CLEGG* & WESLEY BEAMER

Departments of Psychology and Physiology, University of California, Davis

This paper is the first in a series of three devoted to the description and analysis of some determinants of copulatory behaviour in male domestic sheep. Its primary purpose is to delineate the quantitative characteristics of the several components of the copulatory sequence as they were observed under a single set of experimental conditions. Close attention is paid to changes in behaviour produced by removal of the initial female and introduction of a second oestrous ewe. The normative baselines presented here will be utilized in the other papers in the series.

Several other descriptions of sheep copulation and related behaviour have been published (Banks 1964; Hafez & Scott 1962; Hulet *et al.* 1962a, b, c; Pepelko & Clegg 1965a, b). The present experiment differs from these earlier reports by employing either a larger sample size or more highly controlled testing procedures, or both. Results of this study will ·be compared with those from earlier experiments.

A brief description of the copulatory sequence will aid understanding of the quantitative treatment that follows.

If a receptive ewe is introduced into a pen containing a ram with prior copulatory experience, the ram will approach her within several seconds and begin the preliminaries to copulation. There are several more or less stereotyped manoeuvres that the ram may engage in prior to his initial mount. These include pawing repeatedly the ewe's flank with one foreleg while standing slightly behind and at a small angle to her (cf. 'nudging', Banks 1964); nuzzling, licking, and nibbling at her flank and ano-genital areas; elevating the head and retracting the upper lip in response to the odour or taste of the ewe's urine (the *Flehmen*); and, while standing behind the ewe just prior to mounting, achieving brief partial extrusion of the penis and making what appear to be copulation intention movements. Some males additionally emit low-pitched 'gargling' vocalizations before and while pawing the ewe.

*Present address: Delta Regional Primate Research Center, Covington, Louisiana.

There is considerable variation within and among males in the frequency and duration of these precopulatory responses. It is not altogether clear that they play a functional role in inducing the ewe to stand for copulation. A fully receptive ewe stands quite still after the initial approach of the ram. When the ram is behind her she will often turn her head to one side and appear to watch him. There is also a characteristic wagging or twitching of the tail that often accompanies full receptivity.

A partially receptive ewe typically moves away from the approaching ram. If he attempts to mount she may move out from underneath him. She may also position herself along the walls of the pen in ways that make mounting attempts difficult or impossible. In these cases a vigorous experienced ram will increase the intensity of his pawing and nuzzling and occasionally succeeds in butting the female into the centre of the pen.

The ram begins a series of shallow thrusting movements as soon as he has mounted the ewe. As in the cases of the male rat (Adler & Bermant 1966; Bermant 1965), cat (Cooper 1965), and perhaps other species as well, these shallow thrusts appear to orient the penis with respect to the vaginal opening. One factor that seems to be influential in determining the number of shallow thrusts prior to insertion is the relative sizes of ram and ewe. Small rams have difficulty penetrating large ewes. An experienced ram may achieve insertion during his first thrust. Rams do not invariably succeed in penetrating during a single mount. After several (3 to 10) unsuccessful shallow thrusts the ram may dismount or the ewe may walk out from underneath him.

Successful penetration is almost always accompanied by ejaculation. During the single deep lunge that characterizes the intromission/ejaculation response, there is a characteristic elevation of the ram's head. There is evidence that the ejaculatory reflex is triggered by the stimulation of the glans penis (Rodin 1940). The ram usually withdraws and dismounts immediately after ejaculating.

Rams are capable of multiple ejaculations

with a single ewe. The intervals between successive ejaculations grow progressively longer (Pepelko & Clegg 1965a). As in the cases of several domestic and laboratory species, introduction of a new oestrous female restores the male's copulatory behaviour to one degree or another (Almquist & Hale 1956; Bermant, Lott & Anderson 1968; Pepelko & Clegg 1965a; Schein & Hale 1965; Wilson, Kuehn & Beach 1963). We deal with the extent of this restoration in detail below.

A recent study by Pepelko & Clegg (1965b) demonstrated small but reliable seasonal differences in the ejaculation frequencies of Targhee rams tested over a 12-month period in Davis. For other breeds and other localities varying degrees of male seasonality have been reported (Fraser 1968; Hafez & Scott 1962). In the present study, using Targhee rams primarily, testing was conducted in all months of the year (see Table I).

Methods
Subjects

Twenty-eight whitefaced crossbred Targhee type and two Corriedale rams were obtained from the Hopland Field Station of the Davis campus. Two rams were 6 years old, while the others were between 3 and 4 years. All were demonstrated copulators. The rams weighed between 91 to 113 kg. Additional non-experimental rams were used to check female receptivity.

Approximately seventy ewes of various breeds were used. None was castrated. A ewe was brought into behavioural receptivity once approximately every 7 weeks by intramuscular injections of progesterone (20 mg per 48 hr for 2 weeks) followed 24 hr later by a single injection of oestradiol cyclopentylpropionate (20 mg).* Receptivity was usually evident within 24 hr of

*The order of gonadal hormone administration is reversed from that required to induce oestrus in female rodents. For quantitative comparisons, we note the following: on a bodyweight basis, the ewes received less progesterone (1·5 to 2·0 mg per kg) and less oestrogen (0·2 to 0·3 mg per kg) than do intact female rats under a typical regimen for experiments of this sort (progesterone: 4·0 to 5·0 mg per kg; oestrogen: 0·4 to 0·5 mg per kg). In both cases the dosages far exceed physiological levels.

the oestrogen injection, and lasted 2 to 4 days.

Apparatus

The indoor testing pens measured 3·1 m × 4·3 m. Natural light entering the barn was sometimes supplemented in the winter by overhead lamps.

Living pens for pairs of rams were situated inside the barn, while ewes normally occupied a large outdoor pasture separated by approximately 15 m from the rams. Oestrous ewes were penned as a group inside the barn, visually isolated and approximately 10 m distant from the rams.

Procedure

Test sessions were usually begun between 9.00 a.m. and 11.00 a.m. A ram was led from his living pen to a test pen and left alone there for 5 to 10 min before a receptive ewe was introduced. Scoring of behaviour began when the gate of the test pen was closed behind the ewe. The ram had uninterrupted access to the ewe for 1 hr or until 20 min had elapsed without a mount, whichever occurred sooner. When the criterion had been reached, an experimenter entered the pen, removed the ewe, and introduced a second receptive ewe. The switching procedure required approximately 5 min. Uninterrupted exposure to the second ewe was terminated by the same criteria as to the first. Maximum duration of a single test session was thus approximately 2 hr.

Four response categories were recorded: (1) *Teases*, an aggregated measure of premounting responses including the pawing, nuzzling, and 'gargling' described above; (2) *Flehmen*; (3) *Mounts*, in which the ram mounted the ewe completely and exhibited shallow thrusting but did not achieve full insertion; and (4) *Intromission/ejaculation*. In the majority of cases ejaculation could be verified by visible dripping of semen from the ewe or ram following dismount. Records were kept of the time intervals between occurrences of these categories. Data on teasing and the *Flehmen* will not be reported here, primarily because we were not totally convinced of the validity of our measurement of these courtship activities.

Table I presents the distribution of tests over

Table I. Number of Copulatory Behaviour Tests Administered in Each Quarter of the Year

	Jan–Mar	Apr–Jun	Jul–Sep	Oct–Dec	Total
Number of tests	43	36	39	34	152

Table II. Rams Receiving a Given Number of Copulatory Behaviour Tests

Number of Tests	Number of Rams
2	7
3	3
4	3
5	11
6	2
7	1
8	1
17	1
18	1

the quarters of the year. Table II is the frequency distribution of the number of tests given to each ram.

A total of 152 tests were administered during the 16-month period beginning November 1964, and ending February 1966. The average interval between successive tests for a single ram was approximately 7 days. In no case were two tests separated by less than 5 days. A single ram encountered a given female no more often than once every 7 weeks.

Results

Ejaculation Frequency

Mean ejaculation frequencies were computed by averaging across all tests for each ram before taking the average for the sample of 30. Mean ejaculation frequency with the first female was 3.91 ± 2.27 SD, while with the second female it was 2.38 ± 1.22. Twenty-eight of the thirty rams showed reduced frequency with the second female ($P<0.01$, sign test).

The scatterplot of mean ejaculation frequencies in Fig. 1 shows that ejaculatory performance with the first female was a good predictor of performance with the second ($r = +0.83$, $df = 28$, $P<0.005$).

Table III is the matrix of ejaculation frequencies for all 152 tests. Reduction in ejaculations with the second female occurred in 104 tests. In thirty-one tests there were equal frequencies, and in seventeen tests the frequency was greater with the second female. Modal performance (seventeen cases) was four ejaculations with the first female and two with the

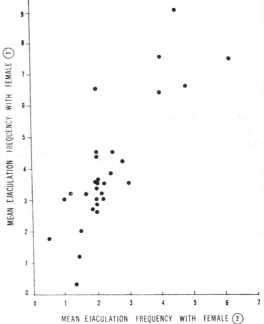

Fig. 1. Scatterplot of ejaculation frequencies achieved with female 1 and female 2.

Table III. Distribution of Ejaculation Frequencies with First and Second Females (N = 152 Tests)

		Ejaculation Frequency, Female 2							
		0	1	2	3	4	5	6	7
	0	0	1	1					
	1	2	8	5					
	2	2	6	12	7	1			
Ejaculation	3	4	15	14*	7	1			
frequency,	4	2	1	17	3	1	1		
female 1	5	1	0	8	4*	2	2		
	6	0	1	2	4	3	0	1	
	7	0	0	0	0	2	1		
	8	0	0	1	1	2	0	2*	2
	9	0	0	0	0	1			
	10	0	0	0	1				

*Data from these cells are used in the analysis of post-ejaculatory intervals (Fig. 2).

second. These values approximate the weighted averages reported above.

Mount Frequencies

There are two ways that the frequencies of non-penetrating mounts need to be analysed. We consider first the mean number of mounts per session, ignoring differences in ejaculation frequencies. With female 1 the rams averaged $11 \cdot 36 \pm 6 \cdot 67$ mounts per session, while with female 2 they averaged $7 \cdot 49 \pm 4 \cdot 36$ mounts per session. Twenty-three of the thirty rams showed reduced mean mounts per session with female 2 ($P < 0 \cdot 01$, sign test).

We consider next the mean number of mounts per ejaculation per session. With female 1 the rams averaged $3 \cdot 83 \pm 3 \cdot 23$ mounts per ejaculation, while with female 2 they averaged $3 \cdot 45 \pm 1 \cdot 66$ mounts per ejaculation. Fourteen of the thirty rams showed reduced mounts per ejaculation with female 2.

These two analyses are interpreted to mean that total mount frequency per session is a correlate of total ejaculation frequency: rams that ejaculate more also mount more. But these high-performance rams do not mount more (or less) for each of their ejaculations than do less active animals. Further evidence for this interpretation is seen in the correlation coefficients presented in Table IV.

Table IV. Correlation Coefficients for Mount Frequencies and Mount Frequencies v. Ejaculation Frequencies

Variables	r
MF female 1 × MF female 2	$+ 0 \cdot 34*$
MF/ejac. female 1 × MF/ejac. female 2	$+ 0 \cdot 05$
MF × EF, female 1	$+ 0 \cdot 43†$
MF × EF, female 2	$+ 0 \cdot 66‡$
MF/ejac. × EF, female 1	$- 0 \cdot 05$
MF/ejac. × EF, female 2	$- 0 \cdot 21$

*$P < 0 \cdot 05$ 1-tailed; †$P < 0 \cdot 01$ 1-tailed; ‡$P < 0 \cdot 005$ 1-tailed.

Total mount frequency with female 1 was positively correlated with total mount frequency with female 2; this is to be expected because mount frequency correlates with ejaculation frequency for each female, and the two ejaculation frequencies are correlated. But there was no significant correlation for either female between mounts per ejaculation and ejaculation frequency, nor between mounts per ejaculation

with female 1 and mounts per ejaculation with female 2.

Postejaculatory Intervals

The postejaculatory interval (PEI) is defined as the time between an ejaculation and the next copulatory event (mount or ejaculation). In order to obtain an accurate averaged estimate of the changes in PEI with successive ejaculations, it is necessary to average across only those sessions with identical ejaculation frequencies (Larsson 1956; Pepelko & Clegg 1965a).

Three sets of PEI curves are shown in Fig. 2. Data from the cells (3, 2), (5, 3), and (8, 6) of Table III were chosen to represent the PEI changes of slow, medium, and fast responders, respectively.* Note that for the (8, 6) group the terminal PEI is not recorded. This is because these rams did not reach the 20-min PEI criterion within the 1-hr period. Their terminal PEI was therefore indeterminable. Also, because of the 20-min criterion, the terminal PEI values shown for the other groups represent minimal estimates.

Introduction of the second female produced a dramatic decrease in the ram's subsequent postejaculatory intervals. The distribution of terminal PEI with female 1 was totally non-overlapping with the distribution of initial PEI with female 2. But the restorative effect on PEI produced by introducing the second female was not complete: on the average, rams recovered more slowly with female 2 than with female 1. For example, the mean completed first postejaculatory interval with female 1 was $5 \cdot 09 \pm 3 \cdot 01$ min, while with female 2 it was $7 \cdot 89 \pm 3 \cdot 35$ min. Twenty-six of the thirty rams showed greater first postejaculatory intervals with the second female ($P < 0 \cdot 01$, sign test). Some rams during some sessions recovered from each ejaculation as rapidly with female 2 as they had with female 1. These were inevitably high performance animals whose ejaculation frequencies were consistently above the group average with both females (Fig. 2).

Introductory Ejaculation Latency

The introductory ejaculation latency is defined as the time from the introduction of a female into the pen to the first ejaculation. The mean latency to ejaculation following introduction of female 1 was $1 \cdot 97 \pm 1 \cdot 79$ min, while with female 2 it was $2 \cdot 09 \pm 1 \cdot 49$ min. Thirteen of the thirty rams showed reduced introductory

*If a single ram was represented more than once in a given cell, his data were averaged before the cell average was obtained.

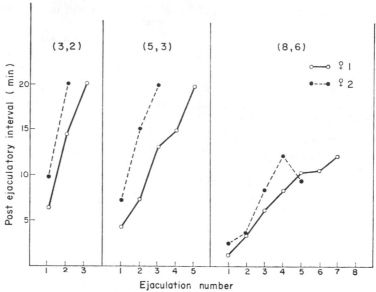

Fig. 2. Mean postejaculatory intervals for slow, medium, and fast responders. The numbers in parentheses refer to ejaculation frequencies with female 1 and female 2, respectively.

latencies with female 2. The two latency measures were not significantly correlated ($r = +0.17$). Thus, the introduction of a second female into the pen led to the occurrence of ejaculation in as brief a period of time as did introduction of the first female. There was not a significant correlation between the rams' ejaculation frequencies with female 1 and their introductory ejaculation latencies with female 2 ($r = -0.13$).

Discussion

The testing procedures employed in this study were quite similar to those used by Pepelko & Clegg (1965a). Using a sample of 10 Targhee rams, Pepelko & Clegg concluded that ejaculation frequency with the second female did not differ significantly from the frequency obtained with the first. As they put it, '... sexual recovery following return of a different not recently mated female did not differ significantly from 100 per cent'. We did not confirm this finding in the present study. The thirty rams in this study equalled or surpassed their performance with female 1 in only forty-eight out of the 152 tests (32 per cent). Only two rams displayed average second female ejaculation frequencies that were equal to or greater than their frequencies with the first female. Moreover, the first postejaculatory interval with female 2 was sig-

nificantly greater than the first interval with female 1. Only four rams showed a shorter average PEI 1 with female 2.

One procedural difference between the two studies might have accounted for the difference in results. Pepelko & Clegg terminated the ram's exposure to a female only when 20 min had elapsed without a mount. In the present study we imposed an additional 1-hr criterion. This means that the rams in Pepelko & Clegg's study potentially had the opportunity to remain with the females for longer periods of time. Conceivably, this could have increased the numbers of ejaculations shown with female 2 to values approximating those shown with female 1. In the present study the rams were not given this opportunity. However, 88 per cent of the tests with female 2, and 66 per cent of the tests with female 1, were in fact terminated by the 20-min criterion. This means that the additionally imposed 1-hr criterion did not act to maximize the difference between ejaculation frequencies with the two females.

An interesting contrast to the present findings is found in the results of Hulet and his colleagues (1962a, b, c), who observed the behaviour of rams continuously for 7 days in pens containing eighteen to thirty-eight ewes. Under these conditions rams were observed to copulate at all hours of the day and night, with maximum rates per hr occurring between 8.00 a.m. and 10.00 a.m. But even these maximum rates were far less than those observed in the present experiment, in which access to receptive ewes was strictly limited. When four ewes were in heat simultaneously (the largest number for which data were reported; Hulet *et al*. 1962a) the rams averaged approximately one ejaculation per hr. In addition, it was shown (Hulet *et al*. 1962b) that successful copulatory performance can be hindered by the presence of other rams in the pen. Aggressive rams were observed actively to prevent successful matings by younger or mature but submissive animals.

The partial recovery of the rams' postejaculatory interval produced by the introduction of female 2 stands in contrast to the behaviour of male rats tested under similar conditions (Bermant *et al*. 1968). Although the rats in the cited experiment did show some performance increments when female 2 was introduced, PEI was not decreased. Further contrasts between rats and rams will be pointed out in the following paper.

We conclude that the quantitative characteristics of copulatory performance in the ram vary substantially as functions of the details of the testing procedures. The results presented here serve primarily as baseline data with which to compare the effects of endocrine and experiential manipulation that are presented in the subsequent experiments in this series.

Summary

Twenty-eight Targhee and two Corriedale rams, ranging in age from 3 to 6 years, were observed in 152 copulatory behaviour tests. In each test a ram was permitted to remain with a single ewe for 1 hr or until 20 min had elapsed without a mount. The first female was then removed and replaced by a second fresh receptive ewe. Rams averaged 3·91 ejaculations with the first female and 2·38 with the second. Ejaculation frequencies with the two females were highly correlated. Mounts per ejaculation averaged 3·83 for female 1 and 3·45 for female 2. Ejaculation frequencies were not significantly correlated with the frequencies of mounts per ejaculation. Initial postejaculatory intervals with the second female were significantly less than terminal intervals with the first female, but the restoration of ability to recover from ejaculation was not complete. Initial ejaculation latencies with the two females were essentially identical.

Acknowledgments

This research was supported by National Science Foundation Grant GB-3328 and United States Public Health Service Grant NB-3977-09.

REFERENCES

Adler, N. & Bermant, G. (1966). Sexual behavior of male rats: effects of reduced sensory feedback. *J. comp. physiol. Psychol.*, **61**, 240–243.

Almquist, J. & Hale, E. (1956). An approach to the measurement of sexual behaviour and semen production of dairy bulls. *Proc. 3rd Int. Congr. Anim. Reprod.* (Cambridge: England), plenary papers, pp 50–59.

Banks, E. (1964). Some aspects of sexual behaviour in domestic sheep *Ovis aries*. *Behaviour*, **23**, 249–279.

Bermant, G. (1965). Rat sexual behavior: photographic analysis of the intromission response. *Psychon. Sci.*, **2**, 65–66.

Bermant, G., Lott, D. & Anderson, L. (1968). Temporal characteristics of the Coolidge Effect in male rat copulatory behavior. *J. comp. physiol. Psychol.*, **65**, 447–452.

Cooper, M. (1965). Questions and group discussion. In *Sex and Behavior* (ed. F. Beach), pp 476–478. New York: Wiley.

Fraser, A. (1968). *Reproductive Behaviour in Ungulates*. London: Academic Press.

Hafez, E. & Scott, J. (1962). The behaviour of sheep and goats. In *The Behaviour of Domestic Animals* (ed. E. Hafez), pp. 297–333. London: Baillière, Tindall & Cassell.

Hulet, C., Ercanbrack, S., Price, D., Blackwell, R. & Wilson, L. (1962a). Mating behavior of the ram in the one-sire pen. *J. Anim. Sci.*, **21**, 857–864.

Hulet, C., Ercanbrack, S., Blackwell, R., Price, D. & Wilson, L. (1962b). Mating behavior of the ram in the multi-sire pen. *J. Anim. Sci.*, **21**, 865–869.

Hulet, C., Blackwell, R., Ercanbrack, S., Price, D. & Wilson, L. (1962c). Mating behavior of the ewe. *J. Anim. Sci.*, **21**, 870–874.

Larsson, K. (1956). *Conditioning and Sexual Behavior in the Male Albino Rat*. Stockholm: Almqist & Wiksell.

Pepelko, W. & Clegg, M. (1965a). Studies of mating behaviour and some factors influencing the sexual response in the male sheep *Ovis aries*. *Anim. Behav.*, **13**, 249–258.

Pepelko, W. & Clegg, M. (1965b). Influence of season of the year upon patterns of sexual behavior in male sheep. *J. Anim. Sci.*, **24**, 633–637.

Rodin, I. (1940). The influence of thermal and mechanical stimuli on ejaculation into the artificial vagina in rams. *Trud. Lab. inkusst. Osemen. Zivotn.* (*Mosk.*), **1**, 46–53.

Schein, M. & Hale, E. (1965). Stimuli eliciting sexual behavior. In *Sex and Behavior* (ed. F. Beach), pp 440–482. New York: Wiley.

Wilson, J., Kuehn, R. & Beach, F. (1963). Modification in the sexual behavior of male rats produced by changing the stimulus female. *J. comp. physiol. Psychol.*, **56**, 636–644.

(*Received* 24 *December* 1968; *revised* 29 *April* 1969; *Ms. number:* A783)

Editor's Comments
on Papers 6 Through 9

VARIATIONS ON HETEROSEXUAL COPULATION

Although sucessful reproduction is generally contingent upon male-female heterosexual copulation, both males and females often engage in other behavior that falls within the rubric of sexual behavior. A major reason for the impact on society of the work of Kinsey and his associates (Kinsey et al., 1949; Kinsey et al., 1953) was the evidence that such behavioral patterns had a much higher incidence among humans in the United States than many people had previously thought.

Variations on heterosexual copulation occur with respect to both the motor patterns displayed and the sex of the stimulus object toward which the behavior is directed. Beach (1979) has suggested a taxonomy of forms of sexual behavior based on the sex of individuals, motor patterns, and stimuli eliciting the behavior. There are reports in the literature of females from a variety of species displaying mounting, intromission/like, and ejaculation/like behavior—patterns normally thought of as characterizing males. Similarly, there are reports, although quite rare in nonprimate mammalian species, of males displaying receptive behavioral patterns that normally are displayed by females. When two like-sexed individuals engage in sexual activity, the interaction is generally termed homosexual. Clearly, patterns of sexual behavior are less rigidly limited to individuals of the designated sex than had been previously thought.

Autoerotic, self-stimulative, or masturbatory behavior often occurs in the absence of any partner. Ford and Beach (1951) defined masturbation as "any sort of bodily stimulation that results in excitation of the genitals."

Because similar or identical motor patterns can function in different ways in different functional contexts, one must use care in interpreting all instances of sex-related behavior as being truly sexual. For example, in primates the "present" posture characteristic of receptive females can be given by males or females in a social context. It may indicate social submission when directed at a dominant male, may function in appeasement after a disagreement, or may serve as a greeting. Both mounting and presenting are quite common in primate social interactions, and these behavioral patterns need not have sexual connotations (Beach, 1976; Wickler, 1967). In guinea pigs and other mammals, the receptive posture of females includes lordosis, a concave arching of the back. A posture identical to the adult lordotic pattern is displayed in some neonates and functions as an aid to the mother in licking the perineal region to stimulate urination and defecation (Beach, 1966).

Major impetus of much early work on the sexual behavior of nonhuman primates stemmed from interest in the possible similarities of human and nonhuman homosexual and autoerotic behavior. Hamilton (Paper 6) conducted one of the earliest systematic studies of variants on heterosexual copulation in nonhuman primates. He found evidence for a wide range of sexual "tendencies" and summarized them in a comprehensive table. Although both heterosexual and homosexual behavior appeared common, the tendency to masturbate appeared to be developed only under "abnormal" conditions. Three years later, Kempf (1917) published another study of primate sexual tendencies. Like Hamilton, Kempf observed much homosexual activity, including anal intercourse. Although both Hamilton and Kempf appeared motivated by a wish to relate nonhuman to human behavior, such an orientation is excessive in Kempf's writing. He concluded, "The phylogenetic constitution of man, as we find it completely exposed in the infrahuman primate, obsesses him with what he feels to be perverse tendencies as he strives to behave in an ideally civilized manner and plunges him into the depths of despair when he fails" (p. 154). Later Bingham (Paper 12) and Carpenter (1942) provided additional observations of homosexual and autoerotic behavior in nonhuman primates. More recent studies have been published by Chevalier-Skolnikoff (1976) and Akers and Conaway (1979).

The next two papers are by C. P. Stone and F. A. Beach, two names appearing frequently throughout the book. Stone and Beach conducted careful investigations of a wide range of sex-related questions with a rigor remarkable for the time at which many of them were done. Stone (Paper 7) reports two cases of the rare phenomenon of apparently normal male rats displaying female receptive patterns.

Beach (Paper 8), in one of the eleven important papers he published in 1942, describes complete mountlike and intromission/like behavior in female rats. Beach provides a clear description of these relatively complete motor patterns in untreated females. The occurrence of mounting by female guinea pigs is described by Goy and Jakway in Paper 11. Mounting by females occurs in a wide range of species and varies as a function of many factors, including the hormonal state of the animals (see the review by Beach, 1968). As such heterotypical behavior occurs in normal animals, it is important that adequate controls be run before one concludes that high levels of heterotypical mounting are the result of an experimental treatment.

A large body of research in the 1960s and 1970s has been directed at the role of the perinatal hormonal milieu in determining the tendencies of males and females to display mounting and lordotic behavior in response to various hormonal treatments and appropriate stimuli in the adult (Beach, 1971; Carter, 1974). The presence or absence of certain gonadal hormones during particular developmental stages is a major determinant of the occurrence of behavior that is either homotypical or heterotypical for genetic males and females. In essence, if the developing young animal is exposed to any of a variety of androgens or estrogens during a sensitive period in development, the animal will tend to display a minimal tendency toward femalelike lordotic behavior and a substantial tendency for malelike mounting behavior when tested under the appropriate conditions regardless of genetic sex. In rats and hamsters the sensitive period occurs a few days after birth; in guinea pigs and primates it is prenatal. Even prenatal stress on the mother can have feminizing and demasculinizing effects on the developing male (Ward, 1972). The sexual behavior displayed by androgenized female rats is characterized by normal malelike temporal patterning, even when fine-grain analyses are performed (Sachs et al. 1973). Perinatally untreated females have also been shown to display the full male pattern including ejaculationlike behavior, especially after certain drug treatments (Emory and Sachs, 1976.)

In a model parametric study Pfaff (1970) assessed the effectiveness of both male and female hormones in eliciting both male and female sexual behavior in both adult males and females. Although mammalian females generally appear more "bisexual" than males, the opposite is true of birds (Adkins, 1975; Adkins and Adler, 1972).

Various authors have described masturbatory acitivity in mammals (for example, Paper 6; Kempf, 1917). Some examples of such behavior in dolphins are described by McBride and Hebb (Paper 9). Various forms of sex play and homosexual behavior also are described in Paper 9.

REFERENCES

Adkins, E. K., 1975, Hormonal basis of sexual differentiation in the Japanese quail, *J. Comp. Physiol. Psychol.* **89**:61–71.

Adkins, E. K., and N. T. Adler, 1972, Hormonal control of behavior in the Japanese quail, *J. Comp. Physiol. Psychol.* **81**:27–36.

Akers, J. S., and C. H. Conaway, 1979, Female homosexual behavior in *Macaca mulatta, Arch. Sex. Behav.* **8**:63–80.

Beach, F. A., 1966, Ontogeny of "Coitus-related" reflexes in the female guinea pig, *Proc. Nat. Acad. Sci.* **56**:526–533.

Beach, F. A., 1968, Factors involved in the control of mounting behavior by female mammals, in *Perspectives in Reproduction and Sexual Behavior,* M. Diamond, ed., Indiana University Press, Bloomington, Ind., pp. 83–131.

Beach, F. A., 1971, Hormonal factors controlling the differentiation, development, and display of copulatory behavior in the *ramstergig* and related species, in *Biopsychology of Development,* E. Tobach, L. R. Aronson, and E. Shaw, eds., Academic, New York, pp. 249–296.

Beach, F. A., 1976, Cross-species comparisons and the human heritage, *Arch. Sex. Behav.* **5**:469–485.

Beach, F. A., 1979, Animal models for human sexuality, *Ciba Foundation Symposium 62 (new series): Sex, Hormones and Behaviour,* Excerpta Medica, Amsterdam, pp. 113–143.

Carpenter, C. R., 1942, Sexual behavior of free-ranging rhesus monkeys (*Macaca mulatta*). II. Periodicity of estrus, homosexual, autoerotic and non-conformist behavior, *J. Comp. Psychol.* **33**:143–162.

Carter, C. S., 1974, *Hormones and Sexual Behavior,* Dowden, Hutchinson & Ross, Stroudsburg, Penn.

Chevalier-Skolnikoff, S., 1976, Homosexual behavior in a laboratory group of stumptail monkeys *(Macaca arctoides):* Forms, contexts, and possible social functions. *Arch. Sex. Behav.* **5:**511–527.

Emory, D. E., and B. D. Sachs, 1976, Hormonal and monoaminergic influences on masculine copulatory behavior in the female rat, *Horm. Behav.* **7:**341–352.

Ford, C. S., and F. A. Beach, 1951, *Patterns of Sexual Behavior,* Harper, New York.

Kempf, E. J., 1917, The social and sexual behavior of infra-primates with some comparable facts in human behavior, *Psychoanal. Rev.* **4:**127–154.

Kinsey, A. C., W. B. Pomeroy, C. E. Martin, and P. H. Gebhard, 1953, *Sexual Behavior in the Human Female,* Saunders, Philadelphia.

Kinsey, A. C., W. B. Pomeroy, C. E. Marin, and P. H. Gebhard, 1953, *Sexual Behavior in the Human Female,* Saunders, Philadelphia.

Pfaff, D., 1970, Nature of sex hormone effects on rat sex behavior: Specificity of effects and individual patterns of response, *J. Comp. Physiol. Psychol.* **73:**349–358.

Sachs, B. D., E. I. Pollack, M. S. Krieger, and R. J. Barfield, 1973, Sexual Behavior: Normal male patterning in androgenized female rats, *Science* **181:**770–772.

Ward, I. L., 1972, Prenatal stress feminizes and demasculinizes the behavior of males, *Science* **174:**82–84.

Wickler, W., 1967, Socio-sexual signals and their intra-specific imitation among primates, in *Primate Ethology,* D. Morris, ed., Aldine, Chicago.

6

Reprinted from pp. 295–301, 313–318 of *J. Anim. Behav.* **4**:295–318 (1914), by permission of the publisher, Holt, Rinehart and Winston

A STUDY OF SEXUAL TENDENCIES IN MONKEYS AND BABOONS

G. V. HAMILTON
Montecito, California

In spite of the considerable advance that has been made in our knowledge of sexual life since the appearance of Freud's (1) *Drei Abhandlungen zur Sexualtheorie*, we still lack that knowledge of infra-human sexual life without which we may scarcely hope to arrive at adequately comprehensive conceptions of abnormal human sexual behavior. For example, the possibility that the types of sexual behavior to which the term " perverted " is usually applied may be of normal manifestation and biologically appropriate somewhere in the phyletic scale has not been sufficiently explored. Homosexual tendencies come to frequent expression in adolescent boys and girls, thereby presenting to the mental hygienist a problem, the solution of which awaits, first of all, biological knowledge of homosexuality which only the behaviorist can supply. It is unnecessary to multiply examples in illustration of the fact that both the theoretical interests of the science of behavior and the practical needs of what we may regard as a group of applied sciences of human behavior (viz., mental hygiene, criminology, psychopathology) place upon the animal behaviorist an obligation to lay the necessary foundations for a scientific and thoroughly comprehensive investigation of sexual life.

The above considerations, of which I have been almost daily reminded by clinical contacts with human sexual problems, have led me to formulate the following problems in animal behavior:

(1) Are there any types of infra-human primate behavior

which cannot be regarded as expressions of a tendency to seek sexual satisfaction, but which have the essential objective characteristics of sexual activity ?

(2) Do such sexual reaction-types as homosexual intercourse, efforts to copulate with non-primate animals and masturbation normally occur among any of the primates, and if so, what is their biological significance ?

It is always a difficult matter to collect scientific data which shall be specifically relevant to behavior problems when such problems do not lend themselves to strictly experimental methods of investigation. Even under carefully prearranged experimental conditions one cannot always be sure that adventitious stimuli may not have played a part in bringing about a given response. This uncertainty is much greater when the animal subject is either at large with his fellows or confined with them in a cage of sufficient size to allow a reasonable approximation to natural conditions. Under such circumstances the best that one can do is to supply a set of conditions which are apt to lead to a fairly definite and uncomplicated development of the desired situation.

The monkey's marked variability of response presents a further difficulty, for in seeking to identify a definite situation-response sequence the observer is called upon to distinguish activities that are essential components of a given reaction-type from purely fortuitous activities. For example, in a given case a monkey's manipulation of his genitalia may be nothing more than reflex scratching of a momentarily irritated area, and not at all a part of his response to the situation of which the observer wishes to determine the reactive value. In many cases I have been unable to ascertain the essential components of a given reaction-type until prolonged contact with my subjects has enabled me to predict with reasonable certainty that whenever the appropriate situation developed a sequence of activities composed of such and such members would be manifested.

A difficulty of another kind is encountered when one seeks to present results that have been obtained by non-experimental methods. Such results can seldom be indicated by tables of figures, habit-formation curves, etc., because reaction to " natural " situations are usually complexes of activities which call for detailed description. When, as in the present instance, the observer arrives at conclusions which have no value unless they are found to be justified by the facts upon which they are based, and when the presentation of all the facts involved would require the space of several printed volumes, some method of abridgment must be adopted.   An abridged journal is not apt to be satisfactory where an extensive program of observation has been followed unless, as rarely happens, the most convenient and logical order in which the facts can be presented coincides with the order of their occurrence.   I have followed the order of presentation outlined below for the sake of effecting a satisfactory abridgment without resorting to the awkward expedient of publishing extracts from my note books in journal form:

I.   List of subjects.
II.   Description of environmental conditions.
III.   A list of the types of situations that were arranged by the observer or encountered by the subjects in consequence of their spontaneous activities; and under each description of a typical situation one or more detailed descriptions of typical responses thereto.
IV.   Classification of reactions as expressions of reactive tendencies.

I. DESCRIPTION OF SUBJECTS

The estimated or known age of each subject is given for January, 1914.   Inability to identify an animal as to species is indicated by a dash after the generic name.   The " pet " name of each animal is given to facilitate the reader's identification of subjects in subsequent descriptions of behavior.   An animal's sex is indicated by its laboratory number—even numbers for males, odd for females.   In reporting the behavior of an animal the first reference to it will include, in the order given, its laboratory number, pet name and initial letters of the genus and species to which it belongs.   E.g., " 7-Becky-M-r " refers to Monkey 7 of the list, and indicates that she is a female *M. rhesus*.

Monkey 1. Bridget. *M. rhesus*. Adult.
Monkey 2. Mike. M. adult. About 1/4 larger than adult male *M. rhesus*. Fur grey and luxuriant. Tail about 8 centimeters long and furred to the tip. Body thick, face broad. Readily identifiable as a macaque.

Monkey 3. Kate. *M. rhesus*. Adult. Mother of Monkey 9.

Monkey 4. Pat. *M. rhesus*. Adult. Vision defective.

Monkey 5. Maud. *M. rhesus*. Young adult.

Monkey 6. Jocko. *M. cynomolgus*. Adult.

Monkey 7. Becky. *M. rhesus*. Adult. Mother of Monkeys 13 and 24.

Monkey 8. Jimmy I. *M. cynomolgus*. Adult.

Monkey 9. Gertie. *M. cynomolgus-rhesus*. Age, 3 years, 2 months. Daughter of monkeys 3 and 10. First pregnancy began September, 1913.

Monkey 10. Timmy. *M. cynomolgus*. Adult. Father of Monkeys 9, 13 and 24.

Baboon 11. Grace. *Papio*—. Adult. A small black baboon, about 1-3 larger than an adult female *M. rhesus*. Tail absent, other anatomical features similar to those of typical members of genus *Papio*. Became pregnant when bred to male baboon (Monkey 12) but frequent previous copulation with male macaques was without result.

Baboon 12. Sandy. *Papio*—. Adult. Fur black over back and forehead, grey elsewhere. " Pig " tail, long muzzle. Legs and body much shorter and thicker and those of the chacma. A very powerful animal, almost equal to the chacma in weight.

Monkey 13. Tiny. *M. cynomolgus-rhesus*. Age, 5 months. Daughter of Monkeys 7 and 10.

Monkey 14. Jimmy II. *M. cynomolgus*. Young adult.

Monkey 16. Sobke. *M. rhesus*. Young adult.

Monkey 18. Baby. *M. cynomolgus*. About 1-2 adult size. Castrated before sexual maturity.

Monkey 20. Chatters. *M. cynomolgus*. About 3-4 adult size. Castrated before sexual maturity.

Monkey 22. Daddy. *M. cynomolgus*. Adult. Castrated— date unknown.

Monkey 26. Skirrel. *M. cynomolgus*. Adult.

Monkey 28. Scotty. M—. Young adult. Probably belongs to the cynomolgus group.

II. ENVIRONMENTAL CONDITIONS

My laboratory is in the midst of a live oak woods in Montecito, California, about five miles from Santa Barbara. Like all of

Southern California, Montecito is not entirely free from frost, but the winters are so mild that when the orange and lemon growers far to the south of us are compelled to use artificial heat in their orchards to prevent damage by frost, our local growers find it unnecessary to take any precautions whatsoever against the cold.   The climate here is therefore exceptionally mild, even for Southern California, and at no times seems to reduce the activities of macaques—except, of course, when rain drives them to shelter.   My subjects have always been in excellent physical condition, the only deaths having been due to accident or to pathological causes that were operative at the time of an animal's purchase.

In front of the laboratory is a quadrangular yard, 16.7 meters long by 7.4 meters wide.   The laboratory encloses one end of this yard, and the cage one side.   The other end and side of the yard is enclosed by a solid board fence which is 1.9 meters high. The entire enclosure is surrounded by live oak trees, and one tree is contained within the yard.   The animals, when at large, could wander to an indefinite distance from the laboratory by passing from tree to tree, but they rarely wander out of sight of the yard.

The cage is 6 meters high, 16.7 meters long and 1.8 meters wide.   The front, top and upper half of the rear and ends are covered with wire netting, the meshes of which are 1.4 centimeters square.   The lower half of the rear and ends is solidly boarded, to give stability to such a tall, narrow structure.   The cage is subdivided into eleven compartments by partitions of which the lower one third is wood and the upper two thirds wire netting (1.4 cm. mesh).   Within 72 centimeters of the top of each compartment is a horizontal shelf, 30 centimeters wide.   Each compartment is also equipped with a sleeping box, a food drawer, and proper drainage for the concrete floor.   A door at the rear gives access to the man who cleans the cage.

An important accessory to the cage is a wooden alley, which extends along the entire rear of the structure, midway between the top and bottom.   This alley is 75 centimeters high by 60 centimeters wide.   Each compartment opens into the alley by means of a sliding door arrangement, which enables the observer to make it accessible to the occupants of one or more compartments, according to the demands of a given experiment.

83

III. TYPES OF SITUATIONS AND OF RESPONSES THERETO

The outdoor conditions described above have enabled me to liberate most of the subjects in selected pairs or groups. 13-Tiny-M-cr was too young, and 10-Timmy-M-c, 12-Sandy-Papio—, 26-Skirrel-M-c and 28-Scotty M— were too large to be safely liberated, but an excellent substitute for outdoor freedom was obtained for them by giving them access to the long alley and three or four of the cages. I have been able to form a fairly accurate estimate of the effects of cage-life on sexual behavior by comparing the activities of animals that have been at large for several months with the activities of those that have been continuously confined. At this point it may be said that, provided the macaque or baboon have a sufficiently wide range for pursuing and fleeing his enemies, playing with his fellows, etc., confinement will be found to have no perverting effects in this climate.

In the following list of observations I have endeavored to spare the reader the inconvenience of being compelled to make frequent reference to the list of subjects by giving in each instance the age of the animals under discussion, rather than the date of the observation. Wherever reference is made to an animal as " sexually immature " it is to be understood, in the case of females, that menstruation has not yet appeared; males are regarded as sexually immature until they have assumed the characteristic strut that I have found to be coincident in appearance with seminal discharge on copulating with females. Sexual maturity, in this sense, antedates the attainment of adult size by at least a year.

Situation 1. Male and female separated from one another (but not from fellows of the opposite sex) for at least one week, then given access to one another.

Observation 1. 10-Timmy-M-c and 5-Maud-M-r, both sexually mature, occupied separate cages, each with a mate, for several weeks. Each animal had been copulating freely. They were then given access to one another, their respective mates having been tolled into an empty cage and confined therein. Timmy rushed from the alley into Maud's cage as soon as her door was opened. She observed his approach from her shelf, and as he ascended toward her, smacking his lips, she, too, smacked her lips. As soon as the male clambered upon the shelf the female assumed the sexual position, viz.: Hind legs fully extended to an almost vertical position; forelegs sharply flexed; tail erect; body inclined forward and downward from the hips; head sufficiently extended and rotated to enable the female to direct her gaze upward and backward. The male grasped the female at the angles formed by the juncture of hips and body with a hand on either side, and in mounting her, he clasped her legs just above the knees with his feet. He leaned forward and downward during copulation, smacking his lips violently. The female seemed to invite contact with his mouth, for she persistently thrust her smacking lips towards the male, until he leaned still further downward and touched her lips with his own. Shortly before copulation ceased, the male uttered a succession of shrill little cries, and greatly increased the vigor of his copulatory movements. As soon as he dismounted the female, he took her tail in one hand and elevated it, then with his free hand examined her vaginal labia, at the same time closely inspecting them with nose and eyes. Then he lay down and the female examined his fur " flea-hunting."

Observation 2. 11-Grace-Papio— (sexually mature), 14-Jimmy-M-c (sexually mature), the three eunuchs and a number of immature monkeys had been at large for several weeks. Grace had freely copulated with all of the males, including the eunuchs. 16-Sobke-M-r (sexually mature) had occupied a cage with a mature female for several weeks. He was now liberated. As soon as he and Grace caught sight of one another they began smacking their lips. Their subsequent behavior was essentially similar to that of the animals described in observation 1.

[*Editor's Note:* Material has been omitted at this point.]

IV. CLASSIFICATION OF SEXUAL REACTIONS AS EXPRESSIONS OF REACTIVE TENDENCIES

1. *Tendencies to Seek Sexual Satisfaction.*

A. *Male Tendencies*

(a) Tendency to engage in typical sexual intercourse with females.

(b) Tendency to increase sexual excitement by preliminary examination of the female's genitalia, or by chasing and biting the female.

(c) Tendency to use a younger or weaker male as a female.

(d) Tendency to play the rôle of female to a copulating male.

(e) Tendency to attempt copulation with non-primates and humans.

(f) Tendency to masturbate (probably developed only under abnormal conditions).

B. *Female Tendencies*

(a) Tendency to engage in typical sexual intercourse with males.

(b) Tendency to play the rôle of male to younger or weaker female.

(c) Tendency to play the rôle of female to friendly female.

(d) Tendency to solicit copulation with non-primates.

2. *Tendencies to Assume the Female Sexual Position as a Defensive Measure.*

A. *Male Tendencies*

(a) Tendency to assume the female sexual position when attacked by a more powerful fellow of either sex.

B. *Female Tendencies*

(a) Tendency to assume the female sexual position when attacked by a more powerful fellow of either sex.

3. *Tendencies to Seek to Lure an Enemy to Attack by Assuming the Female Sexual Position.*

A. *Male Tendencies*

(a) Tendency to lure a male enemy to attack by assuming the female sexual position.

B. *Female Tendencies*

(a) Tendency to lure a female enemy to attack by assuming the female sexual position.

My analysis of the material from which the thirty observations recorded in the preceding pages were taken, at first inclined me toward a classification in which there would appear only the three general tendencies that appear in the above table, viz.: (1) A tendency to seek sexual satisfaction, (2) A tendency to assume the female sexual position as a defensive measure, (3) A tendency to lure enemies to attack by assuming the female sexual position. Had this scheme of classification been adopted, lists of the typical expressions of each of these three tendencies would have been given instead of the lists of specific tendencies that appear in the table. But the viewpoint from which I have come to regard animal and human behavior, taken inclusively, as material for a separate branch of natural science which may be made to serve as an important foundation for various applied sciences finally induced me to adopt the method of classification that appears above. According to my view, the behavior of an organism is the expression of reactive tendencies which have specific representation in its structure. Some of these tendencies have an *inherent* structural representation, such as, e.g., the tendency that finds expression in the kitten's spit and slap when it first encounters the dog-odor—behavior which may be observed in a kitten before its eyes are opened, and which cannot be attributed to the modifying effects of any previous experiences. Other tendencies owe their existence to two factors, viz.: (1) An inherent capacity for post-natal structural modifications by experience; these modifications by their appearance, add the tendencies of which they are the appropriate bases to the list of the organism's properties. (2) The operation of environmental influences that help to produce the necessary structural changes. Any habit-reaction may be regarded as the expression of an acquired tendency of this kind.

My conception of behavior as reducible to expressions of specific reactive tendencies might easily lead to an endless multiplication of such tendencies to account for the apparently innumerable separate modes of organic activity—especially in view of the possibilities afforded by the extreme plasticity of the nervous system—if it did not include something more than the above generalizations. Lasurski (2), whose *Hauptneigungen* bear many points of resemblance to the reactive tendencies of my conception, avoids this danger by stopping short at the point at which he recognizes various relatively separate directions in which a given individual's activities may go in response to a limited number of principal inclinations. He regards these inclinations (tendencies) as artificial abstractions to which, nevertheless, human personality is most profitably and conveniently reducible. McDougall (3) calls attention to the fact that " The activities of each species are directed almost exclusively towards a small number of special ends—reproduction, the securing of food, the escape from danger, the protection of the young, the violent destruction of whatever opposes these great tendencies, and a few others that differ from species to species." To this he adds: " The concentration of the animal upon any of these ends does not depend upon its acquired experience, but upon some feature of its innate constitution; and that feature is what we commonly and properly call an *instinct*, an innate tendency to strive after some end of a particular kind, an *innate conative tendency*."

The reactive tendency of my definition differs from Lasurski's *Hauptneigung* in that it is meant to connote something more than an artificial abstraction, and to refer to specific properties of the organism rather than to the generally inclusive traits of personality that Lasurski has in mind. It differs from McDougall's *innate conative tendency* in its recognition of the fact that the features of an animal's innate constitution are plastic, and capable of modification by experience. I am also inclined to approach the analysis of behavior from a somewhat different viewpoint than that which is implied in McDougall's statements. In dealing with behavior one is apt to be diverted from the most proximate aspect of the phenomena under consideration by estimating the facts solely with reference to the ultimate needs of the individual or, more usually, of the species. It is somewhat artificial, I believe, to assume that a given sequence of activities is set in operation by outer stimuli acting in conjunction with a need which is more apt to be a product of the biologist's analysis than (in other than a rather mystical sense) dynamically a part of the animal's reactive equipment. The preferable course is to identify *individual hungers* which are the product of inner, physiological events and environmental stimulations. The satisfaction of a given hunger may or may not be conducive to the welfare of either the individual or the species—may or may not correspond to a biological need. The total phenomenon with which the behaviorist is concerned in a given case consists of a sequence of events of which the first member is usually, but not always, an external situation. Then come the physiological processes that produce the hunger-impulsion to which the reaction is to be ascribed. The identification of a reactive tendency becomes possible whenever we are able to predict that the operation of a given hunger-impulsion will be directly followed by a series of activities which conform to a known type.

The existence and nature of a given hunger is, of course, arrived at by inference; but this does not necessarily call for a departure from a purely objective attitude toward the facts. If, as Watson (4) suggests, the behaviorist may use the term " consciousness " as it is used by other natural scientists, no insuperable difficulty ought to be encountered in the construction of criteria for the identification of a hunger as a relatively independent dynamic unit. Such criteria would take into account the facts concerning the physiology of the sense-organs to which we already have access; and would recognize the various possibilities for experiencing specifically different satisfactions by employing different modes of stimulating the sense-organs. For example, when I entered the laboratory yard this morning carrying a pail of loquats, the coyote leaned against the large meshes of the wire fence that confines him and whined until I scratched his head; the monkeys gave their characteristic food-calls until I gave them loquats; and after the male monkeys had eaten this agreeable addition to their breakfast they sought the females and copulated with them. I did not need to raise a question as to the contents of each animal's state of consciousness to assist me in the identification of the coyote's dog-like hunger for a mild irritation to his cutaneous sense-organs, or the monkeys' hunger for the various stimulations that are derived from eating food, etc. The coyote may or may not have had a mental picture of the head-scratching or the monkeys of the loquats that usually appear these days whenever I put my hand into the little tin pail; but I am quite sure that these animals were clamoring in response to the particular hungers that I have come to recognize, and that their behavior, the types of which I could have predicted before I entered the yard, was the expression of definite organic properties.

The essential points of the above discussion are these: (1) The essential factors concerned in behavior phenomena are (a) the action of physiological processes usually operating in conjunction with environmental forces, in the production of (b) hungers which impel the individual to manifest (c) activities, the particular types or modes of which are to be ascribed to (d) specific organic properties (reactive tendencies). (2) These reactive tendencies are most conveniently classified with reference to the individual hungers that bring them to expression. (3) The term " reactive tendency," according to my definition, is meant to designate something more specific than an inclination to direct activity toward one of a limited number of general ends, and to include both the innate and the acquired features of an individual's reactive mechanism.

CONCLUSIONS

At least two, and possibly three, different kinds of hunger, or needs of individual satisfaction, normally impel the macaque toward the manifestation of sexual behavior, viz., hunger for sexual satisfaction, hunger for escape from danger and, possibly, hunger for access to an enemy.

Homosexual behavior is normally an expression of tendencies which come to expression even when opportunities for heterosexual intercourse are present. Sexually immature male monkeys appear to be normally impelled toward homosexual behavior by sexual hunger. The fact that homosexual tendencies come to less frequent expression in the mature than in the immature male suggests the possibility that in their native habitat these animals may wholly abandon homosexual behavior (except as a defensive measure), on arriving at sexual maturity.

Homosexual behavior is of relatively frequent occurrence in the female when she is threatened by another female, but it is rarely manifested in response to sexual hunger.

Masturbation does not seem to occur under normal conditions.

The macaque of both sexes is apt to display sexual excitement in the presence of friendly or harmless non-primates.

It is possible that the homosexual behavior of young males is

of the same biological significance as their mock combats. It is
clearly of value as a defensive measure in both sexes. Homosexual alliances between mature and immature males may possess
a defensive value for immature males, since it insures the assistance of an adult defender in the event of an attack.

REFERENCES

(1) FREUD, SIGMUND. Drei Abhandlungen zur Sexualtheorie. Vienna.
      1905.
(2) LASURSKI, A. Ueber das Studium der Individualität. Pädagogische Mono-
      1912.   graphien, XIV Band. Leipzig.
(3) McDOUGALL, WM. The Sources and direction of psycho-physical energy
      1913.   American Journal of Insanity, vol. 69, no. 5, p. 865.
(4) WATSON, JOHN B. Psychology as the behaviorist views it. Psychologica
      1913.   Review, vol. 20, no. 2, pp. 158-178.

7

Reprinted from *Am. J. Physiol.* **68**:39–41 (1924)

A NOTE ON "FEMININE" BEHAVIOR IN ADULT MALE RATS

CALVIN P. STONE

From the Department of Psychology, Stanford University

Received for publication December 1, 1923

Upon rare occasions adult male rats have been observed to mount one another under ordinary conditions of cage confinement. This behavior occurs most frequently, however, when strange males are introduced into a cage of males, or when several males, still in a state of sexual excitement following prolonged copulation, are again brought together in their home cage.

Ordinarily a male thus mounted immediately effects his release from the copulatory clasp by a series of movements such as jumping forward, rearing upward, twisting aside, half turning the body or engaging in combat. Repeated mounting attempts, so far as the writer's limited observations go, do not eventually cause a male thus mounted to assume a pose or attitude characteristic of the receptive female or to modify in other ways his masculine behavior in the direction of femininity. Even in instances wherein a male has been repeatedly mounted by males that habitually dominate him in the home cage, resistance to the copulatory clasp persists.

Within the past year two cases of atypical behavior on the part of adult rats have been observed. Two males were discovered which did not resist the mounting attempts of other males. Furthermore, they responded to the latter's copulatory attempts in a manner characteristic of normal females in the receptive phase of the oestrual cycle. In view of the rarity of such behavior in what seemed to be normal males and because of the possible bearing it may have on reports of hermaphroditism experimentally produced in animals, these cases are herein reported. It is hoped that similar communications from other experimenters will serve to indicate the frequency with which such behavior occurs.

Case I. A medium-sized adult rat, weighing approximately 250 grams. Prior to the first bisexual behavior detected, this male had been used as a "tester" to determine whether female rats were in heat. His particular qualifications for this service were unusual sexual vigor and aggressiveness. Several litters were sired by him.

Males regularly employed as "testers" ordinarily become greatly excited in the course of an evening's work. Because of this excitement they frequently continue in copulatory attempts, one upon the other, when returned to their home cages.

Since such behavior had often been noted in other rats, no particular attention to it was given in the present instance until a chance observation revealed atypical behavior on the part of one male.   It was noticed that one male was repeatedly mounted by two other males with which he was quartered and that he did not in turn attempt to mount either of them.   Contrary to expectation, he did not resist the copulatory attempts of cage-mates; furthermore, he gave in response to their palpation of his sides a typical feminine response.   He depressed his back in the lumbar region, elevated the coccygeal region, arched the tail, and threw back the head in a manner exactly duplicating the behavior of the receptive female in copulo.   From three to four nights per week over a period of thirty days the behavior of this male was observed under various experimental conditions.   Throughout that time he played the rôle of male when in the company of a receptive female, but in the presence of a sexually excited male permitted mounting and responded in a manner typical of a receptive female.

Eventually this male was sacrificed in order that his reproductive tract might be examined.   Active spermatozoa in great numbers were found in the ductus deferentes and the testes.   Nothing resembling ovarian tissue was discovered.

Case II.   A large adult male weighing approximately 300 grams.   His responses to receptive females were normal in pattern and in vigor.   Evidence of ability to beget young was at hand.   Prior to these observations he had been used as a "tester" of sexual receptivity of females.   To the mounting attempts of other males he responded in a manner similar to that described in the foregoing case history.   In the presence of receptive females he played the rôle of a normal, aggressive male, but immediately thereafter, when mounted by other males, responded in a manner typical of the receptive female.   Systematic observations were carried on for three weeks with this male and in that time no variation from this dual form of sexual behavior was noticed.   Unfortunately, during a temporary lapse of experimental work, this rat was lost without opportunity for autopsy.   Hence nothing can be said concerning the reproductive apparatus beyond what may be inferred from his ability to beget young.

Feminization of male rats has been described by various investigators (1), (3), in connection with somatic changes resulting from ovarian transplantation.   So far as one may ascertain from the published accounts of these experiments only partial transformation of masculine behavior was effected.   A feminine "tail reflex" was added to the male's repertoire of responses, his attitude toward other males became less aggressive, or a solicitude for suckling young was acquired.

Such is the condition obtaining in the two cases herein reported.   Responses which have hitherto been considered peculiar to the female,[1] *and to her alone*, were exhibited by males that seemed to be perfectly normal. Naturally the question has arisen as to whether the behavior of these two males was really *very* exceptional as the writer has hitherto been led to suspect.   Has the sexual behavior of the male rat been studied with sufficient thoroughness to enable one to judge with what frequency the "feminine" behavior exhibited by the males herein described might be found

[1] For a more detailed account of the female's sexual behavior, see (4), pp. 101–105; (2), pp. 70–72.

in any random sampling of adult males? What per cent of male rats, if any, will exhibit under ordinary laboratory conditions, and apart from training or ovarian treatment, the feminine sexual responses which experimenters claim to have induced through the influence of ovarian transplants? What per cent of normal male rats can, with the aid of special environmental settings and modes of stimulation, be made to display elements of feminine sexual activity?

Until these questions are answered more satisfactorily than available data now permit, there is no adequate basis for harmonization and evaluation of the published accounts of hermaphroditic behavior experimentally produced in rats.

BIBLIOGRAPHY

(1) Moore: Journ. Exper. Zoöl., 1918, xxviii, 137.
(2) Long and Evans: Univ. of Cal. Pub., 1922, vi, 1.
(3) Steinach: Arch. f. d. gesammt. Physiol., 1912, cxliv, 71.
(4) Stone: Journ. Comp. Psychol., 1922, ii, 95.

8

Reprinted by permission from *J. Genet. Psychol.* **60**:137–142 (1942)

EXECUTION OF THE COMPLETE MASCULINE COPULATORY PATTERN BY SEXUALLY RECEPTIVE FEMALE RATS*

Laboratory of Experimental Biology, American Museum of Natural History

FRANK A. BEACH

Female rats in estrous occasionally respond to other sexually-receptive females with the execution of portions of the masculine copulatory pattern (Long and Evans, 1922; Hemmingsen, 1933; Stone, 1938; Ball, 1940). One published report describes sexually-receptive female rats which directed masculine mating reactions toward sluggish males that had failed to respond to the female's sexual posturings (Beach, 1938).

The complete sex pattern of the male rat first described by Stone (1922) and verified by other workers (Ball, 1939; Anderson, 1936; Beach, 1940) consists of five elements. (*a*) The male mounts the female from the rear, clasping her sides between his forepaws. (*b*) Moving the forepaws inward, backward and upward, and then releasing the pressure, the male palpates the female's sides with extreme rapidity. (*c*) Moving the hind quarters in and out in a piston-like fashion the male delivers a number of vigorous pelvic thrusts which presumably direct the penis into the vagina. (*d*) After the final pelvic thrust the male's forepaws release the female and he dismounts with a forceful backward lunge. The backward lunge often carries the male 6 to 10 inches from the female, and an extremely vigorous copulator may fall over backward or carom off the cage wall. (*e*) Coming to rest in a sitting position the male licks the penis and scrotum for several seconds, often manipulating the external genitals with the forepaws.

Observers who have described in detail the masculine sexual behavior of normal females have emphasized the absence of the fourth and fifth elements characteristic of the copulating male's mating pattern. No account has come to our attention in which normal females are described as displaying the backward lunge or the postcopulatory licking of the genitals.

*Received in the Editorial Office on October 10, 1940.

To induce masculine sexual behavior in female rats Ball (1940) injected 10 adult females with testosterone propionate and studied their reactions to normal females in heat. Females receiving the male hormone showed masculine mating reactions more frequently than did untreated females. Further, injected females performed the complete male pattern including the backward lunge and postcopulatory genital cleaning.

The purpose of the present paper is to report the execution of the complete masculine pattern by untreated female rats. This behavior was noted in the course of a current experiment dealing with the copulatory behavior of male rats.[1] At the beginning of each day's testing three to six sexually-receptive females, to be used as incentive animals with experimental males, are selected from stock. A female's receptivity is tested by placing her with a nonexperimental male of known sexual vigor. Each female may be used in several tests with different experimental males during the day's work. Copulation with a vigorous male frequently appears to increase a female's excitability so that when she is returned to the resting cage containing other estrous females she resorts to the masculine mating pattern.

Recently we were testing a series of sexually active males, using three highly receptive females as incentive animals. These females were multiparous rats 10 to 13 months old. During the intertest rests we noted that two females were alternately mounting the third female. One of the aggressive females was seen to display the backward lunge and postcopulatory genital cleaning; and when this occurred an immediate attempt was made to induce repetition of the behavior and to record it in motion pictures.

The aggressive female which had executed the complete male pattern was placed in the regular observation cage and allowed a few minutes' rest. The submissive female was then placed in the observation cage, and in the next 30 minutes the aggressive female mounted her cage mate nine times. Each time that mounting occurred the bright lights were turned on for movie work; but this stimulus disrupted the behavior in question and the mounting female immediately released her partner to retire to a far corner of the cage. For the next 10 minutes the camera was put aside and

[1]Supported by a grant from the Committee for Research in Problems of Sex, National Research Council.

brilliant lights were not used. In this period eight partial mountings and three complete acts were observed.

The completeness of the act was unquestionable, and the backward lunge was quite pronounced, carrying the mounting female several inches from the animal she had just released. Each time the backward lunge occurred the copulating female assumed a sitting posture and assiduously licked the genital region in a manner characteristic of the copulating male. In a final attempt to obtain a permanent record we were able to film one complete act in which the backward lunge, though present, was somewhat weaker than usual.

At the conclusion of these observations the submissive female was returned to the resting cage, and immediately the second aggressive female mounted and displayed the complete pattern twice. Although we watched the two animals for a half hour this behavior did not recur. Mounting and palpation were frequent, but the backward lunge and postcopulatory genital cleaning did not reappear.

In an earlier study estrous females exhibiting the male sexual pattern were found to possess macroscopically normal ovaries (Beach, 1938). That estrogenic activity in the ovaries of the females herein described was normal is indicated by the animals' marked sexual receptivity when confronted with a male. A partial explanation for the behavior observed appears to lie in a suggestion previously advanced:

> . . . the specificity of the mating patterns for the two sexes, although probably inherited, is not rigidly dictated by the innately organized substratum. Although there may be a strong preference for the normal copulatory response it is obvious that in a few individuals at least, there exists the innate organization essential to the mediation of the mating pattern of either sex (Beach, 1938, p. 334).

In describing the effects of testosterone upon female rats Ball wrote as follows:

> It is concluded that the male copulatory pattern in more or less rudimentary form is part of the equipment of the normal female rat. The threshold of this behavior is very high normally, but it can be lowered by testosterone administration (1940, p. 164).

Although Ball drew no such conclusion, one might have added the conjecture that in addition to lowering the threshold for the rudimentary male pattern in the female, the testosterone contributed an exclusively masculine factor embodied in the two final elements of the copulatory act. Present results indicate that this is not the case. The complete male pattern is present in some untreated females; and occasionally may be evoked in response to a second receptive female.

In this connection it is significant to note that Noble and Wurm (1940) have recorded one instance of a female black-crowned night heron (with histologically normal ovaries) which exhibited elements of the mating pattern normally restricted to the behavior of the courting male.

The reaction of male and female herons to injections of testosterone propionate lead these authors to the following conclusions:

> Differences in the sexual behavior of the adults seem regulated only by differences in the amounts of male hormone normally found in these birds. Estrogens alone fail to stimulate any breeding behavior in either sex (1940, pp. 849-850).

Since castrate birds failed to develop any secondary sex characters it was concluded that the androgens responsible for both male and female courtship are produced in the ovary.

The production of androgenic substances by the mammalian ovary has been demonstrated in the case of the mouse (Hill, 1937, *a, b*), pig (Parkes, 1937), and rat (Deansley, 1938). An alternative source of androgens in the female may be the suprarenal cortex. Allen and Vespignani (1938) found active testicular epithelium in the connective tissue surrounding the adrenal. Virilism in the human female is often related to excessive androgenic material revealed by urine assay, and the masculine characters accompanying such a condition may be reduced or eliminated by unilateral adrenalectomy (Broster, Allen, Vines, Patterson, Greenwood, Marrian, and Butler, 1938). Derivatives of the adrenal cortical hormone exert an androgenic effect (Mason and Myers, 1936); and administration of adrenal tissue effects masculinization of the female guinea pig (Hodler, 1937).

Female rats herein described may have reacted to excessive androgenic activity on the part of the ovaries or adrenals.

The literature includes two reports of virile male rats which displayed portions of the female copulatory pattern when mounted by other sexually-active males (Stone, 1924; Beach, 1938). These animals may have reacted under the influence of excessive amounts of estrogens secreted by the testes. Urine assays reveal that the human testis produces estrogens (Witschi, 1939); and the urine of male homosexuals has been found to contain a higher proportion of estrogenic material than the urine of normal men (Glass, Duel, and Wright, 1940).

Temporary sex reversals in the mating pattern of male and female rats cannot be fully explained on the basis of estrogens in the male and androgens in the female. Such hypotheses, even if they could be fully substantiated in each case with endocrinological evidence, fail to take account of the neural basis of the behavior involved. When an adult male rat, raised in isolation, executes perfect copulatory behavior upon his first contact with a receptive female, we are forced to assume the existence of an inherited neural organization mediating the complex concatenation of reflexes. The same is true of sexually receptive behavior displayed by the inexperienced female during the first estrous.

Findings cited in the present report suggest that some individuals may inherit or acquire the neural organization mediating the mating pattern of the opposite sex. The appearance of this pattern is apparently dependent upon its threshold, and the stimulation offered by the behavior of the incentive animal. No complete explanation for the observed facts is yet available, but they appear to be important to a complete understanding of sexual behavior, and must be taken into account in the interpretation of "sex reversals" in mammals subjected to hormone administration.

REFERENCES

1. ALLEN, E., & VESPIGNANI, P. M. Active testicular epithelium in the connective tissue surrounding a human suprarenal gland. *Anat. Rec.,* 1938, **72**, 293.

2. ANDERSON, E. Consistency of tests of copulatory frequency in the male albino rat. *J. Comp. Psychol.,* 1936, **21**, 447-459.

3. BALL, J. Male and female mating behavior in prepubertally castrated male rats receiving estrogens. *J. Comp. Psychol.,* 1939, **28**, 273-283.

4. ————. The effect of testosterone on the sex behavior of female rats. *J. Comp. Psychol.,* 1940, **29**, 151-165.

5. BEACH, F. A. Sex reversals in the mating pattern of the rat. *J. Genet. Psychol.,* 1938, **53**, 329-334.

6. ————. Effects of cortical lesions upon the copulatory behavior of male rats. *J. Comp. Psychol.,* 1940, **29**, 193-245.

7. Broster, L. R., Allen, C., Vines, H. W. C., Patterson, J., Green-wood, A. W., Marrian, G. F., & Butler, G. C. The adrenal cortex and intersexuality. London: Chapman & Hall, 1938.

8. Deansley, R. The androgenic activity of ovarian grafts in castrated male rats. *Proc. Roy. Soc. London*, 1938, **126**, 122.

9. Glass, S. J., Duel, H. J., & Wright, C. A. Sex studies in male homosexuality. *Endocrinology*, 1940, **26**, 590-594.

10. Hemmingsen, A. Studies on the oestrous-producing hormone (Oestrin). *Skand. Arch. f. Physiol.*, 1933, **65**, 97-250.

11. Hill, R. T. Ovaries secrete male hormone. I. Restoration of the castrate type of seminal vesicle and prostate glands to normal by grafts of ovaries in mice. *Endocrinology*, 1937a, **21**, 495.

12. ————. Ovaries secrete male hormone. III. Temperature control of male hormone output by grafted ovaries. *Endocrinology*, 1937b, **21**, 633.

13. Hodler, D. Surrénales et masculinisation. *Arch. Anat., Hist. & Embr.*, 1937, **24**, 1.

14. Long, J. A., & Evans, H. M. The oestrous cycle in the rat and its associated phenomena. *Mem. Univ. Calif.*, 1922, **6**.

15. Myers, C. S., & Kendall, E. C. Chemical studies of the suprarenal cortex. *J. Biol. Chem.*, 1936, **116**, 267.

16. Noble, G. K., & Wurm, M. The effect of testosterone propionate on the black-crowned night heron. *Endocrinology*, 1940, **26**, 837-850.

17. Parkes, A. C. Androgenic activity of ovarian extracts. *Nature*, **139**, 965.

18. Stone, C. P. The congenital sexual behavior of the young male albino rat. *J. Comp. Psychol.*, 1922, **2**, 95-153.

19. ————. A note on "feminine" behavior in adult male rats. *Am. J. Physiol.*, 1924, **88**, 39-41.

20. ————. Personal communication to the author (1938).

21. ————. Sex drive. In *Sex and Internal Secretions*. (Eds. Allen, Danforth, & Doisy.) Baltimore: Williams & Wilkins, 1939.

22. Witschi, E. Modification of development of sex in lower vertebrates and in mammals. In *Sex and Internal Secretions*. (Eds. Allen, Danforth, & Doisy.) Baltimore: Williams and Wilkins, 1939.

Laboratory of Experimental Biology
American Museum of Natural History
Central Park West & 79th Street
New York City

Reprinted from pp. 111, 114–115, 123 of *J. Comp. Physiol. Psychol.* **41**:111–123 (1948)

BEHAVIOR OF THE CAPTIVE BOTTLE-NOSE DOLPHIN, *TURSIOPS TRUNCATUS*

ARTHUR F. McBRIDE

Marine Studios, Marineland, Fla.

AND

D. O. HEBB

McGill University and the Yerkes Laboratories of Primate Biology, Orange Park, Fla.

Received November 20, 1947

The very large and convoluted brain of the porpoise, or dolphin (*Tursiops truncatus*), and the striking deviations of its structure from that found in other mammals, give a considerable interest to porpoise behavior. The total mass of the brain appears to be considerably larger than man's. A specimen removed by Dr. George Clark from a sick animal that died at the Marine Studios weighed 1480 gm. (the average weight of the human brain is in the neighborhood of 1300 gm.). Von Bonin gives an average brain weight for Tursiops of 1886 gm. The cerebellum is large but not altogether out of proportion to man's, and far the greatest part of the total mass is in the large cerebrum.

Langworthy (7) has reported that there is little or no frontal association area in the porpoise brain, no anterior commissure, and a relatively small corpus callosum. Also of interest is the sensory development. There is of course no sense of smell at all. Tactile and kinesthetic afferents (including the vestibular portion of the VIIIth nerve) do not appear large in number. On the other hand, the optic nerve is large; and the auditory part of the VIIIth nerve, and its central connections, form a system that is even larger than the visual one. Langworthy also suggests (7, p. 463) that the pulvinar—which in primates appears to act as an adjunct to the visual system—has been taken over by the auditory system in the porpoise.

Size of the cerebrum as such, or size of a particular sensory nucleus, cannot tell us that psychological development is at a corresponding level. But an essential problem of comparative psychology is to determine what correlations may exist between structure and function, even though they must be imperfect. One method of doing so is, of course, by the use of experimental lesions of the brain; but an equally important method is the comparison of species, and for this it is clearly desirable to study species that are widely different both in bodily and in neural structure.

Observations of the porpoise in captivity (9, 12) have shown that a systematic study, when it can be made, will be very valuable. For the present, we are able only to offer incidental observations. The object of this report is to make available to comparative psychologists what is known of porpoise behavior, and to attempt to define the problem for further investigation.

[*Editor's Note:* Material has been omitted at this point.]

SEXUAL BEHAVIOR

Townsend observed repeated matings by his captive porpoises, but has given no details of the behavior. Matings have not been observed at the Marine Studios, though they are known to have occurred. The period of gestation is probably about 11 months. There appears to be a mating season of several months in the spring (February to April), but males show evidence of sexual excitation all through the year. This fact is related to the sexual behavior between males ("homosexual" behavior) that has been observed repeatedly.

The sequence of events leading to copulation is not known. During much of the sexual activity observed at the Marine Studios, the females have repeatedly sought to escape the males. On the other hand it has been observed on several occasions that a female has apparently been sexually excited and has persistently approached a male, rubbing her body against his, without eliciting any response (or in a number of instances, arousing the male to leave the female and attempt copulation with another male, as described below). But initiative by the male is much commoner. After a period of days or weeks the female who is at first evasive becomes less so. Nevertheless, in spite of much time spent watching the animals during periods of sexual activity, no actual copulation has been observed.

In the spring of 1939 there were two adult males and one adult female in the tank. The males persistently courted the female and she was quite receptive. The two males were constantly alert, and one never permitted the other to approach the female without competing. As the female had a (still-born) calf about 11 months later, it is apparent that mating eventually took place.

Among males there is a good deal of masturbation, on the floor of the tank and against other males. One male had the habit of holding his erect penis in the jet of the water intake for prolonged intervals. The males also show a good deal of sex play with sharks and turtles, with the appearance of attempted copulation. With the turtle as sex object, the penis is inserted into the soft tissues at the rear of the shell. Ejaculation has not been observed in any of the sex play.

The penis is erected very suddenly, with equally sudden collapse. Erect, it forms an efficient hook, and a young male was seen on several occasions to swim down at considerable speed, scoop up an eel at the bottom of the tank, and swim zigzag all the way across the bottom of the tank with the eel wrapped about its penis. The same animal was also seen swimming upside down at the top of the tank, catching and towing a feather with its erect penis.

In the spring of 1940 there were two large males in the tank and two younger males, smaller and subordinate. One of the younger ones was a long-snouted dolphin (*Stenella plagiodon*). The two larger males repeatedly attempted intromission with the younger ones. In the homosexual activity the active (or "male") male swims upside down under the passive (or "female") male.

In all the notes on behavior at earlier still-births in the tanks of the Marine Studios, mention is made of persistent erection of the penis on the part of the dominant male. None of the observers of the recent live birth saw this in any of the males now in the tanks.

[*Editor's Note:* Material has been omitted at this point.]

REFERENCES

1. BEACH, F. A.: A review of physiological and psychological studies of sexual behavior in mammals. *Physiol. Rev.*, 1947, **27**, 240–307.
2. BONIN, G. v.: Brain-weight and body-weight of mammals. *J. gen. Psychol.*, 1937, **16**, 379–389.
3. HEBB, D. O.: Studies of the organization of behavior: II. *J. comp. Psychol.*, 1938, **26**, 427–444.
4. ——: On the nature of fear. *Psychol. Rev.*, 1946, **53**, 259–276.
5. HOWELL, A. B.: *Aquatic mammals: their adaptations to life in the water.* Springfield, Ill.: Thomas, 1930.
6. KÖHLER, W.: *The mentality of apes.* New York: Harcourt, Brace, 1925.
7. LANGWORTHY, O. R.: A description of the central nervous system of the porpoise (*Tursiops truncatus*). *J. comp. Neurol.*, 1932, **54**, 437–499.
8. LASHLEY, K. S.: Nervous mechanisms in learning. *In* Murchison, C., *The Foundations of Experimental Psychology.* Worcester: Clark Univ. Press, 1929, Pp. 524–563.
9. McBRIDE, A. F.: Meet Mister Porpoise. *Nat. Hist.*, 1940, **45**, 16–29.
10. MAIER, N. R. F., AND SCHNEIRLA, T. C.: *Principles of animal psychology.* New York: McGraw-Hill, 1935.
11. RIESEN, A. H.: The development of visual perception in man and chimpanzee. *Science*, 1947, **106**, 107–108.
11a. TINBERGEN, N.: An objectivistic study of the innate behavior of mammals. Bibliotheca Biotheoretica, series D, 1942, **1**, 39–98.
12. TOWNSEND, C. H.: The porpoise in captivity. *Zool.*, 1914, **1**, 289–299.
13. YERKES, R. M.: *Chimpanzees: a laboratory colony.* New Haven: Yale Univ. Press, 1943.

Part II

DEVELOPMENT

Editor's Comments
on Papers 10 and 11

10   **RASMUSSEN**

   The Relation Between Strength of Sexual Drive and Fertility in Rats, Cocks, and Mice

11   **GOY** and **JAKWAY**
   The Inheritance of Patterns of Sexual Behaviour in Female Guinea Pigs

GENETIC FACTORS

In studying the development of behavior, one studies events occurring over relatively long periods of time in the lives of individual organisms. This may include prenatal development, development of the young organism and the first occurrence of particular behavioral patterns, and changes in adulthood and in old age.

Behavioral development is the result of the mutual and continuous interaction of genetic and environmental factors (Dewsbury, 1978). There is a long history of debate regarding the relative importance of genetic and enviromental factors, "nature" and "nurture," in the display of various behavioral patterns and of the appropriate way in which to conceptualize their actions. Suffice it to say here that there are methods available for the study of both genetic and environmental factors; the relative importance of each in a given instance is an empirical question. Although it is necessary to isolate various factors for analysis, it must be remembered that development is a continuous process and the study of the study of the interaction of the elements is critical. Noting that copulatory behavior is coordinated from constituent elements, Stone (Paper 3) wrote, "No more fundamental problem confronts the student of animal behavior than that of discovering the process by which these constituent elements are integrated into a complex unitary system" (pp. 100–101).

Stone (Paper 3) concluded, "The results of this investigation substantiate the generally accepted view that the copulatory response is the action of an hereditary mechanism" (pp. 146–147).

Stone's primary evidence for this conclusion was the completeness of the copulatory pattern in males experiencing their first interactions with adult receptive females.

There are more powerful techniques available for the demonstration and analysis of genetic determinants of sexual behavior. Goy and Jakway (1962) published an excellent review of the available literature at that time. Most of this literature stemmed from work on nonmammalian species. Among the most convincing lines of evidence of genetic influences are studies of interspecific hybrids. Often animals produced by crossing two species display identifiable elements from the behavioral repertoire of both parental species. This was exactly the result found by Clark et al., (1954) in a significant study of hybrids between platyfish and swordtails. Robert Hinde (1956) found similar results in a study of crosses of different species of finches.

Most genetic studies utilize genetic variability within a single species as the basis for analysis. Either genetically homogeneous or heterogeneous populations can be used. Presumptive evidence of genetic influence stems from the stability of individual differences within heterogenous populations. More powerful conclusions can be drawn when correlations are calculated in a heterogeneous population between the same behavioral measures taken in parents and offspring.

Persuasive evidence of a genetic influence stems from artificial-selection studies in which animals are chosen for mating on the basis of their phenotypic scores on particular behavioral tests. Ideally, a "high" line, a "low" line, and a random-bred line are developed. Wood-Gush (1960) obtained appreciable responses to artificial selection for high and low mating frequencies in cockerels. Manning (1961) completed a study of selection for mating speed in fruit flies.

Rasmussen (*Paper 10*) conducted an early selection study of mammalian as well as nonmammalian sexual behavior. He used the obstruction-box method in which animals had to cross an electrified grid to reach an opposite-sexed partner (see Paper 29). Animals selected for high "sexual drive" crossed the grid six times as frequently as animals selected for low drive. This was an important pioneering study of the genetic bases of rodent copulatory behavior. That a study published in 1953 is labeled "pioneering" is indicative of the youth of the field of behavior genetics. There appear to have been no other published studies of genetic selection for sexual behavior in rodents to this day, probably because such studies are both tedious and time-consuming. Selection for mating frequencies has been successfully conducted using chickens (Siegel, 1972) and Japanese quail (Sefton and Siegel, 1975).

Powerful analytic techniques are available for use with genetically homogeneous populations. Inbred strains are the product of at least twenty generations of brother-sister mating; in theory, each individual is homozygous and identical to other members of the strain at nearly all genetic loci. When consistent differences are observed among different inbred strains raised under identical conditions (for example, Valenstein et al., 1954), one has presumptive evidence of the operation of genetic factors in copulatory behavior.

Even more convincing evidence stems from studies in which inbred lines are crossed to produce F_1, F_2, and backcross generations. Among the first such analytic genetic studies on laboratory mammals were the tandem papers of Goy and Jakway (Paper 11) and Jakway (1959). By studying two inbred strains of guinea pigs plus their F_1, F_2, and backcross generations, Goy and Jakway were able to uncover different patterns of genetic determination for different measures of sexual behavior in female guinea pigs. They emphasized this complexity of patterns of inheritance and explored its implications.

More recent studies have been concentrated on house mice and laboratory rats, largely because a greater variety of inbred strains is available for study. McGill (1962) described strain differences in copulatory behavior among three inbred lines of house mice. He then went on to conduct a sustained and important long-term research program on mammalian sexual behavior (McGill, 1970). More recently McGill has studied genotype-hormone interactions in the determination of sexual behavior (McGill, 1978).

Strain differences in the copulatory behavior of rats were reported by Whalen (1961). Dewsbury (1975) conducted a diallel cross analysis of rat copulatory behavior in which all possible F_1 crosses among four inbred strains were compared.

Together these various lines of evidence provide a convincing demonstration of the important role of genetic determinants as one factor in the development of sexual behavior.

REFERENCES

Clark, E., L. R. Aronson, and M. Gordon, 1954, Mating behavior patterns in two sympatric species of xiphophorin fishes: their inheritance and significance in sexual isolation, *Bull. Am. Mus. Nat. Hist.* **103**: 135–226.

Dewsbury, D. A., 1975, A diallel cross analysis of genetic determinants of copulatory behavior in rats. *J. Comp. Physiol. Psychol.* **88**:713–722.

Dewsbury, D. A., 1978, *Comparative Animal Behavior*, McGraw-Hill, New York.

Goy, R. W., and J. S. Jakway, 1962, Role of inheritance in determination of sexual behavior patterns, in *Roots of Behavior*, E. L. Bliss, ed., Harper, New York, pp. 96–112.

Hinde, R. A., 1956, The behavior of certain cardueline F$_1$ inter-species hybrids, *Behaviour* **9**:202–213.

Jakway, J. S., 1959, Inheritance of patterns of mating behavior in the male guinea pig, *Animal Behav.* **7**:150–162.

McGill, T. E., 1962, Sexual behavior in three inbred strains of mice, *Behaviour* **19**:341–350.

McGill, T. E., 1970, Genetic analysis of male sexual behavior, in *Contributions to Behavior-Genetic Analysis: The Mouse as a Prototype*, G. Lindzey and D. D. Thiessen, eds., Appleton-Century-Crofts, New York, pp. 57–88.

McGill, T. E., 1978, Genotype-hormone interactions, in *Sex and Behavior: Status and Prospectus*, T. E. McGill, D. A. Dewsbury, and B. D. Sachs, eds., Plenum, New York, pp. 161–187.

Manning, A., 1961, The effects of artificial selection for mating speed in *Drosophila melanogaster, Animal Behav.* **9**:82–92.

Sefton, A. E., and P. B. Siegel, 1975, Selection for mating ability in Japanese quail, *Poultry Sci.* **54**:788–794.

Siegel, P. B., 1972, Genetic analysis of male mating behavior in chickens (*Gallus domesticus*). I. Artificial selection, *Animal Behav.* **20**:464–470.

Valenstein, E. S., W. Riss, and W. C. Young, 1954, Sex drive in genetically heterogeneous and highly inbred strains of male guinea pigs, *J. Comp. Physiol. Psychol.* **47**:162–165.

Whalen, R. E., 1961, Strain differences in sexual behavior of the male rat, *Behaviour* **18**:199–204.

Wood-Gush, D. G. M., 1960, A study of sex drive of two strains of cockerels through three generations, *Animal Behav.* **8**:43–53.

10

THE RELATION BETWEEN STRENGTH OF SEXUAL DRIVE AND FERTILITY IN RATS, COCKS AND MICE

E. WULFF RASMUSSEN, STIPENDIARY, BIOLOGICAL INSTITUTE, UNIVERSITY OF OSLO, NORWAY.

Because of the common experience with several species of animals that male potency as well as female willingness to copulate, vary considerably, some years ago we advanced the hypothesis that these individual differences might depend on genetic factors. For possible verification I have measured the strenght of the sexual drive in albino rats of both sexes in a modified "Columbia Obstruction Apparatus", and conducted selective breeding for five generations.

In the apparatus the animals have to cross an electrified grill in order to reach an animal of the opposite sex. The test animal is then removed and must cross the grill anew to achieve contact with the stimulus animal. The number of times it crosses in a fixed period of observation, is used as a measure of the strenght of the drive.

In the first generation of selection, the difference in number of crossings between the two lines was not so great, but in F_2 already, both the male and female offspring of parents with high sex drive (many crossings) crossed the grill twice as often as offspring of parents with low (few crossings).

The selection was then continued by putting together 8 pairs who had crossed the grill most often and 8 pairs who had crossed the smallest number of times. The difference between the two lines increases further in the third and fourth generation. In F_5 the male and female offspring of parents with high sexual drive cross about 6 times as frequently as offspring of parents with low.

From a statistical viewpoint the difference between the means for the number of crossings of the two lines is in F_2 4.2 times the standard error and increases to 9.3 times the standard error in the last generation.

Individual differences in skin resistance among the rats are eliminated by the fact that we had an external resistance of 23,750,000 ohms (the voltage being 475) in circuit with the grill. Otherwise, in order to accomplish 6 obligatory crossings *before* the current is switched on the grill, the offspring of parents with high sex drive performed the task from 2—3 times as quickly as offspring of parents with low. Further, the former offspring copulate sooner after male and female are put together than the latter offspring.

Unexpectedly, a difference infertility between the two groups became evident. In the first two generations of selection the difference was small, but in F_3, 8 pairs of rats with low sexual drive produced 71 young as compared to 8 pairs with high sexual drive producing 57. In the fifth generation, the corresponding figures are 68 and 22. In all, 49 pairs with low sexual drive produced 411 young in comparison to 50 pairs with high drive producing 301.

The difference in fertility manifested itself in 3 different ways:

a) 22.5 per cent of the couples with high drive produced no young at all — while that was the case in only 2.6

per cent in the couples with low drive. In "normal" females breeding at this age the percentage is 15.

b) The breeding couples with high drive produced a lesser number of young per litter than did those with low. In the fourth and fifth generation this difference was as great as 3 1/10 young per litter.

c) More days elapsed from the time males and females were put together until delivery of the young among animals with high drive. Practically this means that rats with *low* drive manage to produce more litters during the breeding period of their lives (for females from about 2½ to about 15—18 months of age) than animals with high sex drive.

Calculating with all generations, the difference in the final rate of fertility was 2.54. The standard error of this difference was 0.81.

As a basis of comparison a birth record of 9457 rats from "normal" parents was compiled. In this material, the fertility manifested itself as a function of the age of the breeding female. The rate of fertility is highest the first time the females breed (at an age of about 3 months). The rate then decreases slowly until an age of about 11—12 months. From this age it declines rapidly until this little animal at an age of about 18 months totally ceases to contribute to the preservation of its species. Our group with low sex drive had a higher rate of fertility than even the females breeding for the first time when 3 months of age — while the group with high sex drive had a fertility corresponding to "normal" females breeding at an age of 13 months. The rats from our two groups were about 9—10 months when they gave birth.

As no factor previously known to influence fertility seems to be involved, the conclusion is drawn that in the rat there is *an inverse ratio between strength of sexual drive and fertility*.

As a probable explanation of the observed facts I will say that what is inherited is likely partly a higher or lower degree of hormonal production, partly a greater or lesser sensitivity of the nervous system to these hormones. On the basis of several other experiments in which about 14,000 successful copulations have been registered in rats, poultry and mice, we have advanced a "two-component charge-discharge hypothesis of the copulatory act and orgasm". According to this hypothesis the underlaying physiological mechanism for the observable sexual behavior is partly nervous, partly endocrine.

The rate of hormonal production thus determines to a great extent the strength of the sexual drive and at the same time — together with other factors — the fertility of the organism. From the literature it is known that injections of large doses of sex hormones under certain conditions may be able to reduce the fertility.

On the basis of these data on rats it cannot be said to what extent the phenomena might depend upon the female or male organism. In order to clarify this and at the same time to obtain some data concerning other species I have performed some experiments with White Leghorns.

The number of successful copulations during 4 hours, from 2.30 P. M. to 6.30 P. M. (cocks copulate most frequently during that time of the day) was used as a measure of the strength of the sex drive (the cocks were isolated from hens when not under observation). Those with high and low sex drive were then put together with other hens (which had been isolated from cocks for at least 33 days) and allowed one coitus with a hen before the latter was removed.

To date 9 cocks with high sex drive have fertilised 75 eggs (55 chicks) — while 9 cocks with low, after the same number of copulations, have fertilised 146 eggs (105 chicks). Because of possible methodical errors and the great variability, these figures are presented with great reservations.

I have also constructed an apparatus for measuring the strength of the sex drive in mice and standardized the detailed procedure for testing, but because of many methodical difficulties I can to date unfortunately give no data concerning fertility for this species. At the time the congress meets, some results will be available also for mice, I hope.

In rats we have recently conducted more series of tests. The aim was to determine the fertility after *one* ejaculation in animals with high and low sexual drive (in which no selective breeding had been conducted). As the difference in fertility among these animals was not so great, we have started selective breeding anew in order to study different aspects of the fertility in animals which are more homogenous for the hereditary factors which must be supposed to determine the strength of the sexual drive.

La relation entre l'intensité de l'instinct génésique et la fertilité démontrée par des recherches expérimentales.

Résumé

1. Au moyen d'une méthode dérivée d'un modèle américain, nous avons mesuré la force du besoin sexuel chez les rats blancs des deux sexes. Par l'étude de cinq générations, comprenant en tout 356 animaux, que nous avons examinés, nous avons pu prouver par sélection que la force du besoin sexuel est génétiquement déterminée.

2. Durant les investigations, une différence très nette de fertilité entre les deux groupes devint évidente. Les couples de rats avec un fort besoin sexuel se montraient peu fertiles — tandis que les couples avec un faible besoin sexuel produisaient un grand nombre de rejetons. La différence était due au nombre de couples qui ne produisaient pas de rejetons aussi bien qu'au nombre de rejetons par portée.

3. Au cours d'expériences postérieures avec d'autres espèces, 9 coqs avec un grand besoin sexuel qu'on laissait copuler *ad libitum* fécondèrent 75 oeufs alors que 9 coqs avec un faible besoin sexuel fécondèrent 147 oeufs. Tenant compte de sources possibles d'erreur et d'une variabilité très grande, on ne peut considérer ces expériences avec des coqs comme définitives.

Experimentelle Untersuchungen über den Geschlechtstrieb im Verhältnis zur Fertilität.

Zusammenfassung

1. Mit einer geeigneten Methode nach amerikanischem Muster modifiziert, haben wir den Sexualtrieb bei Albinoratten beider Geschlechter gemessen. Mit Selektion in fünf Generationen haben wir mit 356 Tieren gezeigt, dass die Stärke des Sexualtriebes sich vererbt.

2. Im Laufe des Versuches, liess sich ein statistisch haltbarer Unterschied in der Fruchtbarkeit erkennen. Rattenpaare mit *starkem* Triebe hatten eine *geringe* Fruchtbarkeit, während Rattenpaare mit *schwachen* Triebe eine auffällig *grosse* Nachkommenschaft hatten . Der Unterschied hat seine Ursache sowohl in der Anzahl der erzeugenden Paare als auch in der Zahl der Jungen per Wurf.

3. In weiteren Versuchen mit anderen Tierarten, befruchteten 9 Hähne mit starkem Triebe, die sich *ad libitum* paaren durften. 75 Eier während 9 Hähne mit schwachem Triebe bei der nämlichen Anzahl von Bedeckungsakten 147 Eier befruchteten. Auf Grund möglicher Fehlerquellen sowie der grossen Variabilität können indessen die Versuche mit den Hähnen einstweilen nicht als definitiv betrachtet werden.

11

Copyright © 1959 by Animal Behaviour

Reprinted from *Animal Behav.* 7:142–149 (1959)

THE INHERITANCE OF PATTERNS OF SEXUAL BEHAVIOUR IN FEMALE GUINEA PIGS*

By ROBERT W. GOY and JACQUELINE S. JAKWAY†

Department of Anatomy, University of Kansas, Lawrence, Kansas

In a previous study, patterns of sexual behaviour were found to differ significantly between inbred female guinea pigs from Strains 2 and 13 (Goy & Young, 1957). In that report a distinction was made between (1) vigour (the frequency or intensity of the œstrous reactions, and (2) responsiveness (the ease with which œstrus is induced by exogenous hormone). Briefly, females from Strain 2 were found to be very responsive to the hormonal treatment, but low in the vigour of their œstrous reactions. In contrast, females from Strain 13 were relatively low in responsiveness but high in vigour. The genetical basis for such a relationship between traits cannot be understood from a study of inbred strains alone. The mode of inheritance of the traits under consideration, and the stability of the relationship under various conditions of hybridization need to be examined. The present experiments were performed as part of a general programme to study the interaction of genetic and hormonal factors in the determination of patterns of sexual behaviour. These experiments have given emphasis to the problem of whether strictly genetic factors or selection factors would account for the inverse relationship between vigour and responsiveness in female sexual behaviour.

As in the study with the male guinea pig (Jakway, 1959), it has been possible to study the inheritance of different elements or components of female sexual behaviour. The observation has been made repeatedly that the female guinea pig displays both a receptive component (the lordosis reflex) and an active component (male-like mounting behaviour) while in œstrus. The mode of inheritance of these two components forms the main body of the present work.

Material and Methods

Previous work indicates that behaviour of spayed females treated with suitable quantities of œstrogen and progesterone is similar to that exhibited prior to ovariectomy (Boling, Young & Dempsey, 1938; Boling & Blandau, 1939; Young & Rundlett, 1939). Largely for convenience, but also to rule out differences in amount of endogenous hormone, only spayed females were used in the present study.

Two hundred and twenty-eight females were used. Twenty-six were from highly inbred Strain 2, 29 from Strain 13, 23 from the F_1 generation, 47 from the F_2, 48 from the backcrosses between F_1 and Strain 2 (B_2), and 55 from the backcrosses between F_1 and Strain 13 (B_{13}). Reciprocal matings were represented in all hybrid populations.

The median age at the time of ovariectomy was 3·5 months in each genetic group. The distributions were not skewed. Eight strain 2 and 8 strain 13 females were spayed at relatively extreme ages (6 to 9 months) and less than 1 month). Previous work has shown that age at the time of ovariectomy does not significantly alter the measures currently being studied (Goy & Young, 1957b). Tests of reproductive performance began one month later on the average. Except where indicated each animal was tested 5 times at intervals of approximately 21 days. For the first 3 tests, each animal was injected with 100 I.U. of œstradiol benzoate followed 36 hours later with 0·2 I.U. of progesterone.‡ For the remaining tests, each animal was injected with 50 I.U. of œstradiol benzoate followed 36 hours later with 0·2 I.U. of progesterone. The volume of all injections was constant (0·5 cc.), and injections were given subcutaneously in the left axilla. Immediately after injection with progesterone, the animals were placed in a standard observation cage (in groups consisting of 6 to 12 individuals) and observed continuously for 14 hours. Each animal was tested once every hour to determine the time of appearance of the lordosis reflex, and its duration was measured with a stop watch on every occurrence. The first

*This investigation was supported by research grant M-504 (C₅) from the National Institute of Mental Health of the National Institutes of Health, Public Health.

†Present address: 1400 South 21st Street, Lincoln, Nebraska.

‡Œstradiol benzoate (Progynon-B) and progesterone (Proluton) were supplied by the Schering Corporation, Bloomfield, New Jersey.

lordosis obtained was regarded as the onset of œstrus, and animals failing to respond on any of the 14 hourly tests were viewed as not in œstrus. For those animals responding on at least one hourly test, œstrus was regarded as terminated when they failed to lordose on two successive hourly tests.

Control animals from the inbred strains were studied intercurrently with the hybrid generations throughout the course of the study from June 1955, until June 1957.

The measures employed in the present study were the same as those described in detail previously (Goy & Young, 1957a). Briefly, they are as follows: (1) latency of heat; (2) duration of heat; (3) duration of the maximum lordosis; (4) frequency of male-like mounting behaviour; and (5) the percentage of tests on which œstrus was observed (per cent. response). Only the latter measure, in the present study, was employed in those tests following the 50 I.U. of œstradiol benzoate. Eight females from Strain 2 and 10 females from Strain 13 were not tested at this dosage.

The data from the present study were not normally distributed and the variances of the different genetic groups were unequal. Because of these characteristics, conventional parametric analysis was not feasible. Therefore, only X^2 and non-parametric statistics were employed in the analysis.

Results

1. Quantitative Inheritance of the Basic Measures

Strain 13 females displayed a long latency and a relatively short duration of œstrus, a long duration of maximum lordosis, and a large number of male-like mounts during œstrus. Strain 2

females are characterised by a short latency and a long duration of œstrus, a short duration of maximum lordosis, and a general lack of male-like mounting behaviour (Table I). The differences between the inbred strains for these measures is highly significant ($P < ·001$) for every measure.

The values obtained for the F_1 resemble Strain 2 for latency and duration of œstrus. For both the duration of the maximum lordosis and the frequency of male-like mounting, the values for the F_1 females are intermediate to the values found for the parent strains. Since no significant differences were found between the reciprocals the results were combined.

In the F_2 generation values highly similar to those obtained for the F_1 generation were found. Again, latency and duration of œstrus resemble those values obtained for the inbred Strain 2. Values for duration of maximum lordosis and frequency of male-like mounting are intermediate to those of the inbred strains.

Within the B_2 population, the values obtained for latency and duration of œstrus and the duration of the maximum lordosis closely approximate those found for inbred Strain 2. On the other hand, mounting behaviour is not reduced to the low level characteristic of the inbred parent strain.

Backcrossing from F_1 to parent Strain 13 (B_{13}) produced females with shorter latencies and longer durations than those characteristic of Strain 13. The values for maximum lordosis and number of mounts were essentially restored to those obtained for the parent strain.

The mode of inheritance of the duration of the maximum lordosis deserves special attention. Table I presents the obtained means from the in-

Table I. Responses of Inbred and Hybrid Spayed Female Guinea Pigs Injected with 100 I.U. Œstradiol Benzoate and 0·2 I.U. Progesterone.

	Number ♀♀	tests	% of tests + for œstrus	Latency of œstrus (hours)	Duration of œstrus (hours)	Duration of max. lordosis (seconds)	Number of mounts per œstrus
Strain 2	26	67	94·0	4·2±·19	7·9±·28	13·6±·69	1·0±·27
Strain 13	29	87	79·3	6·6±·26	4·8±·25	24·5±1·34	19·5±3·34
F_1	23	69	98·6	4·8±·16	7·2±·30	19·4±1·29	5·9±1·19
F_2	47	130	95·4	4·7±·13	7·2±·20	19·5±·78	7·1±1·25
B_2	48	144	94·4	4·3±·10	7·9±·22	13·8±·22	5·1±1·31
B_{13}	55	165	86·7	5·7±·13	5·9±·18	22·4±·93	11·5±1·84

breds and crosses. For comparison, means predicted from the assumption of a single gene without dominance are as follows: F_1, 19·1 sec.; F_2, 19·1; B_2, 16·3; and B_{13}, 22·7. The close correspondence of obtained and predicted values strongly supports the assumption of a single gene in the determination of this trait.

Percentage response at 100 I.U. of estradiol, like latency and duration, displays phenotypic dominance of the strain 2 values (Table I). When 50 I.U. of estradiol are injected the same pattern of inheritance is demonstrated: strain 2 (63·6 per cent.), F_1 (62·8 per cent.), F_2 (63·0 per cent.), and B_2 (65·6 per cent); in contrast, strain 13 and B_{13} displayed only 41·9 and 37·9 per cent. response respectively.

2. Independence of the Basic Measures in Mode of Inheritance

Within the parent population as a whole (Strains 2 and 13 combined), duration of œstrus and per cent. response were positively associated and latency and duration inversely related. These associations exist not only between inbred strains, but also within each inbred strain. In addition, the relationship remains essentially unaltered for every genetic group (Table II). Retention of the association in populations of

Table II. Product Moment Correlations Between Latency and Duration of Œstrus within each Genetic Group.

Strain 2	Strain 13	F_1	F_2	B_2	B_{13}
—·53	—·79	—·76	—·52	—·64	—·56

increased heterogeneity suggests pleitropy (multiple manifestations of the same alleles). The common mode of inheritance displayed by these three measures (phenotypic dominance of the strain 2 values) supports the interpretation of a single underlying character. Since latency, per cent. response, and duration of œstrus fail to segregate independently, only the latter measure is used in the remainder of the analysis.

Significant negative associations exist between duration of heat and duration of maximum lordosis within the parent population ($X^2 = 27·69$, P<·001). However, genetic independence of these measures is established by the disappearance of any trace of association within the F_1 population and all subsequent crosses. Even within each pure strain the measures are not associated.

3. Analysis of Phenotypic Distributions

The distributions of duration of œstrus, maximum lordosis, and mounting within each genetic group are presented in Table III. The significance of the overall differences among the groups was evaluated by the use of X^2. On this and subsequent X^2 comparisons, expected frequencies of suitable magnitude were obtained by combining adjacent caetgories. For duration of œstrus, X^2 equals 74·02 (10 d.f., P<·001). For duration of maximum lordosis, X^2 equals 77·49 (10 d.f., P<·001). For frequency of mounting, X^2 equals 63·14 (10 d.f., P<·001).

The significance of the differences obtained from relevant between groups comparisons is presented in Table IV. From these comparisons the phenotypic dominance of the strain 2 duration of œstrus is evident. Among the crosses, only the distribution of duration within B_{13} differs significantly from that obtained for the inbred Strain 2. The distribution within B_{13} also differs significantly from that within Strain 13 indicating a retention of the strain 2 value by a significant proportion of the individuals (34·6 per cent.).

For both maximum lordosis and mounting, the distributions within B_2 do not differ from that of the parent strain but do differ from Strain 13. Correspondingly, these distributions within B_{13} do not differ from parent Strain 13 but do differ from Strain 2. Within the other crosses (F_1 and F_2) the distributions differ significantly from both parent strains due to an increase in the proportion of individuals displaying intermediate values.

All of the F_1 individuals have the same genetic constitution. The comparisons between distributions of F_1 and other hybrid groups will therefore serve as an indication of the extent of recombination achieved. Due to the phenotypic dominance of Strain 2 with respect to duration of œstrus, statistically significant reconstitution of the strain 13 type does not appear in the F_2. Instead, the reappearance of the "recessive" parental values for this measure is achieved only among the B_{13} individuals. It should be pointed out that a larger number of individuals would show a statistically significant difference between the F_1 and F_2 hybrids with respect to this trait.

In line with the hypothesis of a single pair of alleles and an intermediate mode of inheritance, the distribution of maximum lordosis ought to display bimodality in the F_2 generation. The failure to achieve this is probably due to the proximity of the modal categories of the parent

Table III. Proportions of Individuals within the Categories Listed for the Various Traits.

	Duration of œstrus in hours									
	2	3	4	5	6	7	8	9	10	11 or more
2	—	—	—	—	—	·423	·269	·115	·154	·038
13	·034	·137	·207	·207	·344	·069	—	—	—	—
F_1	—	—	·087	—	·131	·391	·261	·087	·043	—
F_2	—	—	·021	·064	·255	·212	·255	·149	·021	·021
B_2	—	—	—	·043	·146	·229	·250	·208	·064	·064
B_{13}	—	·018	·127	·237	·272	·218	·072	·036	·018	—

	Duration of maximum lordosis in seconds					
	5 to 9·98	10 to 14·9	15 to 19·9	20 to 24·9	25 to 29·9	30 and above
2	·192	·346	·423	·038	—	—
13	—	·069	·172	·379	·172	·206
F_1	—	·261	·304	·174	·174	·087
F_2	·021	·170	·426	·234	·064	·085
B_2	·104	·604	·250	·043	—	—
B_{13}	—	·127	·272	·237	·200	·163

	Frequency of mounting during œstrus				
	0 to 2·4	2·5 to 4·9	5·0 to 7·4	7·5 to 12·4	above 12·5
2	·804	·154	·038	—	—
13	·034	·137	·172	·137	·158
F_1	·217	·261	·304	·174	·043
F_2	·409	·170	·085	·149	·191
B_2	·583	·146	·083	·041	·147
B_{13}	·218	·127	·182	·164	·309

strains. For mounting activity, where the modal categories of the parent strains are widely separated, bimodality of the trait within the F_2 is achieved.

4. Combinations of Behavioural Components and the Recovery of a New Type

The existence of three independently assorting traits or measures permits the classification of sexual behaviour in a novel manner. Either a high or a low value of any measure may be dis- played by a given female. A "high" value is defined as any individual value above the median of the parental population. For the purpose of determining a parental median, the Ns of the two inbred strains were first equalised by random elimination of 3 strain 13 females. With two alternatives for each of three phenotypic elements, 8 distinct combinations are possible. The inbreds and crosses may then be compared on the basis of the frequency with which the different phenotypic combinations are observed.

116

Table IV. Significance of Differences between Trait Distributions of Inbreds and Crosses.

		2	13	F_1	F_2	B_2	B_{13}
Duration of œstrus	Strain 2		·001	—	—	—	·001
	Strain 13	·001		·001	·001	·001	·05
	F_1	—	·001		—	—	·001
Maximum lordosis	Strain 2		·001	·01	·001	—	·001
	Strain 13	·001		·05	·05	·001	—
	F_1	·01	·05		—	·05	—
Mounting activity	Strain 2		·001	·001	·01	—	·001
	Strain 13	·001		·01	·01	·001	—
	F_1	·001	·01		·02	·001	·07

These sexual behaviour "profiles" for the female are presented in Fig. 1. The behaviour profile most characteristic of Strain 13 falls within the categorical combination labelled "high maximum lordosis (HML), low duration of heat (LD), and high mounts (HMts); for strain 2 females, the modal category comprises the elements of low maximum lordosis (LML), high duration of heat (HD) and low number of mounts (LMts). Each inbred strain occupies relatively few of the 8 possible types. Within the parent population as a whole (P, in Fig. 1) 7 out of 8 possible types are represented. The F_1 crosses neither lose nor add types to the total array, but a significant increase, suggestive of heterosis, occurs in the proportion of females displaying high values on all measures (C.R.= 2·02, P < ·05). Within the F_2 crosses, the array of phenotypic combinations is much flatter, and a small number of cases appear with low values on every measure. This proportion, though small, is reliably greater than zero. The identification of females with a phenotypic combination not found in the parent population is expected on the basis of independently assorting genes. The deviation of the obtained proportion below that expected by chance is due to the phenotypic dominance of high values for duration of œstrus.

For the B_2 crosses, a significant proportion of females still display the new phenotypic combination. None of the females in B_2 group display the combination most characteristic of strain 13. Correspondingly, the new phenotypic combination is retained in females from B_{13}, but none displays the combination typical of pure strain 2.

Discussion

For the inbred strains, the findings from the present study agree substantially with those reported earlier (Goy & Young, 1957). Latency of heat is considerably lengthened, but both strains show the same increase in latency and the relative difference between the strains remains the same. We are unable to account for the observed increase at the present time but the change is unimportant, regardless of its origin, to the interpretation of the results.

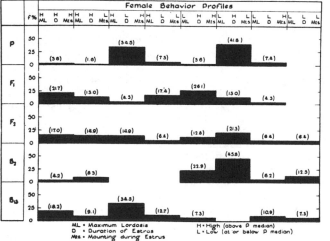

Fig. 1. Frequency distribution of combination of Sexual behaviour traits.

For the present, the precise mode of inheritance of the various components is regarded as significant only in so far as it establishes the independence of the components. There seems little reason to doubt the genetic independence of lordosis and mounting even though each displays intermediate inheritance. For the maximum duration of lordosis only a single or relatively few genetic factors are involved. This is clearly indicated first by the close agreement of means obtained from the hybrids with those predicted on the basis of a single gene without dominance. In addition, the restoration of the values characteristic of the pure strains in the first backcross generation is also indicative of relatively few genes in the determination of the trait. In contrast, mounting does not attain the values characteristic of the parent strains in either backcross group.

The association between a large amount of mounting and a long lordosis, observed within the parent population as a whole, is not maintained in the hybrid populations, again indicating genetic independence of the two components. Thompson (1957) has indicated that, for behavioural traits, covariations may result from four separate sources. These he has called genetic communality, chromosomal cummunality, selection communality, and environmental communality. Chromosomal and environmental communality are excluded as bases for the association since a long lordosis segregates independently of mounting in the hybrid generations.

From such considerations it seems likely that the association resulted from the artificial "selection" of a particular sub-family, or alternatively, only those lines in which the traits were associated continued to reproduce. Supporting the latter possibility is the informal observation that mounting by the female has a facilitating effect on the mating activity of the male. The value of such facilitation for the perpetuation of a strain in which the males are repeatedly characterised as sluggish (Jakway, in press; Valenstein, Riss & Young, 1954; Riss, 1955) is apparent. On the other hand, the ease with which hybrids were obtained from matings between males from Strain 13 and females from Strain 2 indicates that mounting by females is not crucial to mating even though it may be facilitating.

The separateness of the genetic bases for the quantitative aspects of mounting and lordosis is supported by physiological findings. Young, Dempsey, Myers & Hagquist (1938) have shown that the two components are differently related to the ovarian condition at the time of œstrus. The mounting component appears later in ontogeny than the lordosis and mounting is displayed earlier than the lordosis in the cycle of the mature female (Young & Rundlett, 1939). And a few cases have been reported of intact female guinea pigs who mount at cyclic intervals without ever displaying lordosis (Young, 1941).

The present finding provides evidence of separate genetic bases only for the quantitative aspects of mounting and lordosis; whether genetically separate hormonal thresholds exist cannot be determined. The finding has direct implications for our concept of vigour since it shows clearly that the vigour of one component is completely independent of the vigour of the other. This principle, conclusively demonstrated for the female, has been demonstrated in studies on the male also. In a study dealing with the inheritance of the male mating behaviour pattern, different modes of inheritance were demonstrated for ejaculation frequency and rate of intromission on the one hand and the rate of occurrence of the lower components on the other hand (Jakway, in press). Thus vigour of the mating behaviour pattern, as we have come to view it, is not inherited as a unitary trait (i.e. a set of genes regulating the vigour of all aspects of mating behaviour), but rather as a separate set of genetic factors for each of the genetically independent behavioural elements.

In the present study we have been able to record a number of measures which reflect the character of the lordosis reflex. This has permitted us to evaluate both the strength or vigour of that reflex (as measured by the duration in seconds of the maximum lordosis displayed in response to fingering) and also its latency of appearance, the number of successive hours it is displayed, and the per cent. of the tests on which it is displayed. According to the genetic analysis, these four aspects of the lordosis component can be accounted for by two independent genetic factors. The vigour of the lordosis reflex is determined by one of these genetic factors; latency of heat, duration of heat, and per cent. response are jointly determined by the second set of factors as indicated by their common mode of inheritance (phenotypic dominance of Strain 2) and their failure to assort independently during hybridization. Previously (Goy & Young, 1957) "responsiveness" was defined as "a measure of the effectiveness of the hormone in inducing œstrus, regardless of the character of the induced œstrus." Operationally, the principal criterion was percentage response. In view of the present findings it seems necessary to conclude that latency and duration of heat are also measures of responsiveness to the hormone.

We are unable to account for the findings with regard to mounting behaviour on a strictly genetical basis. When the data for the male (Jakway, 1959) are compared to those for the female, the inheritance of mounting appears quite complex. For in the males, those from Strain 2 exhibit more mounting (Valenstein, Riss & Young, 1954) and a higher rate of mounting (Jakway, 1959) than males from Strain 13. Exactly the opposite quantitative relationships are found for females from these same strains. Possibly alpha-œstradiol benzoate and progesterone produce mounting by acting upon tissues completely different from those acted upon by male gonadal hormone. But the kind of hormone does not appear to be the only factor involved in the low expressivity of this trait in females from Strain 2 (Goy & Young, 1958). For the present, the possibility of modifiers associated with the sex of the individual must be entertained.

Particular attention is called to the finding that the genetic factors controlling responsiveness, maximum duration of lordosis, and frequency of mounting recombine in the F2 population in such a way as to produce a phenotypic combination which is completely new (i.e. without prototype in the parent population). This principle, although well known for morphological traits has never been completely documented for behaviour. Recently, Lorenz (1958) has pointed out the significance of such a principle to the evolution of behaviour and some evidence for the novel recombination of behavioural elements has been obtained by him in studying interspecific crosses. Our findings extend the concept advanced by Lorenz in so far as recombination occurs not only with respect to the presence and arrangement of elements within a pattern of behaviour, but also with respect to the quantitative expression or vigour of the various elements.

Summary and Conclusions

Two hundred and twenty-eight spayed female guinea pigs were used. Twenty-six were from highly inbred Strain 2, 29 from Strain 13, 23 from the F_1 generation, 47 from the F_2, 48 from backcrosses of F_1 to Strain 2, and 55 from back-

crosses between F_1 and Strain 13. The animals were injected three times at monthly intervals with 100 I.U. of œstradiol benzoate and twice with 50 I.U. of œstradiol. In every case, the estradiol was followed 36 hours later with 0·2 I.U. of progesterone. The females were placed in unisexual groups, observed continuously for 14 hours, and their behavioural responses recorded. The data included individual determinations for latency and duration of heat, duration of the maximum lordosis displayed during heat and the frequency of male-like mounting. Percentage response was determined by the number of females displaying behavioural œstrus divided by the total number tested.

The results indicated that three independent genetic factors determine the character of œstrus as we have measured it. Latency of heat, duration of heat and per cent. response all show correlated inheritance and may be multiple manifestations of a single set of factors which determine the responsiveness of the individual to œstradiol benzoate. These traits display phenotypic dominance of the strain 2 values.

A completely independent genetic mechanism determines the duration of the maximum lordosis. The mode of inheritance is intermediate and the results may be interpreted in terms of a single genetic factor without dominance.

The genetic basis for male-like mounting activity appears completely independent of the genetic mechanisms for maximum lordosis. Inheritance is of the intermediate type but more than a few genetic factors are indicated and the possibility of modifiers (associated with the sex of the individual) exists.

For the female guinea pig, the pattern of behaviour displayed during œstrus is not inherited as a unitary trait. It is concluded that the negative association between mounting and duration of heat, as well as that between maximum lordosis and duration of heat, are a product of selection pressures. These associations found among the inbred strains are not attributable to genetic linkage of the traits. In addition, it is concluded that sensitivity to a hormone is genetically independent of the factors determining the quantitative character of the response.

Acknowledgments

The authors wish to express their gratitude to Dr. John A. Weir of the Department of Zoology, University of Kansas, for his help and discussion during the planning and execution of this work.

We are particularly indebted to him for a critical evaluation of the genetical problems in this work.

REFERENCES

Boling, J. L. & Blandau, R. J. (1939). The estrogen-progesterone induction of mating responses in the spayed female rat. *Endrocrinology*, **24**, 359-364.

Boling, J. L., Young, W. C. & Dempsey, E. M. (1938). Miscellaneous experiments on the estrogen-progesterone induction of heat in the spayed guinea pig. *Endocrinology*, **23**, 182-187.

Goy, R. W. & Young, W. C. (1957a). Strain differences in the behavioural responses of female guinea pigs to alpha-œstradiol benzoate and progesterone. *Behaviour*, **10**, 340-354.

Goy, R. W. & Young, W. C. (1957b). Somatic basis of sexual behaviour patterns in guinea pigs: Factors involved in the determination of the character of the soma in the female. *Psychosom. Med.*, **29**, 144-151.

Jakway, J. S. (1959). The inheritance of patterns of mating behaviour in the male guinea pig. *Anim. Behav.*, **7**, 150-162.

Lorenz, K. (1958). The evolution of behaviour. *Sci. Amer.*, **199**, 67-82.

Riss, W. (1955). Sex drive, oxygen consumption and heart rate in genetically different strains of male guinea pigs. *Amer. J. Physiol.*, **180**, 530-534.

Thompson, W. R. (1957). Traits, factors, and genes. *Eugen. Quart.*, **4**, 8-16.

Valenstein, E. S., Riss, W. & Young, W. C. (1954). Sex drive in genetically heterogeneous and highly inbred strains of male guinea pigs. *J. comp. physiol. Psychol.*, **47**, 162-165.

Young, W. C., Dempsey, E. W., Myers, H. J. & Hagquist, C. W. (1938). The ovarian condition and sexual behaviour in the female guinea pig. *Amer. J. Anat.*, **63**, 457-483.

Young, W. C. & Rundlett, B. (1939). The hormonal induction of homosexual behaviour in the spayed female guinea pig. *Psychosom. Med.*, **1**, 449-460.

Young, W. C. (1941). Observations and experiments on mating behaviour in female mammals. *Quart. Rev. Biol.*, **16**, 135-156.

Accepted for publication 23rd March, 1959.

Editor's Comments
on Papers 12 Through 15

DEVELOPMENT AND EFFECTS OF EXPERIENCE

The behavioral patterns and capacities of organisms change radically during their lifetimes. Over the range of youth, maturity, and old age, new patterns appear, old patterns disappear, and the relative frequencies of occurrence of behavioral patterns change. In analyzing behavioral development, one is confronted with both a baseline that changes over an animal's lifetime thus affecting its responses to experience and with the need to assess the effects of experience that, in turn, act on a changing baseline. We shall first consider the changes that occur as a function of age, under relatively typical environmental conditions, and then proceed to consider the effects of particular kinds of environmental input.

The study by Bingham (Paper 12) deals with the development of sexual behavior in young chimpanzees. Bingham stresses the fact that copulation cannot be studied in isolation, as events and experience occurring prior to the first overt copulatory activity are of overwhelming importance for the occurrence of copulation. Like other students of primate sexuality, Bingham stresses the role of experience as well as the variability that characterizes primate behavior. More recently, Michael and Wilson (1973) summarized the current status of the literature on the ontogeny of primate sexual behavior.

Stone (Paper 3) studied the ontogeny of copulation in male laboratory rats, concluding that the earliest age at which copulation was observed was sixty-four days. Larsson (1956) refined the behavioral measures of rat copulation and found that mounts and intromissions occurred at an age earlier than did ejaculation. Larsson found copulation in young males to differ quantitatively from that of older males. For example, young males required more intromissions prior to reaching ejaculation. Beach (1966) traced the course of lordotic behavior in guinea pigs. Lordosis is seen in young males and females but disappears as the animals mature, only to reappear under hormonal control in the adult female.

Sexual behavior also changes during the period of an animal's mature life, from its first occurrence through the remainder of its adult life. Much of this change is attributable to experience. Michael and Wilson (1973) reported that whereas rhesus macaques displayed poorly oriented copulatory behavior prior to their first-ever ejaculation, both orientation and quantitative aspects were changed in older, experienced animals. Larsson (1967) reported that the developmental changes that occur in the copulatory behavior of maturing rats as discussed above depend on the actions of androgens but not on sexual activity during this time.

Following up the earlier work of Larsson (1956), Larsson and Essberg (Paper 13) described the course of the changes in sexual activity in aging male rats. They conclude that, whereas the aging male has a heightened threshold for the arousal of the copulatory response, the capactiy of males to attain multiple ejaculations increased with age within the range tested. By periodically handling old males, Larsson (1963) accelerated their rate of copulation to a level comparable to that of younger males. The female reproductive system becomes less responsive with age. For example, old, multiparous females require more stimulation for the initiation of pregnancy (Paper 38) than younger females (Davis et al., 1977).

Numerous researchers have attempted to manipulate the rate of develpment of the capacity for and occurrence of sexual behavior (see Beach, 1948). In Paper 14 Stone reports precocious mating in male rats given subcutaneous injections of testosterone propionate. Stone's results were further refined by Beach (1942a), Baum (1972), and Södersten et al. (1977). Similar results were reported for young chicks by Noble and Zitrin (1942). Precocious development in female rats given hormone administration has been reported by Beach (1942a), Södersten (1975), and Williams (1979). In the latter study, the complete pattern of lordosis and "ear wiggling" was observed in six-day-old females. Various other treatments affect sexual maturation and the age at which sexual activity is observed, including delivery of peripheral electric shock (Goldfoot and Baum, 1972), pheromones (Vandenbergh, 1971, 1973), and nutritional state (Larsson, et al., 1974).

There is a large literature on the effects of early experience on adult sexual behavior, perhaps stimulated in part by Freudian theories. In Paper 15 Valenstein, Riss, and Young studied the consequences of early experience on the develpment of sexual behavior in guinea pigs. They found that contact with other animals was important for the develpment of normal sexual behavior and that different genotypes were differentially affected by early experience. Males reared in social isolation were deficient in their sexual behavior relative to socially reared animals. This effect was further refined by A. A. Gerall (1963) and H. D. Gerall (1965).

The effects of early experiental deprivation in primates have generated much interest. An effect of experience was implied in Bingham's work (Paper 12). An important program of research on chimpanzees was conducted by H. W. Nissen. Nissen died before publishing a complete report, and his results were summarized by Riesen (1971). Nissen studied a group of chimpanzees deprived of normal maternal care and peer interaction from chimpanzees but given much human handling. Although these deprived animals did not copulate with each other at maturity, most females eventually mated when paired with experienced males; pairing naive males with experienced females was slightly less effective. Rogers and Davenport (1969) found that most chimpanzees reared in more complete isloation learned to copulate with experienced partners and suggested that human interaction is more detrimental to the young chimpanzee than isolation. In a series of very important studies of the effects of early deprivation in rhesus monkeys, Mason (1960), Harlow (1965), and Missakian (1969) found substantial deficits in sexual behavior to follow early social deprivation in rhesus monkeys. These defecits appear more refractory to tutoring by experienced animals than those of chimpanzees (Harlow 1965).

The literature on the effects of early social deprivation on sexual behavior in rats is somewhat more confusing than that in other species. Stone (Paper 3) reported normal sexual behavior in males reared in isolation from the time they were weaned. Beach and his associates have repeatedly examined this question and found no detrimental effect of early social deprivation on mating behavior in male rats (Beach, 1942b, 1958; Kagan and Beach, 1953). Indeed, there was a suggestion that prepubertal social experience of some kinds may be detrimental, perhaps by establishing behavioral tendencies that compete with the tendency to mate. However, there is also a considerable body of more recent literature that suggests that early social isolation is detrimental to male rats (for example, Gerall et al., 1967; Gruendel and Arnold, 1969; Hård and Larsson, 1968). The reasons for these discrepancies are not presently clear.

As indicated above, adult mating experience also affects copulatory behavior. Various measures of copulatory behavior in male rats have been found to show relatively small, quantitative alterations as a function of copulatory experience (Dewsbury, 1969; Larsson, 1959; Rabedeau and Whalen, 1959). Beach (1947, 1948) suggested that the copulatory behavior of "higher forms" is more dependent on the cerebral cortex and on experiential factors and less dependent on hormones than that in "lower forms" (see also Paper 34). Much comparative research has been conducted on the effects of copulatory experience on the retention of copulatory behavior after castration in an attempt to assess this proposal. Facilitative effects of experience have been found in cats (Rosenblatt, 1965), hamsters (Bunnell and Kimmel, 1965), house mice (Manning and Thompson, 1976), and in some strains of rabbits (Ågmo, 1976). By contrast, no facilitative effects were found in laboratory rats (Rabedeau and Whalen, 1959; Bloch and Davidson, 1968) or dogs (Hart, 1968). Rhesus monkeys are quite variable with respect to the rate at which copulation disappears after castration (Phoenix et al., 1973). There is no obvious pattern to the comparative data on the effects of experience on the loss of copulatory behavior after castration.

REFERENCES

Ågmo, A., 1976, Sexual behavior following castration in experienced and inexperienced male rabbits, *Z. Tierpsychol.* **40**:390–395.

Baum, M. J., 1972, Precocious mating in male rats following treatment with androgen or estrogen, *J. Comp. Physiol. Psychol.* **78**:356–367.

Beach, F. A., 1942a, Sexual behavior of prepuberal male and female rats treated with gonadal hormones, *J. Comp. Pysiol. Psychol.* **34**:285–292.

Beach, F. A., 1942b, Comparison of copulatory behavior of male rats raised in isolation, cohabitation, and segregation, *J. Genet. Psychol.* **60**:121–136.

Beach, F. A., 1947, Evolutionary changes in the physiological control of mating behavior in mammals, *Psychol. Rev.* **54**:297–315.

Beach, F. A., 1948, *Hormones and Behavior*, Harper, New York.

Beach, F. A., 1958, Normal sexual behavior in male rats isolated at fourteen days of age, *J. Comp. Physiol. Psychol.* **51**:37–38.

Beach, F. A., 1966, Ontogeny of "coitus-related" reflexes in the female guinea pig, *Proc. Nat. Acad. Sci. U.S.A.* **56**:526–533.

Bloch, G. J., and J. M. Davidson, 1968, Effects of adrenalectomy and experience on postcastration sex behavior in the male rat, *Physiol. Behav.* **3**:461–465.

Bunnell, B. N., and M. E. Kimmel, 1965, Some effects of copulatory experience on postcastration mating behavior in the male hamster, *Psychonomic Sci.* **3**:179–180.

Davis, H. N., G. D. Gray, and D. A. Dewsbury, 1972, Maternal age and male behavior in relation to successful reproduction by female rats (*Rattus norvegicus*), *J. Comp. Physiol. Psychol.* **91**:281–289.

Dewsbury, D. A., 1969, Copulatory behavior of rats (*Rattus norvegicus*) as a function of prior copulatory experience, *Animal Behav.* **17**:217–223.

Gerall, A. A., 1963, An exploratory study of the effect of social isolation variables on the sexual behavior of·male guinea pigs, *Animal Behav.* **11**:274–282.

Gerall, H. D., 1965, Effect of social isolation and physical confinement on motor and sexual behavior of guinea pigs, *J. Pers. Soc. Psychol.* **2**:460–464.

Gerall, H. D., I. L. Ward, and A. A. Gerall, 1967, Disruption of the male rat's sexual behavior induced by social isolation, *Animal Behav.* **15**:54–58.

Goldfoot, D. A., and M. J. Baum, 1972, Initiation of mating behavior in developing male rats following peripheral electric shock, *Physiol. Behav.* **8**:857–863.

Gruendel, A. D., and W. J. Arnold, 1969, Effects of early social deprivation on reproductive behavior of male rats, *J. Comp. Physiol. Psychol.* **60**:123–128.

Hård, E., and K. Larsson, 1968, Dependence of adult mating behavior on the presence of littermates in infancy, *Brain, Behav., Evol.* **1**:405–419.

Harlow, H. F., 1965, Sexual behavior in the rhesus monkey, in *Sex and Behavior*, F. A. Beach ed., Wiley, New York, pp. 234–265.

Hart, B. L., 1968, Role of prior experience in the effects of castration on sexual behavior of male dogs, *J. Comp. Physiol. Psychol.* **66**:719–725.

Kagan, J., and F. A. Beach, 1953, Effects of early experience on mating behavior in male rats, *J. Comp. Physiol. Psychol.* **46**:204–208.

Larsson, K., 1956, Conditioning and sexual behavior in the male albino rat, *Acta. Psychol. Gothoburg.* **1**:1–269.

Larsson, K., 1959, Experience and maturation in the development of sexual behavior in male puberty rat, *Behaviour* **14**:101–107.

Larsson, K., 1963, Non-specific stimulation and sexual behavior in the male rat. *Behaviour* **20**:110–114.

Larsson, K., 1967, Testicular hormone and development changes in mating behavior of the male rat, *J. Comp. Physiol. Psychol.* **63**:223–230.

Larsson, K., S. G. Carlsson, P. Sourander, B. Forsstrom, S. Hansen, B. Henriksson, and A. Lindquist, 1974, Delayed onset of sexual activity of male rats subjected to pre- and postnatal undernutrition, *Physiol. Behav.* **13**:307–311.

Manning, A., and M. L. Thompson, 1976, Postcastration retention of sexual behavior in the male BDF$_1$ mouse: The role of experience. *Animal Behav.* **24**:523–533.

Mason, W. A., 1960, The effects of social restriction on the behavior of rhesus monkeys. I. Free social behavior, *J. Comp. Physiol. Psychol.* **53**:582–589.

Michael, R. P., and M. Wilson, 1973, Changes in the sexual behavior of male rhesus monkeys (*M. mulatta*) at puberty, *Folia Primatol.* **19**: 384–403.

Missakian, E. A., 1969, Reproductive behavior of socially deprived male rhesus monkeys (*Macaca mulatta*), *J. Comp. Physiol. Psychol.* **69**: 403–407.

Noble, G. K., and A. Zitrin, 1942, Induction of mating behavior in male and female chicks following injection of sex hormones; including notations on body weight and comb growth, *Endocrinology* **30**:327–334.

Phoenix, C. H., A. K. Slob, and R. W. Goy, 1973, Effects of castration and replacement therapy on sexual behavior of adult male rhesuses, *J. Comp. Physiol. Psychol.* **84**:472–481.

Rabedeau, R. G., and R. E. Whalen, 1959, Effects of copulatory experience on mating behavior in the male rat, *J. Comp. Physiol. Psychol.* **52**:482–484.

Riesen, A. H., 1971, Nissen's observation on the development of sexual behavior in captive-born, nursery-reared chimpanzees, in *The Chimpanzee*, G. H. Bourne, ed., Karger, Basel, pp. 1–18.

Rogers, C. M. and R. K. Davenport, 1969, Effects of restricted rearing on sexual behavior of chimpanzees, *Dev. Psychol.* **1**:200–204.

Rosenblatt, J. S., 1965, Effects of experience on sexual behavior in male cats, in *Sex and Behavior*, F. A. Beach, ed., Wiley, New York, pp. 416–439.

Södersten, P., 1975, Receptive behavior in developing female rats, *Horm. Behav.* **6**:307–317.

Södersten, P., D. A. Damassa, and E. R. Smith, 1977, Sexual behavior in developing male rats, *Horm. Behav.* **8**:320–341.

Vandenbergh, J. G., 1971, The influence of the social environment on sexual maturation in male mice, *J. Reprod. Fert.* **24**:383–390.

Vandenbergh, J. G., 1973, Acceleration and inhibition of puberty in female mice by pheromones, *J. Reprod. Fert.* **19**:411–419.

Williams, C. L., 1979, Steroids induce lordosis and ear wiggling in infant rats, paper presented at the Eastern Conference on Reproductive Behavior, New Orleans, La.

12

Reprinted from pp. 1–3, 151–161 of *Comp. Psychol. Monog.* 5:1–165 (1928)

SEX DEVELOPMENT IN APES

HAROLD C. BINGHAM

Institute of Psychology, Yale University

INTRODUCTION

Naturalistic accounts of anthropoid behavior have long emphasized social relations. In the early records of explorers and hunters, including reports of natives, there are many stories of aggression by the manlike apes. Little in the way of detailed behavior between the sexes, however, can be obtained from such sources. Yet it is not uncommon to find references to sexual excitement observed in free living apes and attributed to the sight of humans. Among the reports of such responses there is emphasized with surprising frequency differential behavior towards the opposite sex.

Stories of the ravishing of native women by powerful male

apes, often reputed to be half human, appeared long before there were recognized differences among orang utans, chimpanzees, and gorillas.   In various though similar forms these abduction stories seem to have been in the possession of many tribes.   Just what foundation there was for the legend has been the subject of considerable speculation.   Several explanations have been suggested including religious superstitions and ceremonies, partially witnessed tragedies, and the lies of unfaithful wives.

Native accounts are also reported in which it is maintained that anthropoid and human matings occur voluntarily in a state of freedom and that such crossing produces fertile offspring. Such reports seem never to have been verified.   In the reports of natives, authors, or publishers there is such probability of a mixture of observation and imagination that the accounts cannot be given a place in objective studies of behavior.

For the present study, selections from the literature have been restricted to reported observations made under such conditions that reliability is reasonably possible and generally probable. A large body of references is thus excluded from consideration. Publications on the orang utan and gibbon are omitted because my original data are selected primarily from the chimpanzee. Incomplete records of sexual activities in lower primates and a few references to the gorilla are included chiefly for purposes of comparison.

It was in my initial plan to include sections on menstruation, gestation, and parturition.   References to and incomplete descriptions of these aspects of reproduction appear in the literature, but the lead of my own observations compelled me, at least temporarily, to ignore such topics.   The sexual behavior that I have described is preceded by a section on the external genitalia of the chimpanzee and gorilla.[1]   These structural facts will not be equally acceptable to all readers, but their introductory value is by no means negligible at a time when knowledge of the anthropoids is rapidly growing.

Throughout the report there is emphasized the significance of

[1] The structures of the orang utan and gibbon are omitted because the behavior reported in succeeding chapters does not involve these organisms.

genetic continuity. Random records from the lower primates, designated as cross sections of behavior, emphasize by contrast the importance of developmental history. This point of view demands consideration of social behavior, in itself non-sexual, that contributes to the developing sexual function. In several observations I have been confronted with indications of sexual insight[2] but, for convenience in presenting here the facts about development and enlarging experience, the topic is not discussed. It seems more appropriate to reserve such discussion for a separate paper.

Four stages of postnatal development in the chimpanzee— infancy, childhood, adolescence, and maturity[3]—have come within my observations. Following a summer's work in 1924 at Havana, Cuba, with the Abreu colony of primates, there was inaugurated the following year in the Yale Institute of Psychology a more extended period of study on chimpanzee childhood. From chimpanzees in these early years of development the most valuable records of sexual genesis have been obtained, but the origins of this behavior clearly reach back to, if not beyond, the beginnings of postnatal life.

Financial assistance has come from the Carnegie Institution of Washington, the Laura Spellman Rockefeller Memorial, the Rockefeller Foundation, and the Committee for Research in Problems of Sex, National Research Council. Madame Rosalía Abreu generously gave me access to her collection of animals. It is quite impossible to acknowledge all of my personal obligations. Highest appreciation goes to Professor Robert M. Yerkes, Professor Raymond Dodge, and Doctor Donald K. Adams. In various ways I have also had valuable assistance from Doctor Leon S. Stone of the Department of Anatomy, Yale School of Medicine; Mr. Chauncey McK. Louttit, Mr. Carleton F. Scofield, Mrs. Margaret Child Lewis, Doctor O. L. Tinklepaugh, and Mr. Morgan Upton, all of the Yale Institute of Psychology.

[Editor's Note: Material has been omitted at this point.]

[2] I am inclined to recognize insight in behavior involving sexual as well as nutritional adjustments.
[3] Cf. page 39.

viously described in mutual relations. The variety of methods employed by the female chimpanzee is especially significant. Even the behavior of the male when he was obviously irritated revealed variety. The change from pulling with the hand to rubbing with the toe, and the alternations in the use of both great toes, however, lack the versatility displayed by the female. That masturbatory behavior occurs under different emotional circumstances, the facts tend to indicate. That experience may significantly condition the behavior, there can be little doubt. Whether the conditioning is so sudden and so variable as in mutual relations, the data are too meager to reveal.

XI. SUMMARY AND CONCLUSIONS

The facts presented in the foregoing sections may be reviewed under the following topics:

1. Cross section versus genetic records of sexual behavior.
2. Mating behavior of adolescent and mature chimpanzees.
3. Ontogenetic data from chimpanzees in infancy and childhood.
4. Non-reproductive sexual behavior.
5. Universality of sexual excitability.
6. Methods of studying sexual behavior in man and apes.

1. Former studies of sexual behavior in monkeys and apes, whether they include field or laboratory data, have rested primarily on random or cross section observations. Such records have been used in extensive discussion of their phylogenetic significance without evaluating them first in their ontogenetic relations. My observations have been presented primarily from the standpoint of individual development. I have assembled various kinds of evidence showing that the fundamental features of copulatory adjustments between chimpanzees appear and develop years before the individual reaches reproductive maturity.

The importance of knowing the ontogeny of primate representatives when dealing with phylogenetic problems was demonstrated in investigations on the genital structures a half century or more ago. The applicability of this genetic approach to an explanation of sexual behavior has been emphasized by my work on the rôle of experience in reproductive relations. This evidence convinces me that sexual experience, as such, is initially derived from non-sexual adjustments, and that such adjustments, after the appearance of a sexual focus, continue to supplement the developing organization of reproductive behavior.

In sexual behavior, as in genital structure, human and anthropoid resemblances have been more noticeable in the early stages of ontogeny. Until the behavior of reproductively mature apes has been observed in detail, however, as I have done with young animals, and until more subjects have come under observation throughout their growth and development, this suggestion of decreasing resemblance with increasing age should be accepted only tentatively. In support of the suggestion, and in disagreement with previous reports and assumptions, my demonstration that ventroventral adjustments are commonly made in the copulatory play of young chimpanzees is particularly significant. With young animals, also, it has been observed that intromission is possible from the ventroventral position.

In this report there have also been presented for comparison, descriptions of copulatory behavior between various pairs of monkeys. In such behavior the patterns of response seemed to be well perfected, but the assumed development had reached a stage where little in sexual genesis could be observed. Descriptions that are now available suggest greater uniformity and precision in the reproductive behavior of monkeys than of apes. However, such a conclusion can scarcely be supported until more is known about sexual development in various representatives of the primates.

Existing evidence probably justifies the conclusion that variability in the copulatory adjustments of monkeys may be expected in different individuals, species, and colonies. Observed differences in copulation have centered more on secondary behavior than on the copulatory act itself. Relatively stereotyped copulatory behavior has been observed in monkeys under varying social situations, probably representing defense, attack, and exhibitionism, as well as reproduction. It is not improbable, however, that further study of sexual ontogeny will reveal increasing variability in mutual adjustments as the subjects respectively represent monkeys, apes, and man.

2. My first clue to the importance of experience in the sexual behavior of chimpanzees was revealed when Anumá and Malapulga were initially mated. Both had probably reached reproductive maturity, yet their initial attempts to copulate were failures. The overt essentials for copulation were present, but their organization was incomplete. The male seemed to proceed no farther with any of these copulatory factors than he had been observed to do previously in situations probably having for him no sexual significance. During the mating experiment, when these factors were assembled in behavior having a probable sexual focus, the female appeared much better prepared to consummate the mutual adjustments than the male. The conspicuous factors in her behavior also had been observed before she was with the male.

Few opportunities have existed for scientists to observe the reproductive behavior of adult apes, and almost none for studying its origins and development. The records that exist, like those concerning Anumá and Malapulga, are cross sections of behavior. The copulatory relations of Jimmy and Cucusa have been partially reported, but the published account suggests mechanical precision in spite of the reporter's comment about the extraordinary procedure.

Reports indicate that the male chimpanzee has a sex dance which he performs in the presence of the female preparatory to copulation. Prior to my observations of Jimmy, I had assumed that this performance was quite stereotyped and definitely limited to mating situations. It is my present opinion that this dance is variable with individual development and performed with non-sexual as well as sexual significance. Its performance is probably not far removed from the erect swaggering, waddling, and dancing that have been observed in many social situations and reported for pre-adolescent chimpanzees.

3. The rôle of experience and variability in anthropoid reproduction has been further emphasized by following the development in detail of four young chimpanzees, and in cross sections, of other chimpanzee infants. Months before there was any indication of a sexual focus in the behavior of the Yale chimpanzees, I began to recognize sexual factors in other social foci. The accompanying figure represents in schematic outline the relationship between various activities and copulation. The details themselves are subject to revision, but they support as a whole the principle of widespread distribution and genetic significance of copulatory factors.

The development of copulatory play between these young animals was observed in controlled meetings. They were brought together in pairs, and later as a group. The responses between animals of the same sex and of the opposite sex were compared. Their behavior towards non-primates both larger and smaller than themselves, and towards inanimate objects was observed and recorded day after day.

Individual histories prior to my observations are practically unknown, but my records support the judgment that I witnessed initial copulatory play. Further enlightening facts may be obtained if the mating behavior of these same animals at the beginning of reproductive maturity can be similarly recorded for comparison with my early records. Several years must yet elapse, however, before all will have reached such a stage of development.

Following the initiation of mutual sex play, there were many repetitions and variations. Conspicuous differences appeared in ventroventral and dorsoventral adjustments. Other variations of significance were noticeable in the social situations that fostered copulatory play. Moreover, the variability that occurred in the sexual behavior itself appeared to be only a continuation of that which prevailed in the sources of such behavior. A heterogeneous group of variable activities, including both primitive responses and recent acquisitions, seemed to contribute in varying degrees to these sexual foci. Excitement—revealed by romping, teasing, petting, fleeing, eating, fighting, tantrums,

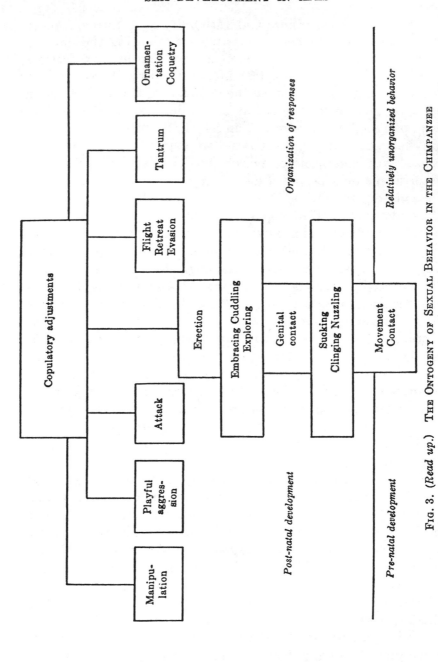

FIG. 3. (*Read up.*)   THE ONTOGENY OF SEXUAL BEHAVIOR IN THE CHIMPANZEE

Schematic presentation showing some of the developing factors that contribute to copulatory adjustments

and commonly by mixtures of these and other activities—was a consistent forerunner of sexual responses.

4. In homosexual and masturbatory responses, the apes have revealed varieties of enlarging sexual experiences that are not directly related to the reproductive function. My evidence is insufficient to decide whether these aspects are to be attributed to the existence of the variable principle or to its relative absence. There seem to be individual differences in such expression, and the behavior is evidently modifiable. To the extent that I have observed it, I suspect it is quite natural, in the open as well as in confinement, among animals that are actively engaged in making varieties of contacts with one another, and in manipulating their environment.

Homosexual activities between male monkeys have been widely reported. Relatively few observations of such relations between female monkeys seem to have been made. Among the chimpanzees I have made many observations of sexual behavior between females. Both ventroventral and dorsoventral adjustments have occurred. The fact that I have no reliable records of sexual play between male chimpanzees should not be accepted as negative evidence, because the opportunities for making such observations have been relatively limited. It is probable that relative age, size, strength, and perhaps other individual characteristics, are significant factors in this behavior.

Variety and exploration in masturbatory behavior have been pronounced. When holding food in the hands, the young chimpanzee regularly sits with elevated knees close against the sides of the abdomen and feet against the buttocks. The great toes often turn in and readily reach the penis which commonly becomes erect at the sight of food. The practice of rubbing the penis while eating may thus be established.

The use of the hands in the masturbatory acts of males is probably less common. In different expressions of excitement, however, it has been observed. It seems to occur in irritation as a part of the common practice, exaggerated in tantrums, of pulling the hair or scratching the body with the fingers. It has been observed in playful activities only when inanimate objects were combined with manual operations.

Erections in the young female chimpanzee have been noted occasionally in response to tactual stimuli. Following urination, copulatory play, and body contacts, enlargements of the clitoris have often occurred. Like the males, the females frequently push their genitalia against inanimate objects. They have also been observed to manipulate their genitals with fingers, and also to use improvised instruments such as sticks, pebbles, and fruit skins.

5. Social stimulation seems to increase the frequency of sexual expression among infrahuman primates. The visible appearance of four monkeys was followed during the next twenty-four hours by more copulatory play in the adjoining cage of chimpanzees than had been witnessed during the preceding two months. Whether aroused by human or infrahuman spectators, such exaggerations, at least among the monkeys, may lead to excesses. Rivalry, it appears, as well as companionship, may encourage these exaggerations.

It was in exhibitionistic activities that some of the most pronounced explorations in sexual behavior appeared. A phase of developing ostentation among the young chimpanzees was preceded by a period of repression in the presence of human visitors. Under similar social conditions the spontaneity and sexual responsiveness of the same animals in the later phase became exaggerated. The direction of the sexual behavior was often towards the human visitor or companion, but it was commonly diverted to one of the chimpanzee companions. There is evidence of preference among the chimpanzees for human individuals of the opposite sex.

In the general excitement which regularly precedes the copulatory responses of the chimpanzee, the perceptions are apparently dominant factors. I believe tactual and visual receptors are most important in the initiation of such behavior. Contact is probably the basic factor in sexual development, although vision may later replace it in relative importance. There is evidence that touch and vision are supplemented somewhat by smell, taste, and possibly hearing.

Thus it appears that a considerable variety of stimuli may set off sexual responses in the apes and monkeys. The versatility of the chimpanzee and the variety of situations that elicit sexual behavior demonstrate the readiness with which excitations spread from non-sexual to sexual areas. There is clearly a wide variation from individual to individual in sexual adaptability. The evidence points to the necessity of a preliminary sexual development before adequate mutual adjustments for reproduction can be made. Even in the absence of anthropoid companions for mutual sex play, the versatility of the apes is such that varieties of sexual expression are revealed.

6. The readiness with which results have been obtained on the sexual life of apes stands out in considerable contrast with the varied and complicated methods employed in such studies of human behavior. There are definite limitations to the use with humans of methods that are applicable without restrictions to the lower primates. Traditions in civilized society are such that continual observance of sexual expression during its entire ontogeny is quite impossible. Unfortunately for direct observation of the developing sex life, man has unexcelled facility in sharing and concealing his own mental content. Direct study of his spontaneous sexual behavior during childhood becomes more and more difficult as his mentality increases. During early adolescence such study probably becomes quite impossible.

It would be helpful if we knew the extent to which sexual shyness in humans is inherent or conditioned. It is conceivable that there are primitive peoples among whom this secretiveness is absent at all stages, as it seems to be in the apes. It is unlikely that human groups exist in which taboos, superstitions, or traditions do not exercise regulating influences. Probably the nearest approach to such influences among the apes is the possibility of surreptitious relations encouraged by the dominance of a stronger male among the younger and weaker animals of the band.

Assuming a society of humans free from sexual restraints, as among the anthropoids, an observer attempting to do what I have done with chimpanzees would still have serious problems. He would probably have difficulty in keeping subjects continually under observation without introducing restraints that the assumption precludes. Human reliance in the observer probably could not be established as with the chimpanzee. On the other hand, I suspect the conditions under which sexual behavior appears and develops would be found long before the human individual begins to conceal his sexual tendencies.

Before direct observation of sexual activities becomes impracticable in humans, the subjects are able to share information, making possible the method of personal report. The value of this supplementary method depends upon the accuracy of the subject in furnishing from memory facts that have been concealed for varying intervals. At its very best the method can only provide cross sections of memory in matters pertaining to sexual genesis. It further demands complete objectivity. Although the method of personal report is more complicated than direct observation, it nevertheless furnishes a possibility of supplementation in human behavior that cannot be used with anthropoids. Unfortunately, the two methods cannot be made fully to supplement one another because the possibilities in the one are waning, while in the other they are increasing.

By contrast with these human complications in problems of sex, the anthropoid subject permits the use of a single method throughout his life history. This consistency in procedure, in addition to experimental controls, should furnish results that supplement significantly our human records. The single method may be used with the apes without limitations. Experiments in anthropoid reproduction have social sanction no less than with birds and mammals. An approaching parallel between human and anthropoid life histories indicates at once that comparisons may help materially in dealing with the sexual problems of man.

Although human and anthropoid similarities might be pointed out, it is by no means a simple matter to evaluate the customs of man in terms of the more primitive sexual behavior of apes.

Differences may be more significant than similarities. The necessity of repeated trials in the heterosexual relations of chimpanzees before adequate reproductive adjustments are possible may be due to limited or wanting means of conveying to one another information about sexual matters. The necessity for individual exploration in sexual relations among humans has partially disappeared. Human individuals can convey by language varieties of knowledge about such matters. Although their sexual background probably goes through a process of development comparable with that of chimpanzees, it scarcely follows that direct experience is necessary or desirable. Human imaginativeness and language may be substituted in large measure for anthropoid exploration.

REFERENCES

(1) ALLESCH, G. J. VON: Bericht uber die drei ersten Lebensmonate eines Schimpansen. Sitz. d. Pr. Akad. d. Wis., Berlin, 1921, 672–685.

(2) ALLESCH, G. J. von: Geburt und erste Lebensmonate eines Schimpansen. Die Naturwiss., 1921, ix, 774–776.

(3) BARKOW, H.: Comparative Morphologie des Menschen und der Menschenähnlichen Tiere. Breslau, ii, 1862, pp. vii + 152.

(4) BINGHAM, H. C.: Parental play of chimpanzees. Jour. Mammal., 1927, viii, 77–89.

(5) BISCHOFF, T. L. W.: Untersuchung der Eingeweide und des Gehirns des Chimpanse. Mitth. K. Zool. Mus. Dresden, 1877, (2), 251–260.

(6) BISCHOFF, T. L. W.: Vergleichend anatomische Untersuchungen über die äusseren weiblichen Geschlechts- und Begattungs-organe des Menschen und der Affen, insbesondere der Anthropoiden (1879). Abh. d. k. Bay. Ak. d. Wiss. Munchen, 1880, xiii, (2), 207–274.

(7) BLAIR, W. REID: Notes on the birth of a chimpanzee. Zool. Soc. Bull. (New York), 1920, xxiii, 105–111.

(8) BOLK, L.: The part played by the endocrine glands in the evolution of man. Lancet, 1921, cci, (2), 588–592.

(9) CHAPMAN, H. C.: On the structure of the chimpanzee. Proc. Acad. Nat. Sci., Philadelphia, 1879, 52–63.

(10) CRISP, E.: On the os penis of the chimpanzee (Troglodytes niger) and of the orang (Simia satyrus). Proc. Zool. Soc., London, 1865, 48–49.

(11) DARWIN, C.: Sexual selection in relation to monkeys. Nature, 1876, xv, 18–19.

(12) DENIKER, J.: Recherches anatomiques et embryologiques sur les singes anthropoïdes. Arch. Zool. Exper., 1885, (series 2, iii), suppl., pp. 265.

(13) DENIKER, J.: Sur les singes anthropoïdes de la ménagerie Bidel. Bull. Soc. Zool., France, 1882, vii, 301–304.

(14) ELLIOT, D. G.: A review of the primates. Monographs Amer. Mus. Nat. Hist., New York, vol. 2, 1913, pp. xviii + 382 + xxvi.

(15) GRATIOLET, P. ET ALIX: Recherches sur l'anatomie du Troglodytes aubryi. Nouv. arch. du. Mus. Hist. Nat., Paris, 1866, 1–263.

(16) HAMILTON, G. V.: A study of sexual tendencies in monkeys and baboons. Jour. Animal Behav., 1914, iv, 295–318.

(17) HARTMANN, R.: Beiträge zur zoologischen und zootomischen Kentniss der sogenannten anthropomorphen Affen. Archiv. f. Anat., Physiol., u. Wiss. Med., 1872, 107–152, 474–502.

(18) HOFFMAN, G. VON: Ueber die weiblichen Genitalien eines Schimpansen. Zeit. f. Geburts. und Gynäk., 1878, ii, 1–8.

(19) HUXLEY, T. H.: Man's Place in Nature. New York, 1893, pp. xv + 329.

(20) KEMPF, E. J.: The social and sexual behavior of infra-human primates with some comparable facts in human behavior. Psychoanal. Rev., 1917, iv, 127–154.

(21) KÖHLER, W.: The Mentality of Apes. Trans. from the German by E. Winter. New York, 1925, pp. viii + 342.

(22) MONTANÉ, L.: Notas sobre un chimpancé nacido en Cuba. Memorias de la Sociedad Cubana de historia natural "Felipe Poey," 1915, i, 259–269. (This report, as it appeared in El Siglo XX, Havana, was translated and published in part by C. S. Rossy, Jour. Animal Behavior, 1916, vi, 330–333.)

(22a) Passemard, E.: Quelques observations sur des chimpanzés. Jour. de Psychol., 1927, xxiv, 243–253.

(23) POCOCK, R. I.: Observations upon a female specimen of the Hainan Gibbon (Hylobates hainanus), now living in the Society's Gardens. Proc. Zool. Soc., London, 1905, ii, 169–180.

(23a) SCHULTZ, A. H.: Studies on the growth of gorilla and of other higher primates. . . ., Mem. Carnegie Mus., 1927, xi, s.n. 134, pp. 87.

(24) SOKOLOWSKY, A.: The sexual life of the anthropoid apes. Urologic and Cutaneous Rev., 1923, xxvii, 612–15.

(25) SONNTAG, C. F.: On the anatomy, physiology, and pathology of the chimpanzee. Proc. Zool. Soc., London, 1923, (1), 323–429.

(26) SONNTAG, C. F.: The morphology and evolution of the apes and man, London, 1924, pp. xi + 364.

(27) SONNTAG, C. F.: On the pelvic muscles and generative organs in the male chimpanzee. Proc. Zool. Soc. Lond., 1923, (2), 1001–1011.

(28) SPERINO, G.: Anatomia del Cimpanzè. Torino, 1897, pp. 487.

(29) SYMINGTON, J.: On the viscera of a female chimpanzee. Proc. Royal. Phys. Soc., Edinburgh, 1888–1890, x, 297–312.

(30) TRAILL, T. S.: Observations on the anatomy of the orang-outang (1817). Mem. Wern. Nat. Hist. Soc., Edinburgh, 1821, iii, 1–49.

(31) YERKES, R. M., AND LEARNED, B. W.: Chimpanzee Intelligence and Its Vocal Expressions. Baltimore, 1925, pp. 157.

(32) YERKES, R. M.: Almost Human. New York, 1925, pp. xxi + 278.

(33) YERKES, R. M. The mind of a gorilla. Genetic Psy. Mon., 1927, ii, pp. 1–193; Part II, pp. 377–551.

13

Reprinted by permission from *Gerontologia Clinica* 6:133–143 (1962)

Effect of Age on the Sexual Behaviour
of the Male Rat

By Knut Larsson and Leif Essberg

Department of Psychology, University of Göteborg

The capacity of the male rat to attain repeated ejaculations is reported to change with age, its peak being at about one year of age and low at puberty and old age (*Larsson*, 1956, 1958a). It has become apparent in recent studies, however, that the observational technique previously used, does not admit an adequate evaluation of the animal's capacity. The mating behaviour has been recorded during a limited period, usually one hour. An old male, however, shows relatively long latencies of the copulatory responses. It may be that the lowered ejaculation frequency observed in old age was caused by the prolongation occurring to the response latencies, and does not reflect a lowered potency of the animal as has earlier been believed.

It has further been noted that a young and an old male does not always react in the same way to a testing situation. Handling of the animal and noise in the environment, which does not influence the sexual activity of the young male, increases the activity of the old male. The lowered responsiveness to the receptive female in old age can be compensated for by non-specific sensory stimulation (*Larsson*, in press).

The present study was conducted in order to investigate the effect of age on the mating behaviour with avoidance of the short-comings of the observational technique previously used. Instead of recording the behaviour during a period of predetermined length equal to all individuals, the experimenter observed the animal until the male reached a specified criterion of exhaustion. Handling and repeated presentation of fresh receptive females provided the subjects with an optimum of sensory stimulation.

Received for publication October 14, 1961.

Methods

Subjects. The study was conducted with males of the stock of albino rats bred and maintained at the Göteborg Laboratory. Four experimental groups were used, the characteristics of which are summarized in table I.

Table I

Groups of Subjects

Group	N	Average age Days	Age range Days	Average weight Grams	Weight range Grams
A	18	112	103–114	128	105–177
B	30	147	125–164	220	170–290
C	14	470	450–510	385	325–440
D	10	613	573–651	325	250–400

In order to select subjects for the youngest group, A, the following procedure was adopted. After a rat had reached 100 days of age, and every day thereafter, he was observed with a female in heat for 10 min. The first time a male achieved an intromission, he was set aside and used as a subject in the standard test procedure on the following day. If an animal did not meet this criterion of potency by its 113th day of age, no further observations were made, and the rat was discarded. In this way rats were excluded which might have reached puberty, but due to conditioned inhibitions or other reasons did not respond to the female.

Subjects for groups B, C and D were selected from animals available at the different age levels. All animals had sexual experience but of unequal amounts. The subjects in group B had one hour of experience prior to the observation, and the subjects in C and D had on various occasions served as breeders. They were all rested for approximately two weeks prior to the tests.

Table II

Observational Period in Minutes

Group	Average	Range
A	106	57–143
B	121	81–202
C	109	72–174
D	145	111–194

The animals were extremely docile, gently and easily handled. They showed no signs of respiratory or other diseases.

They were kept in wooden cages. Each cage, 22 in. wide 16 ½ in. deep and 7 in. high with a wire-mesh top, contained approximately six animals. The floors were covered with sawdust and cleaned twice weekly.

Water and food were continuously available in the cages. The diet consisted of "Anticimex" commercial rat pellets supplemented with fresh vegetables.

Lights in the rooms were clock-controlled, being turned off at 12 and on at 22. The room temperature was 19–21°C.

Apparatus

The experimental cages were semicircular, having a straight glass front and sheet-metal sides. They were 27″ in. wide at the front, 16″ in. deep and 16″ in. high. Four observation cages were fastened to a rack in such a position that they could be watched simultaneously. Each cage was illuminated by a 15 Watts bulb.

The observer sat in front of the cages at a distance of approximately two meters. The laboratory room was dark so that the observer was not visible to the animals. The sexual responses of the animals were recorded by help of a stopwatch.

Stock and experimental cages were kept in separate rooms.

Variables Measured

The mating pattern consists of a series of mounts and pelvic thrusts. The male mounts the female, palpates her flanks with his forepaw, and intromission is made when she raises her perineum in response to his palpation. He then dismounts, licks his penis and the erection disappears. After a brief interval he remounts the female. The series of intromissions is ended by ejaculation. The ejaculatory response is easily distinguishable from the intromission response, by a deep thrust.

During each test the following items were measured:

Mount frequency: number of mounts with penetration,

Intromission latency: time from entrance of the female to the first intromission,

Ejaculatory latency: time from the first intromission until ejaculation,

Intercopulatory interval (ICI): average delay between intromissions,

Postejaculatory interval (PEI): time from an ejaculation to the next intromission.

The term "Series" is used to refer to an ejaculation and the sequence of mounts and intromissions leading up to it. Thus "Series I" indicates the sexual responses preceding and including the first ejaculation; "Series II" refers to those leading up to and including the second, etc.

Procedure

Prior to testing, stimulus females were brought into estrus by hormone treatment. Thirty hours before the test .08 mg. of estradiol benzoate in oil was injected subcutaneously and this was followed six hours before the test by an injection of 1.00 mg. of progesterone.

The tests were begun at approximately 14. Five minutes before the beginning of the test the subject was placed in the observation cage and at the end of this period a female was introduced.

All tests were continued until the criterion of sexual exhaustion was met. A male was judged to be exhausted when he did not mount and penetrate the female 30 min after ejaculation. If he penetrated the female before this period was ended, the behaviour was recorded for an additional 30 min and if failing to ejaculate within this period the test was discontinued.

When a male did not mount the female within ten minutes after ejaculation, he was taken up from the floor of the observation box, handled for a few seconds and again put into the cage. This procedure was repeated every tenth minute until the male had mounted the female or the criterion of exhaustion was met. The female was replaced after each ejaculation.

Results

The ejaculation frequency progressively increased with age, the lowest being observed in Group A and the highest in Group D (table III). Examining the performances during the first hour of observation only, an increase in the ejaculation frequency was observed up to and including Group C. The frequency again decreased in Group D. In the second hour of observation, however, the Group D males made almost four times as many ejaculations as did the younger animals. This shows that the lesser number of ejaculations appearing in the first hour does not reflect a failing capacity to attain repeated ejaculations but is a result only of the prolongation of the response latencies occurring in this age group.

Table III

Median Number of Ejaculations during Different Periods of the Observation

Period of observation	Group A	Group B	Group C	Group D	A–B	B–C	p C–D	B–D	A–D
0–60	3.1	4.9	5.3	4.5	<0.001	–	<0.02	–	<0.02
60–120	0.7	1.4	1.0	3.7	–	–	<0.001	<0.002	<0.002
Total testperiod	4.0	6.1	6.7	7.0	<0.001	–	–	<0.01	<0.002

Tables IV, V and VI give a more detailed picture of the changes occurring in the mating pattern with advancing age. In Group B, C and D all subjects attained four or more ejaculations (table IV). Since in Group A, most subjects failed to achieve more than four ejaculations, it was decided to base group comparisons upon the first four series of copulations.

Table IV

The Percentage of Animals Achieving Ejaculations

Number of ejaculations before exhaustion	Group A	B	C	D
1	6			
2	11			
3				
4	61	6	14	
5	16	14	14	
6	6	53	14	30
7		10	50	50
8		14		10
9		3		10
10				

According to table V the younger rat needed more intromissions to ejaculate than the older one. The first ejaculation, for instance, was preceded in A by 24.5, in B by 8.9, in C by 8.5 and in D by 8.0 intromissions. In all series the same trend of a lowering

Table V

Median Number of Intromissions and Mounts Preceding Ejaculation in the Different Age Groups. Probability Values Based on Two-Tailed Mann-Whitney Test (*Siegel*, 1956).

Behavioural component	Series	Group A	B	C	D	A–B	p B–C	C D
Intromissions to ejaculate	I	24.5	8.9	8.5	8.0	<0.001	<0.03	–
	II	7.7	5.2	4.7	4.5	<0.001	<0.002	–
	III	8.8	5.5	4.5	6.5	<0.001	<0.05	<0.05
	IV	13.9	5.8	5.2	5.2	<0.001	–	–
Mounts to ejaculate	I	1.3	3.0	2.5	6.5	–	–	<0.02
	II	1.8	1.5	0.8	2.2	–	–	–
	III	1.0	2.3	2.3	1.8	–	–	–
	IV	3.3	2.8	3.2	2.5	–	–	–

of the intromission frequency with increasing age was apparent. No systematic changes were seen in the mount frequencies.

The decrease in intromission frequency was accompanied by a shortening of the intercopulatory intervals and the postejaculatory intervals (table VI). In the oldest age group, however, this trend

Table VI

Median Ejaculatory Latencies, Intercopulatory Intervals and Postejaculatory Intervals in the Different Age Groups (in Minutes)

Behavioural component	Series	Group A	B	C	D	A–B	p B–C	C–D
Intromission latency		1.11	0.40	0.18	0.19	<0.001	<0.01	–
Ejaculatory latency	I	13.0	3.9	5.0	11.1	<0.001	–	<0.002
	II	4.9	2.4	1.5	2.9	<0.001	<0.002	<0.02
	III	8.0	2.7	1.6	2.9	<0.001	<0.003	<0.05
	IV	9.8	2.8	2.8	2.7	<0.001	–	–
Intercopulatory interval	I	0.55	0.43	0.48	1.27	–	–	<0.002
	II	0.54	0.44	0.40	0.79	–	–	<0.002
	III	0.73	0.59	0.40	0.47	<0.03	<0.03	–
	IV	0.78	0.54	0.43	0.61	<0.004	–	–
Postejaculatory interval	I	6.7	5.5	5.0	6.5	<0.003	<0.01	<0.02
	II	8.2	6.9	5.9	7.5	<0.001	<0.002	<0.002
	III	11.0	8.1	6.8	9.1	<0.005	<0.001	<0.002
	IV	13.1	10.1	8.8	9.7	<0.001	<0.05	–

was reversed and the latencies became prolonged. Some of these changes are partly due to differences in the amount of sexual experience given to the subjects. It is known from previous investigations that experience selectively influences some measures of sexual activity but not others (*Larsson*, 1959). The intromission latency and the postejaculatory intervals are slightly shortened as result of one hour of sexual experience, while the intromission frequency, the ejaculatory latencies and the ICI's remain uninfluenced by experience. In this study the Group A males had no experience and the Group B males one hour of experience. This is in contrast to the very experienced subjects in Group C and D. It is therefore concluded that while the prolongation of the postejaculatory intervals in the oldest age group is wholly to be attributed to effects of age, the reduction of the postejaculatory intervals in Groups B and C was partly due to experience. Also the reduction of the intromission latencies is probably caused by experience. However, the changes in the intercopulatory intervals, in the ejaculatory latencies and in the intromission frequencies are to be considered as genuine effects of age.

In all groups, the ejaculations following the first one were attained after fewer intromissions. This facilitatory effect of a preceding series of copulations upon the following one is a feature characteristic to the rat mating behaviour. As shown in figure 1 the reduction systematically changed with increasing age. In Group A

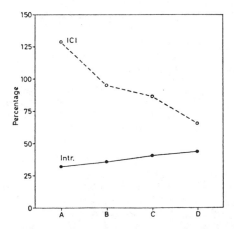

Fig. 1. Showing the relationship between the performances in the first and second series of copulations in the different experimental groups. See text.

the intromission frequency in the second series was reduced to 32% of the frequency recorded in the first series. In Groups B, C, and D the corresponding figures were 36, 41, and 44% respectively (A–B p <0.001, B–C p <0.04 and C–D non-significant, Wilcoxon two-tail test). Also the effect of the first series upon the ICIs of the second series varies with the age of the animal but in opposite direction. The Group A males showed prolonged, not shortened, ICIs in the second series. In the other groups a progressively increasing reduction occurred (A–B p <0.002, B–C p non-significant, C–D p <0.05).

It is of interest to study the effect of growing exhaustion upon performances in the different groups. Table VII shows the intro-

Table VII

Performances in the Second, Last and Last But One Series Median Values. In Group A the Comparison is Based on the Subjects, which Attained Four or More Ejaculations. Probability Values Based on Two-Tailed Wilcoxon Test (*Siegel*, 1956)

Behavioural component	Group	N	Series of copulation II (a)	Second last (b)	Last (c)	p a–b	b–c
Intromissions to ejaculate	A	15	7.4	11.3	13.7	<0.01	<0.01
	B	30	5.2	6.5	6.8	–	<0.02
	C	14	4.7	· 5.5	7.2	<0.01	<0.02
	D	10	4.5	5.8	7.5	–	<0.05
Ejaculatory latency in minutes	A	15	4.8	8.8	9.8	<0.01	<0.01
	B	30	2.4	4.1	5.3	<0.01	<0.01
	C	14	1.5	3.9	7.6	<0.01	<0.01
	D	10	2.9	4.2	7.0	–	–
ICI in minutes	A	15	0.60	0.82	0.85	<0.02	<0.01
	B	30	0.44	0.63	0.96	<0.01	<0.01
	C	14	0.40	0.84	1.48	<0.01	<0.01
	D	10	0.79	0.84	0.73	–	–

mission frequencies, the ejaculatory latencies and the ICIs in the second, the last and last but one series of copulations presented by each individual. Since in all groups with the exception of D, the minimal values of the different measures were found in the second series, this series was selected as the basis for the comparison. As shown by the table, the last ejaculation attained by the animal was preceded by more intromissions and attained after a longer latency than was the second ejaculation. The Group D males did not show any effect on the ICIs in the last series. Also

147

in this group, however, a prolongation of the very last series occurred, as will be seen when the third instead of the second series is selected as the base for the comparison (the difference is significant on the 0.01 p level). It is therefore concluded that in all age groups, a prolongation of the response latencies and an increase in the intromission frequencies occurs when the animal approaches the state of sexual exhaustion.

After the last ejaculation occasional mounts and intromissions were recorded as shown by table VIII. When the observation was

Table VIII

Number of Subjects Responding to the Female after the Last Ejaculation and the Median Number of Mounts and Intromissions Presented before the Interruption of the Observation

Group	N	Number of animals mounting female	penetrating female	Mounts Mdn	Intr. Mdn
A	18	12	11	15.0	6.5
B	30	15	12	7.0	3.3
C	14	8	7	5.3	7.0
D	10	5	7	6.0	5.3

ended, most, although not all individuals, seem to have reached the limit of their capacity. In a number of animals the observation period was prolonged with one or two hours and a few of these animals attained another ejaculation. The behaviour recorded in the prolonged tests are not included in the data presented.

Discussion

It is apparent that while an observation of one hour will give a good indication of the capacity of the young animal to attain repeated ejaculations, such a test is inadequate when studying the capacity of the old male. The low ejaculation frequency in the first hour of observation in the oldest group in comparison with other groups is in complete agreement with findings earlier reported, but contrary to earlier beliefs, this decrease cannot be taken as an indication of a lowered capacity of the old male. This does not exclude the possibility of a decrease at still higher age levels than here investigated.

Not only did age influence the capacity of the male to attain repeated ejaculations but it also profoundly changed the appearance of the mating pattern. The increasing potency was accompanied by a decrease in the number of intromissions necessary to attain ejaculation. The response latencies were prolonged in old age. It is of particular interest to note the reduction effect of the preceding ejaculation upon the length of the average intercopulatory interval of the following series. While the young rat started to copulate at a maximal speed as soon as the receptive female had been presented to him, the old male had to attain one or two ejaculations before being capable of responding at a maximal rate. The young rat, although less potent, seems to be more easily aroused than the old one.

The conclusion that there occurs a rise of the arousal threshold in old age is supported by two other findings. As indicated above, nonspecific sensory stimulation, which does not influence a young rat, increases the sexual activity of the old male. It is also known that the sexual activity is submitted to a diurnal periodicity, being low in the daytime and high at night. This periodicity, however, is more pronounced in the old rat than in the young one (*Larsson*, 1958 b). While a young rat shows a relatively great amount of sexual activity also in the passive phase of the diurnal periodicity, the old male is likely to be completely inactive during this phase. It would appear that the old male's sexual activity is more dependent on environmental stimulation.

It is at present too early for speculation regarding the physiological basis for the changes in the mating behaviour observed. Studies recently completed in this laboratory, however, indicate that some of the changes may have an endocrine basis. The low ejaculation frequency and high intromission frequency at puberty seem to reflect, either a lowered sensitivity to androgen of the nervous mechanisms mediating the behaviour, or a lowered amount of androgen circulating in the body at that age.

Acknowledgment. These experiments were supported by grants from Magnus Bergwall and Hierta Retzius Foundation to the senior author. The hormone preparations were generously supplied by the Hässle Tika Corporation, Göteborg.

Summary

The male rat mating behaviour was studied in four different groups, aged 103–114, 125–164, 450–510 and 573–651 days respectively. The subjects were allowed to mate until they attained a specified criterion of exhaustion. It was found that the animal's capacity to attain repeated ejaculations increased with age, the youngest subjects being most rapidly exhausted and the oldest ones most resistent against exhaustion. There occurred great changes in the mating pattern with increasing age. The number of intromissions preceding ejaculation was highest in the youngest age group and progressively decreased with age. The response latencies were progressively reduced up to 450–510 days of age but become prolonged in the oldest age group. While the young male copulated with maximal speed immediately after having been presented with the receptive female, the subjects in the two oldest age groups responded by minimal response latencies only in the presence of facilitatory effects from one and two series of copulations.

This was interpreted as indicating a heightened threshold in old age for the arousal of the copulatory response.

Zusammenfassung

Die sexuelle Aktivität vier verschiedener Altersgruppen männlicher Ratten wurde untersucht. Die Gruppen umfaßten Tiere folgenden Alters: 103–114 Tage, 125–164 Tage, 450–510 Tage und 573–651 Tage. Die Tiere durften sich paaren, bis sie ein bestimmtes Kriterium der Erschöpfung erreicht hatten. Es wurde festgestellt, daß die Fähigkeit der Tiere, wiederholte Ejakulationen zu bekommen, mit dem Alter zunahm. Die jüngsten Tiere erschöpften sich am schnellsten, und die ältesten Tiere zeigten sich mehr widerstandsfähig gegen Erschöpfung. Mit zunehmendem Alter zeigten sich große Veränderungen im Paarungsverhalten. Die jüngsten Ratten benötigten die höchste Anzahl von Intromissionen, um Ejakulation zu erreichen. Mit zunehmendem Alter nahm die Anzahl ab. Die Latenzzeit verkürzte sich laufend bis zu einem Alter von 450–510 Tagen, stieg jedoch wieder in der ältesten Gruppe. Die jungen Männchen kopulierten mit größter Geschwindigkeit, kurz nachdem sie mit einem empfänglichen Weibchen zusammengeführt worden sind, während die Tiere in den zwei ältesten Gruppen mit minimaler Latenzzeit nur nach Einwirkung verstärkender Effekte einer oder zweier vorgehender Ejakulationen reagierten.

Der Versuch deutet auf eine Erhöhung der Reizschwelle zur Erweckung des Kopulationsreflexes bei höherem Alter hin.

Résumé

Nous avons étudié le comportement copulateur du rat mâle dans quatre groupes d'individus, âgés respectivement de 103 à 114, de 125 à 164, de 450 à 510 et de 573 à 651 jours. On a laissé copuler les animaux jusqu'à ce qu'on ait pu constater un critère spécifique d'épuisement. Nous avons trouvé que la capacité de l'animal d'obtenir des éjaculations répétées augmentait avec l'âge. Les animaux les plus jeunes ont été le plus vite épuisés et les plus vieux ont résisté le mieux à l'épuisement. Il y a eu, avec l'âge, de grands changements dans le comportement copulateur. Le nombre d'intromissions qui précédaient l'éjaculation était plus élevé dans le groupe des plus jeunes et diminuait progressivement avec l'âge. Les latences de réponse décroissaient progressivement jusqu'à l'âge de 450–510 jours, mais se prolongeaient dans le groupe des plus vieux. Alors que le jeune mâle copulait avec une rapidité maximum après être présenté à la femelle, les individus des deux groupes plus âgés ne donnaient de réponse à latence minimum qu'en la présence d'effets facilitants, provenant d'une ou de deux séries de copulations.

Nous avons interprêté cet effet comme l'indice d'un seuil plus élevé, à l'âge avancé, pour la provocation de la réponse copulatrice.

References

Larsson, K.: Conditioning and sexual behaviour in the male albino rat (Almqvist & Wiksell, Stockholm 1956). – The sexual activity in senile male rats. J. Gerontology *13:* 136–139 (1958a). – Age differences in the diurnal periodicity of the sexual behaviour. Gerontologia (Basel) *2:* 64–72 (1958b). – Experience and maturation in the copulatory behaviour. Behaviour *14:* 101–107 (1959). – Non-specific stimulation and sexual behaviour in the male rat. Behaviour (in press).
Siegel, S.: Nonparametric statistics (McGraw Hill, New York 1956).

Authors' address: Dr. K. Larsson and L. Essberg, Department of Psychology, University of Göteborg, *Göteborg* (Sweden)

14

Reprinted from *Endocrinology* **26**:511–515 (1940)

PRECOCIOUS COPULATORY ACTIVITY INDUCED IN MALE RATS BY SUBCUTANEOUS INJECTIONS OF TESTOSTERONE PROPIONATE[1]

CALVIN P. STONE

From the Department of Psychology, Stanford University

STANFORD UNIVERSITY, CALIFORNIA

GONADOTROPIC and gonadal hormones induce many of the somatic and physiological signs of pubertas praecox in immature rats if administered repeatedly in appropriate doses (1, 2). Whether these somatic changes are accompanied by psychosexual changes, however, cannot be definitely concluded from the published studies available to the author. Reports on the masculinization of young chicks suggest that the somatic and psychosexual changes may, to a marked extent, go hand in hand. Domm and v. Dyke (3) reported that male chicks crowed at the age of 9 days, following 6 daily injections of a gonadotropic hormone (hebin), and that treading appeared when the chicks were 13 days old, following 10 daily injections of hebin. Hamilton (4) caused male chicks to crow at the age of 10 days, by injections of testosterone propionate from the second day after hatching. These chicks struck with their feet as though attempting to use spurs (which they did not have) and pecked belligerently at intruding objects. However, treading or other evidence of copulatory aggressiveness was not mentioned by Hamilton in this report or in the report by Breneman (5) who induced crowing in male chicks when they were only 5 days old by daily injections of dihydroandrosterone benzoate. Absence of this element of masculinization suggested to the author the desirability of further experimental studies of the behavioral aspects of pubertas praecox in laboratory animals.

In the present study the chief objectives were to determine whether precocious copulatory behavior could be induced in immature male rats by subcutaneous injections of testosterone propionate, and to compare this activity with that normally appearing at a later time in untreated males.

Animals. The young rats were reared in our own laboratory. Their diet and also that of their parents consisted of the Steenbock mixture (6) supplemented by lettuce one day per week. They were weaned when 25 days of age. No special selection of litters or of individuals was made when forming the control and experimental groups. Because of this, body weights at weaning, sizes of litters and numbers of males per litter are variable. The groups on the whole, however, appear to be representative of the colony with which we have worked on problems of sex during the past decade and a half at Stanford University.

Injections. A close approximation of 0.62 mg. of testosterone propionate dissolved in one-eighth of one cc. of sesame oil was injected beneath the loose skin over the shoulders or at the flank of each experimental animal daily from the beginning of in-

---

[1] This study was financed by the Committee for Research on Sex Problems, National Research Council. My research assistant for the observational part of the study was Mr. Max W. Lund, graduate student in psychology, to whom I am greatly indebted for painstaking effort to adhere to a standard procedure in testing the males.

jections to the day on which the first copulatory activity was recorded. Animals with numbers falling between 1 and 18, inclusive, received their first injections when 25 or 26 days of age; those with numbers between 19 and 51 received theirs when they were 22 days old. Some of the injected substance followed the needle as it was withdrawn or was pressed out by the animal later on as it moved about the cage. Although the small loss of hormone probably was of no consequence to the individual injected, since it would be effective even on the surface of the skin (7), the point is mentioned because the litter-mate controls were kept with the injected animals in order to equalize, so far as possible, the factors of husbandry and testing of litter mates. Anticipating our test results at this point, it may be said that there is no indication that the controls absorbed sufficient hormone to affect their psychosexual development in a measurable degree. Control injections of pure sesame oil into the normal animals were not made. It seemed unnecessary to subject these young to this type of control as no experiments so far reported have suggested that this substance alone has androgenic effects. Repeated injections of the oil might, however, have a retarding effect on sexual development.

Acceleration of growth rate of the external genitals of injected animals (testes excepted) was apparent almost immediately (cf. 1). Within 24 hours after the first injection there was a change in the redness of the prepuce. On the third day elongation of the penis could be detected. By the fifth day the glans was well developed and could be exposed by the animal itself or by manual retraction of the foreskin. At this time the penis resembled that of an untreated male of the age of 45 to 50 days. Although the scrotum became more pendulous than usual the testes appeared not to be accelerated in growth rate. Whether or not there was actually some retardation, as one might expect in accordance with the study of Moore and Price (8), we did not attempt to ascertain. Erection of the penis was noted many times as the young males made copulatory attempts or sat back on tail and hind feet after copulating.

Tests for copulation. The term copulation is used here to embrace the following elements of the complete copulatory act as it is found in adult males: pursuit of the female; mounting and palpation of the female's sides with forepaws; rapid pelvic movements that terminate by the animal's sitting back on its tail and hind feet and licking the penis. Usually the final, slightly more vigorous pelvic thrust is present but in some instances this is absent. In the latter case the act resembles abortive copulatory acts of adults in which they fail to effect intromission. No instance of the prolonged clasp in which the ejaculatory reflex occurs was recorded.

The tests for copulatory behavior were made between 7:30 and 11:00 P.M. A receptive female was placed with one or two of the young males at a time and left there for a period of from 8 to 10 minutes while the experimenter kept close watch for copulatory behavior. From 2 to 3 tests per evening were given and usually from twenty minutes to three-quarters of an hour elapsed between tests. No distinction between experimental and control animals was made and the cage-mates tested together varied from time to time. The females were relatively small adults weighing about 150 to 175 gm. No female that was wild or pugnacious during the preliminary tests was used lest the young males be intimidated. Even so, it was not possible to prevent occasional instances of biting or mauling of young. This type of behavior, seemingly, cannot be prevented when males that are much smaller than the females are being tested.

RESULTS

The net results of the tests for copulation are presented in figures 1 and 2. The first figure is a frequency distribution of the ages of first copulation of the injected animals, their litter-mate controls, and a group of non-litter-mate controls similarly tested. The second figure provides a simple basis for comparing the litters as well as the individuals within litters; it also gives the dates of injections and the time of and number of tests for each animal.

Unpublished and published data (9) at our disposal indicate that 35 to 40 days may be regarded as the typical lower range of ages at first copulation in untreated rats which are reared and tested as we have done in the present experiment. The upper age range cannot be fixed so accurately but probably it terminates in the neighborhood of 70 to 80 days (9). For small groups of animals the median age of first copulation usually falls between 47 and 58 days.

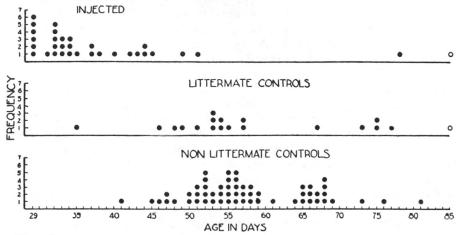

Fig. 1. Frequency distributions of the ages of first copulations of the injected rats, their litter-mate controls, and a group of non-litter-mate controls similarly tested.

Taking the foregoing values as points of reference when considering figure 1, it is clearly apparent that the injected males, as a group, copulated precociously for there is only a small amount of overlapping between them and the controls. For the 9 injected males (*rats 1 to 18*, fig. 2) which received their first treatment when 25 or 26 days of age, the lowest age at first copulation was 32 days and the median age for the 9 animals was 35 days. For the 22 that were injected from their twenty-second day, the earliest age of first copulation is 29 days, and the median age is 32 days. Whether the greater acceleration in the second sub-group is due to earlier injections, to more numerous injections prior to the first tests (cf. fig. 2), or to both of these factors cannot be determined from our present data. Probably earlier testing would have shunted some of the cases now massed at the age of 29 days to lower ages. Massing of cases at the lower end of the range in tests of this kind usually indicates that opportunity for overt sex behavior lagged behind the animal's state of readiness.

That testosterone propionate is not 100% effective in inducing the behavioral signs of pubertas praecox is clear from the instances of relatively late copulation. Animals *11, 21* and *45* (fig. 2) did not at any time display sexual aggressiveness at the early ages when other injected animals began to copulate, and *rat 45* failed to copulate even within the 85-day period of testing. *Rat 41* which copulated at the age of 49 days made attempts at the age of 29 and again at 31 days. Copulatory attempts have never been seen in normal animals at such early ages, hence we are probably on safe grounds in crediting him with precocious sexual activation even though the full copulatory pattern was not elicited at an early age. While observing that the injected hormone did not invariably produce precocious sexual aggressiveness, we must not overlook the possibility that the fault may lie in part in our testing procedures. Some of the receptive females posture before the young males much more aggressively than others and thus provide better lures to elicit sexual behavior. Although it is unlikely that a given male would have a long series of relatively inactive females, the relevancy of this point depends on whether dosage administered by us was inadequate to activate certain males for more than a day or two, and through faulty testing their activation was not detected at that time. The point is deserving of further inquiry.

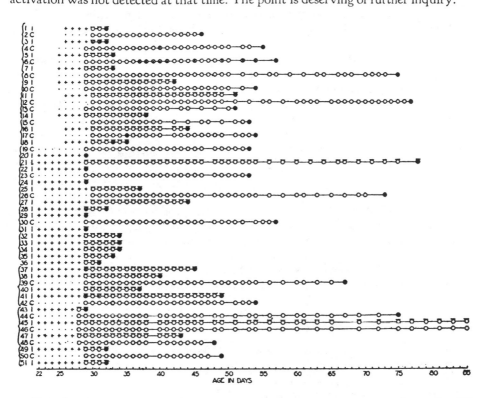

Fig. 2. INDIVIDUAL TEST RECORDS FOR THE INJECTED AND THE LITTER-MATE CONTROLS. Letters I and C refer to injected and control. The animals' ages are given on the abscissa. Circles designate the ages at which tests for copulation were given. Open circles mean that the animals failed to copulate and closed circles (dots) indicate that the males copulated in one or more of the tests given at that age. Plus signs indicate the days on which injections were given prior to the beginning of testing; thereafter a short line that is tangent to the circles indicates this. A cross inside an open circle indicates that the male made copulatory attempts, but failed to complete the act.

With respect to variability of initial copulation in the experimental and the control groups it seems apparent that the greater degree is present in the controls. This statement is based on inspection rather than on computed measures because the varied ages at the time of initial injections of the experimental animals would tend to increase the variability of this group beyond that normally present if all animals had been injected at the same age. The greater variability in the untreated group is explicable if we assume that there is considerable variability in the ages at which an amount of testicular hormone that is adequate to arouse the copulatory impulse normally appears in the blood stream. In the experimental group we should expect less variability because of the synchronous appearance of a relatively large amount of

155

the male hormone in the blood stream as a result of the subcutaneous injections. Moore and Price (8) have found that the beginning of rapid increase in hormonal secretion by the rat testis is between the ages of 35 and 40 days. They based their statement on the trend of means, however, and did not attempt to deal specifically with individual differences. From marked differences that are apparent in time at which the penis of rats undergoes the puberal growth spurt it would seem that we may safely infer marked differences in the time of great acceleration of hormonal output by the testis. A weakness in these a priori deductions, however, is the fact that we lack precise data on the relative amounts of hormone required to induce copulatory behavior on the one hand and accelerated growth of the various elements of the accessory sexual organs on the other.

Considering the data of this study in their entirety the most striking result is the change in the developmental schedule of the injected rats, in the psychosexual sphere. The awakening of the copulatory impulse has been predated, on the average, by approximately 20 days. The precocity was induced without alteration of the pattern of the copulatory response as it normally appears in untreated young males. The ultimate fate of the precociously awakened sexual impulses is a phase of the problem which we are not yet prepared to discuss although the importance of this aspect of the problem is recognized.

SUMMARY

By daily injections of 0.62 mg. of testosterone propionate into male rats on and after the ages of 22 to 26 days, the median age of first copulation was set ahead of that of control males by about 20 days. The earliest age at first copulation by an injected male was 29 days; that of an untreated male, 35 days. Probably still earlier copulation may be induced by beginning the injections and the tests somewhat earlier. The pattern of the copulatory act is like that normally appearing in untreated males.

REFERENCES

1. MOORE, C. R.: Chap. VII, p. 353, Sex and Internal Secretions. 2nd ed. Baltimore, Williams and Wilkins. 1939
2. SMITH, P. E.: Chap. XVI, p. 931, Sex and Internal Secretions. 2nd ed. Baltimore, Williams and Wilkins. 1939.
3. DOMM, L. V., AND H. B. v. DYKE: Proc. Soc. Exper. Biol. & Med. 30: 349. 1932.
4. HAMILTON, J. B.: Endocrinology 23: 53. 1938.
5. BRENEMAN, W. R.: Endocrinology 24: 55. 1939.
6. STONE, C. P., M. I. TOMILIN AND R. G. BARKER: J. Comp. Psychol. 19: 215. 1935.
7. MOORE, C. R., J. K. LAMAR AND N. BECK: J. A. M. A. 111: 11. 1938.
8. MOORE, C. R., AND D. PRICE: Anat. Rec. 71: 59. 1938.
9. STONE, C. P.: Am. J. Physiol. 68: 407. 1924.

15

Reprinted by permission from *J. Comp. Physiol. Psychol.* **48**:397–403 (1955)

EXPERIENTIAL AND GENETIC FACTORS IN THE ORGANIZATION OF SEXUAL BEHAVIOR IN MALE GUINEA PIGS[1]

ELLIOT S. VALENSTEIN, WALTER RISS,[2] AND WILLIAM C. YOUNG

University of Kansas

The relative stability of the patterns of sexual behavior displayed by individual male and female mammals has led to speculation and inquiry into the nature of factors that might be responsible for the establishment of such patterns. The possibility that gonadal hormones have an organizing action has long been questioned (1, 2, 4, 15, 19). Genetic factors, on the other hand, may be influential (18). The suggestion has been made also that experiential factors play an important role in primates (5, 9), but in vertebrates below the primates the motor patterns of copulation are said to be innately organized (5, 9, 10, 13, 17). Inasmuch as strains of guinea pigs exist in which the patterns of sexual behavior are relatively homogeneous (18), these strains seemed well suited for a study that would comprehend not only the genetic contribution, but also the possible effect of experience on the development of sexual behavior in a rodent. The investigation was organized as follows: experiment I, in which the role of experience gained during contact with other animals and the influence of the genetic background were studied; experiments II and III, designed for the purpose of obtaining information about the form in which the genetic differences are manifested; experiment IV, in which the effectiveness of male compared with female cage-mates was studied.

EXPERIMENT I

Method

Males of three groups (the isolated males) were placed alone in cages with their mothers from the day of birth until day 25 and isolated thereafter. One group of 17 males was from the inbred Strain 2 (16). Males of this strain are characteristically the smallest animals in the colony, but in their sexual behavior they are moderately vigorous. The second group was composed of 7 males from Strain 13. The males from this inbred strain are large, but generally sluggish behaviorally. The third group consisted of 7 males from a genetically heterogeneous stock. These males are also large, and they are sexually the most vigorous animals in the colony. Control males (the social males) for each of the three groups were raised with their siblings until day 25 when they were weaned. From then until day 73 when they were isolated, they were confined with from 3 to 5 females of the same age.

Beginning on day 77 the sexual performance of the males was determined by seven weekly tests with estrous females of about the same size. The testing procedure and scoring method described by Valenstein, Riss, and Young (18) were used. The number of times a male mounted a female, the number of intromissions and ejaculations, and the amount of the lower measures of sexual behavior (sniffing, nuzzling, and abortive mounts) were recorded. The sexual behavior score takes into account the quantity and maturity of sexual behavior exhibited during a 10-min. test with a female in estrus as well as the length of the interval between the beginning of the test and ejaculation, when ejaculation occurs.

Results

The data obtained from the Strain 2 males (Table 1) reveal a striking difference between the isolated and social males. The difference in average sexual behavior scores is highly significant ($p < .01$). Similarly, the difference in the higher measures of sexual behavior (mounting, intromission, ejaculation) are also significant ($p < .01$) whether the average frequency per animal or the percentage of animals displaying the behavior is considered. Only one isolated Strain 2 male achieved intromission and ejaculation, whereas 16 of 19 social males ejaculated.

When comparing the Strain 2 males with respect to the lower measures of sexual behavior, a different picture is seen. At this level the isolated males surpassed the social males. The former were exceedingly active, displaying a persistent and vigorous interest in the female, but despite this lively interest the isolated males, with one exception, did not mount the females in such a way that intromission and ejaculation could be accomplished.

---

[1] This investigation was aided in part by research grants M-504 and M-504(C) from the National Institute of Mental Health, Public Health Service, and in part by grants 44 and 180 from the University of Kansas Research Fund.

[2] Present address: Department of Anatomy, State University Medical Center, 350 Henry Street, Brooklyn 2, New York.

TABLE 1

Comparison of Sexual Behavior of Male Guinea Pigs Weaned on Day 25 and Raised in Isolation with that of Males Raised with Females*

Animals		N	Lower Measures		Mountings		Intromissions		Ejaculations		Average Scores
			Average per Animal	Per cent Displaying	Average per Animal	Per cent Displaying	Average per Animal	Per cent Displaying	Average per Animal	Per cent Displaying	
Strain 2	Isolated	17	149.0	100	1.2	35	0.5	6	0.06	6	3.9
	Social Situation	19	97.3	100	18.0	100	18.0	90	3.7	84	6.8
Strain 13	Isolated	7	129.4	100	2.0	71	0.0	0	0.0	0	3.6
	Social Situation	7	96.4	100	6.4	86	7.6	57	0.6	57	3.0
Genetically Heterogeneous	Isolated	7	70.2	100	17.4	71	12.6	71	4.4	71	7.5
	Social Situation	7	36.4	100	19.6	100	20.6	100	6.7	100	9.9

* Data obtained from 7 tests of each male during isolation, days 77–120.

Differences were seen in the behavior of isolated and social Strain 13 males. To be sure, the average sexual behavior score of the social males was not significantly different from that of the isolated males. On the other hand, an analysis of the individual measures contributing to the sexual behavior score reveals that 57 per cent of the social males had intromissions and ejaculations while none of the isolated males exhibited these higher measures of behavior. The differences analyzed statistically yield a χ^2 of 3.15 (corrected for discontinuity), which lies between the .05 and .10 levels of confidence. However, inasmuch as the difference is in the expected direction, there is justification for using a one-tailed test, which would place the probability below the .05 level. In either case the trend is clear. The lack of a significant difference in average sexual behavior scores is explained by the fact that the socially raised males did not exhibit a sufficiently large number of the higher measures of behavior to counteract the greater activity of the isolated males in the lower measures. It was noted further that the periods of inactivity following intromission were unusually long.

A third result was obtained from the heterogeneous males. The differences in average sexual behavior scores and in the separate measures that contribute to the scores of the isolated and social males are not striking. Again, the differences are in favor of the social males, but they are not significant; they are due to the behavior of two isolated males which did not mount, achieve intromission, or ejaculate. The fact that the weight of these two males was less than average suggested an explanation for their poor performance, which was tested in experiment II.

EXPERIMENT II

The hypothesis was tested that the lack of any striking difference in the performance of the genetically heterogeneous males raised in the two caging situations might be explained by the more rapid growth of these animals. It was considered possible that most of the heterogeneous males had acquired the experience they needed for the organization of the complete pattern of sexual behavior before they were weaned, for males of this stock less than 25 days of age have been observed mounting their mothers.

Method

Ten males were left alone with their mothers until day 10, when they were weaned and isolated. Ten control males were weaned on day 10 but kept with three females of the same age until day 73, when they were isolated. As in experiment I, seven weekly tests were begun on day 77. Weaning at 10 days of age seemed to have no ill effect. They were vigorous, and their weights on day 73 were similar to those of the animals weaned on day 25 (558 and 550 gm., respectively).

Results

The differences between the isolated and social males are clearly apparent (Table 2). The differences in average scores and in the percentage of animals displaying behavior in the higher categories are significant ($p < .01$). Again, the isolated animals exhibited a great

TABLE 2

Comparison of Sexual Behavior of Genetically Heterogeneous Male Guinea Pigs Weaned on Day 10 and Raised in Isolation with that of Males Raised with Females*

| | N | Lower Measures | | Mountings | | Intromissions | | Ejaculations | | Average Scores |
		Average per Animal	Per cent Displaying	Average per Animal	Per cent Displaying	Average per Animal	Per cent Displaying	Average per Animal	Per cent Displaying	
Isolated	10	120.8	100	10.6	30	6.5	30	1.5	30	4.9
Social Situation	10	52.9	100	14.3	100	19.6	100	5.3	100	8.1

* Data obtained from 7 tests of each male during isolation, days 77–120.

TABLE 3

Comparison of Sexual Behavior of Isolated and Socially Raised Castrate Strain 13 Male Guinea Pigs Receiving 500γ of Testosterone Propionate per 100 Grams Body Weight per Day*

| | N | Lower Measures | | Mountings | | Intromissions | | Ejaculations | | Average Scores |
		Average per Animal	Per cent Displaying	Average per Animal	Per cent Displaying	Average per Animal	Per cent Displaying	Average per Animal	Per cent Displaying	
Isolated	6	130.8	100	2.8	67	0.7	33	0	0	3.6
Social Situation	7	105.7	100	16.6	100	11.7	86	0.9	57	4.4

* Data obtained from 7 tests of each male during isolation, days 77–120.

deal of interest in the females which is reflected in the greater amount of lower-measure behavior.

EXPERIMENT III

The sexual performance of the Strain 13 males in experiment I was significantly below that of the Strain 2 and heterogeneous males. Perhaps, therefore, the deficiency of some substance such as androgen accounted for the poor performance of these males even in the social situation. If so, large amounts of an androgen might improve the performance.

Method

Thirteen males were castrated the day of birth and injected daily beginning the following day with 500γ of testosterone propionate[3] per 100 gm. body weight. This amount of testosterone propionate is over 20 times that found to be sufficient to restore the precastrational level of sexual behavior in adult castrates (12). Six of the males were isolated from other siblings the day of birth and weaned on day 12. Seven males were left with their siblings until weaning on day 12; from then until day 73 they were caged with three females of the same age. As before, seven weekly tests were begun on day 77.

Results

The average sexual behavior score of the social males was significantly better ($p < .01$)

[3] Testosterone propionate was supplied by Ciba Pharmaceutical Products, Inc.

than that of the isolated males (Table 3). In addition the social males in this experiment performed significantly better ($p < .02$) than the intact social males in experiment I (Table 1). They achieved higher average scores and exhibited more of each measure of behavior. The behavior of the castrated isolated Strain 13 animals (Table 3) was not significantly different from that of the intact isolated males in Experiment I (Table 1). It is to be noted that the performance of the socially reared, injected Strain 13 males was still significantly below that of the untreated, socially reared Strain 2 or heterogeneous males in experiments I and II.

EXPERIMENT IV

The contact with other animals in experiments I, II, and III was with females. This experiment was designed to ascertain the influence of contact with males. The results would reveal whether experience gained during possible copulations had contributed to the differences between the social and isolated males.

Method

Ten heterogeneous and seven Strain 2 males were used. As in experiment II, the heterogeneous males were weaned on day 10. The Strain 2 males were weaned

TABLE 4

Sexual Behavior of Genetically Heterogeneous and Strain 2 Male Guinea Pigs Raised with Males*

| | N | Lower Measures | | Mountings | | Intromissions | | Ejaculations | | Average Scores |
		Average per Animal	Per cent Display-ing	Average per Animal	Per cent Display-ing	Average per Animal	Per cent Display-ing	Average per Animal	Per cent Display-ing	
Strain 2	7	112.7	100	18.0	100	26.6	100	4.6	86	8.2
Genetically Hetero-geneous	10	75.6	100	16.9	90	11.9	90	3.9	80	6.7

* Data obtained from 7 tests of each male during isolation, days 77–120.

on day 25. Both groups were confined with two or three males of the same age until day 73. Seven weekly tests were begun on day 77.

Results

For every measure of behavior the performance of the heterogeneous males raised with males (Table 4) was intermediate between that of the social and isolated males of this stock. On the other hand, the performance of the Strain 2 males raised with males (Table 4) surpassed that of the males raised with females (Table 1). The lower measures of behavior and mounting were about the same, but the average number of intromissions and ejaculations and the average sexual behavior score were significantly higher ($p < .01$). In this experiment the sexual behavior score of the Strain 2 males was even higher than that of the heterogeneous males, although the difference was just short of statistical significance ($t = 2.05$).

DISCUSSION

The data show that males which have a minimum of contact with other animals have difficulty in mating. Although each isolated Strain 2 male was given seven tests with estrous females, only 1 of 17 achieved intromission or ejaculation. In contrast 16 of the 19 socially reared males ejaculated, and 17 of 19 had intromissions. Among 7 isolated Strain 13 males, none achieved intromission or ejaculation, whereas 4 of 7 in the social group displayed intromission and ejaculation. A similar difference was seen when the isolated and social Strain 13 males given exogenous androgen were compared. Of the heterogeneous males weaned at 10 days, only 3 of the 10 isolated animals had intromissions or ejaculations, but all the social males displayed the full pattern of sexual behavior. The isolated

animals gave evidence of being as much aroused by the presence of the female as were those from the social group, and there was no evidence of any emotional disturbance that could have interfered with their display of sexual behavior. The best explanation of these results appears to be that *the sexual behavior of the isolated animals had not been organized into an effective pattern.*

Not all that is involved in the organization of the pattern is clear, but certain observations are suggestive. The socially raised male orients itself rather specifically to the posterior end of the female. If the posterior end is inaccessible, the male will attempt to force the female into a more favorable position. When mounting the female, the experienced male approaches from the rear, placing his chest over her back while simultaneously clasping her sides with his forepaws. The female is held by this clasping grip of the male, and as his genitalia are brought into contact with the female, pelvic thrusts are made. These thrusts usually result in an intromission and sometimes culminate in an ejaculation.

Isolated male guinea pigs exhibit many of the components of sexual behavior, but they are not organized in such a way that copulation is possible. These males pursue the female, but when overtaking her, they frequently circle around, attempting to mount her head or side. The mounts are attempted by leaning one or both forepaws directly against the back or sides of the female whereupon she generally jumps away. The inability of the inexperienced male to properly mount and clasp the female appears to be his greatest handicap in achieving intromission and ejaculation. Once a male has properly mounted a female, pelvic thrusts (coital reflex) are made, and these generally result in insertion of the penis.

The demonstration that in the guinea pig the organization of sexual behavior into an effective pattern depends on contact with other animals, requires a re-examination of the basis for the opinion (6, 9, 10, 13, 14, 17) that in lower mammals such patterns are innately organized. One possible explanation is that the rat, which is the species that has been studied most commonly, does not require learning in order to exhibit the complete sexual act whereas the guinea pig does. If so, the generalization suggested by the experience with the rat was too broad. A second possible explanation is based on the procedure used in the work on the rat; they were weaned at 21 days of age, and isolation was not begun until this time (6, 17). While in neither case is it stated specifically, the implication is given that the animals remained with their siblings and mother up to weaning. The average litter size of eight or nine would afford considerable opportunity for contact with siblings prior to isolation. That such contact may be important is suggested by the fact that rats, despite their immaturity at birth, have been observed to pursue, mount, and clasp other individuals as early as 21 days after birth (10). Perhaps, therefore, the experimental males should have been isolated, at least from their siblings, at an earlier age.

Each of the three strains of guinea pigs has been seen to present a characteristic picture. The heterogeneous animals possessing large size and vigor are ideally suited for a rapid organization of sexual behavior. The Strain 2 males are also vigorous, but their smaller stature may perhaps handicap them in mounting and clasping the female. The inexperienced Strain 2 males in particular tend to place their forepaws directly on the back of the female rather than around her sides. Stone (17) also suggested that size may be important. The Strain 13 males appear to be still different. It is possible that their sluggishness retards both learning and the display of what was learned, although their large size probably aids them in mounting and clasping the female properly. The latter possibility may be the explanation for the large percentage of these males that exhibited mounting even in the isolated group.

An attempt to improve the performance of the Strain 13 males by administering large quantities of exogenous androgen met with only limited success. The performance of the socially raised castrate males receiving androgen was significantly better than that of the untreated males, but it was still significantly below the level of either the Strain 2 or the heterogeneous males raised under comparable conditions. Recently (16) thyroxin was used in an effort to activate males of this strain. The rate of oxygen consumption was elevated, but sexual behavior was not altered appreciably.

The results from experiment IV in which males of Strain 2 and the heterogeneous stock were provided contact with males present a problem. It will be recalled that while the performance of the Strain 2 males raised with males was significantly elevated above that of those raised with females, the performance of the heterogeneous males raised with males was below that of those raised with females. When both groups are considered together, the results indicate that contact with males generally provides sufficient experience for the organization of sexual behavior into a functional pattern. However, it remains to be explained why the heterogeneous and Strain 2 males should be affected differently. A possible explanation is that the sexual behavior of some of the heterogeneous males was suppressed by the inhibiting action of fighting among these heavier and more aggressive animals. At the same time, the Strain 2 males may have been aroused by the great amount of mounting that takes place among males caged together without suffering from the detrimental effects of fighting.

The performance of the males raised with males brings us to the suggestion developed by Beach (5, 8), that sexual behavior has a duality of components or "functions." One of the components is that of "potency" or the "capacity for sexual performance." It would appear to be associated with organized patterns depending on the activity of centers in the brain stem and spinal cord (5). According to one statement, this function includes "the promptness with which mating is initiated, the frequency of copulation, and the rapidity with which ejaculation occurs" (8, p. 288). The other component is "erotic sensitivity" or the "susceptibility to sexual arousal." It would appear to be mediated by a central excitatory mechanism of the forebrain (5).

161

Development of the concept appears to have followed observations on the copulatory behavior of partially decorticate rats (3, 7, 10). There was a proportional decrease in the percentage of tests in which copulatory behavior occurred, but the number of copulations during each positive test was not affected, even in the lesion-group in which the percentage of postoperative copulators was lowest. As Beach interpreted the results, there had been a decrease in the ease of arousal to the point of the practical abolition of copulatory behavior, but without an effect on the actual copulatory pattern.

The work with the guinea pig also provides evidence for the existence of dual components of mating behavior, the *organization of the sexual response*, which would correspond roughly to the "capacity for sexual performance" as used by Beach, and *sexual excitability*, which would correspond to "susceptibility to sexual arousal." But as we have thought of "excitability," the promptness with which mating is initiated, the vigor with which the mating is attempted, and the rapidity with which ejaculation occurs are a part of this function rather than of organization. The suggestion originated from the observation that an occasional male will achieve complete copulation in only one or two of ten tests with an estrous female, while giving the appearance of being generally indifferent in the other tests. The organization of the sexual response exists, but the level of excitability is so low that copulation usually does not occur. Strain 13 males fall in this category. On the other hand, most males raised in isolation exhibit excitability, but no organized pattern is present and they do not know how to copulate. While it is convenient to regard sexual behavior as comprising two components, it should be emphasized that they do not function independently. The data suggest that the greater the sexual excitability and vigor, the more rapidly organization of sexual behavior takes place. Also, it has been suggested that the ability or inability to copulate successfully (5, 11), affects the level of excitability.

SUMMARY AND CONCLUSIONS

Four experiments were performed in an effort to determine whether sexual behavior is innately organized in the male guinea pig or whether contact with other animals plays a role in its organization into an effective pattern.

In the first two experiments genetically heterogeneous males and males from the highly inbred Strains 2 and 13 were raised in a social (with females) or in an isolated situation. Significant differences were found between the sexual behavior of social and isolated Strain 2 males. Less marked but nevertheless significant differences were obtained with Strain 13 males. There were no significant differences between the social and isolated heterogeneous animals weaned at 25 days. However, when weaning was done at ten days, significant differences were obtained.

An attempt was made to improve the performance of the behaviorally sluggish Strain 13 males by administering androgen to castrated animals. Their behavior was improved over that of the intact Strain 13 males, but the characteristic differences between this strain and the other two was not overcome.

In the fourth experiment males were raised in a caging situation which afforded them contact only with other males except for the lactating sow. In general, male cage-mates provided sufficient experience for the organization of the sexual behavior pattern, although the picture was complicated somewhat by factors of inhibition and arousal which resulted from the caging of males together.

The main conclusions follow: (*a*) contact with other animals has an organizing action on the development of the copulatory pattern of the male guinea pig; (*b*) the influence of contact may be exerted very early in the life of the animal; (*c*) genetic differences between strains of guinea pigs are responsible for differences in the age at which organization of sexual behavior may take place and for the amount of sexual excitement exhibited during tests; (*d*) genetic differences in level of sexual excitement are not overcome by the administration of large quantities of exogenous androgen; (*e*) contact with males as well as with females generally provides sufficient experience for the organization of the copulatory pattern.

REFERENCES

1. BALL, J. Sex activity of castrated male rats increased by estrin administration. *J. comp. Psychol.*, 1937, **24**, 135–144

2. BALL, J. Male and female mating behavior in prepubertally castrated male rats receiving estrogens. *J. comp. Psychol.*, 1939, **28**, 273–283.

3. BEACH, F. A. Effects of cortical lesions upon the copulatory behavior of male rats. *J. comp. Psychol.*, 1940, **29**, 193–245.

4. BEACH, F. A. Female mating behavior shown by male rats after administration of testosterone propionate. *Endocrinology*, 1941, **29**, 409–412.

5. BEACH, F. A. Analysis of factors involved in the arousal, maintenance and manifestation of sexual excitement in male animals. *Psychosom. Med.*, 1942, **4**, 173–198.

6. BEACH, F. A. Comparison of copulatory behavior of male rats raised in isolation, cohabitation, and segregation. *J. genet. Psychol.*, 1942, **60**, 121–136.

7. BEACH, F. A. Relative effects of androgen upon the mating behavior of male rats subjected to forebrain injury or castration. *J. exp. Zool.*, 1944, **97**, 249–295.

8. BEACH, F. A. A review of physiological and psychological studies of the sexual behavior in mammals. *Physiol. Rev.*, 1947, **27**, 240–306.

9. BEACH, F. A. Sexual behavior in animals and men. Harvey Lectures, 1947–48, Series 43, 254–280.

10. BEACH, F. A. Instinctive behavior: Reproductive activities. In S. S. Stevens (Ed.), *Handbook of experimental psychology.* New York: Wiley, 1951. Pp. 387–434.

11. GRUNT, J. A. Exogenous androgen and non-directed hyperexcitability in castrated male guinea pig. *Proc. Soc. Exper. Biol. & Med.* 1954, **85**, 540–542

12. GRUNT, J. A., & YOUNG, W. C. Consistency of sexual behavior patterns in individual male guinea pigs following castration and androgen therapy. *J. comp. physiol. Psychol.*, 1953, **46**, 138–144.

13. KAGAN, J., & BEACH, F. A. Effects of early experience on mating behavior in male rats. *J. comp. physiol. Psychol.*, 1953, **46**, 204–208.

14. LOUTTIT, C. M. Reproductive behavior of the guinea pig. II. The ontogenesis of the reproductive behavior pattern. *J. comp. Psychol.*, 1929, **9**, 293–304.

15. NISSEN, H. W. The effects of gonadectomy, vasotomy and injections of placental and orchic extracts on the sex behavior of the white rat. *Genet. Psychol. Monogr.*, 1929, **5**, 451–457.

16. RISS, W. Sex drive, oxygen consumption and heart rate in genetically different strains of male guinea pigs. *Am. J. Physiol.*, 1955, in press.

17. STONE, C. P. The congenital sexual behavior of the young male albino rat. *J. comp. Psychol.*, 1922, **2**, 95–153.

18. VALENSTEIN, E. S., RISS, W., & YOUNG, W. C. Sex drive in genetically heterogeneous and highly inbred strains of male guinea pigs. *J. comp. physiol. Psychol.*, 1954, **47**, 162–165.

19. YOUNG, W. C. The hormonal regulation of reproductive behavior. In E. W. Dempsey (Ed.), *Allen's sex and internal secretions* (3rd Ed.). Baltimore: Williams & Wilkins, in press.

Received October 1, 1954.

Part III

CONTROL

Editor's Comments
on Papers 16 Through 21

SENSORY FACTORS

In studying the control of behavior, we consider the short-term factors that are important in the regulation of behavior. Part III is divided into three sections—sensory factors, neural correlates, and motivational and social factors. The important area of hormonal influences on sexual behavior is covered only peripherally because it has already been treated in a previous volume in this series (Carter, 1974).

As sexual behavior occurs primarily in relation to stimuli from the environment, its coordination with respect to environmental factors must depend on sensory input. Since the initial impetus provided by K. S. Lashley, much effort has been devoted to the study of exactly which senses are involved and how they are co-ordinated. The history of the study of sensory control of sexual behavior follows the familiar pattern of progressive refinement of the questions being asked and of the methods used. Earlier studies were aimed at determining which senses were *necessary* for copulation to occur. More recent literature is directed at determining more precisely the ways in which specific sensory input functions to control sexual behaivor.

As has been true with respect to several other problem areas, one can begin with the 1922 study of Stone (Paper 3). He demonstrated that neither visual, olfactory, nor gustatory stimuli nor the vibrissae are necessary for the elicitation of copulatory behavior in young, naive male rats. Partial elimination of the auditory system in blind or anosmic, experienced animals had little effect. The following year, 1923, Stone (Paper 16) extended these observations on the necessity of various senses for the display of copulatory behavior. Even though he eliminated cutaneous sensitivity of the anterior belly wall, the inguinal region, and the scrotum in the 1923 paper, copulatory behavior was not eliminated.

As part of his study of the role of the cerebral cortex in the control of copulation in rabbits, Brooks (see Paper 22) studied the effects of bilateral destruction of the labyrinths and auditory apparatus, enucleation of the eyes, and removal of the olfactory bulbs without eliminating copulatory behavior.

A similar approach was again used by Beach (Paper 17) with laboratory rats. Beach studied both inexperienced and experienced males, tried several different stimulus objects in attempting to elicit mounting, considered the possibility that behavioral deficits following lesions might be secondary to endocrine deficits, and imposed sensory deficits in combinations on various individual animals. He found no single sensory modality to be essential for the elicitation of copulatory activity. Loss of two modalities had a much greater effect on inexperienced than on experienced males. Beach concluded that stimuli in the various modalities all contribute to sexual arousal in a mutually facilitative manner.

These data were important in the formulation of a hypothetical model of the control of copulatory behavior in rats proposed by Beach in the same year (Beach, 1942). The primary construct of this system was a Central Excitatory Mechanism (c.e.m.) that received input from the various sensory modalities. Excitation of the c.e.m. was viewed as a function of input from the various sensory systems and hormones. Although different afferent systems were recognized as making contributions of differing importance, the differences were of degree, not kind, as all were summed in the nonspecific c.e.m.

Subsequent research has been focused more specifically on single sensory modalities and has generally progressed from consideration of whether or not input from a given source is necessary for copulation to a consideration of the precise ways in which such input modulates copulation, whether or not it is essential. Pauker (Paper 18) reviews various theories of the role of accessory glands in stimulating sexual activity. She notes that Tarchanoff believed that tension from distended seminal vesicles provided a mechanical stimulus for the arousal of a sexual instinct. However, Pauker found no effect from removing the seminal vesicles and prostate from male hamsters. Although she did not divide copulatory behavior into individual series to consider a full range of measures, Pauker did quantify the total number of copulations and the rate at which they occurred. Considering a broader range of measures characterizing individual series, Beach and Wilson (1963) found no effect of removal of the seminal vesicles from rats. Input from the reproductive accessory glands appears to be of little importance for male copulatory behavior.

Ball (Paper 19) found that female rats continued to show female sexual behavior after removal of their uteri and vagina, although they experienced some lowered "degree of sexual excitability" (p. 420). Thus input from these organs is not essential for female sexual activity. However, such input does appear important in other respects such as in stimulating the female to remain immobile and in blocking certain responses to pain (Komisaruk and Wallman, 1977). Female rats with the genital area anesthetized are ready to copulate sooner after previous copulations than intact animals (Bermant and Westbrook, 1966).

Other tactile input from the genital region has been found to be of great importance for copulation. For an excellent review, see the paper by Diakow (1974). Aronson and Cooper (Paper 20) desenitized the glans penis of sexually experienced male cats by sectioning the nerves dorsalis penis. Although the males with reduced sensory feedback experienced difficulty in achieving intromissions, they mounted the female readily. Interestingly, desensitized males showed a seasonality of mating not observed in unoperated animals. Similar results were found in a study of sexually inexperienced males (Aronson and Cooper, 1969). The role of sensory feedback from the genital region in regulating rat copulatory behavior has been analyzed by applying local anesthetics (for example, Carlsson and Larsson, 1964; Adler and Bermant, 1966) and by sectioning various nerves (for example, Dahlöf and Larsson, 1976; Lodder and Zeilmaker, 1976). In most cases, treatments that interfere with erection and vaginal insertion appear to have little effect on mounting behavior.

The more recent data are consistent with earlier conclusions that for tests in small cages, visual stimulation is relatively unimportant in the copulatory behavior of male rats (Hård and Larsson, 1968). It should be noted that virtually all laboratory research on sexual behavior has been done under conditions that would minimize the importance of distance receptors.

Major olfactory effects on copulatory behavior have been found. Lesions of the olfactory bulbs have been shown to interfere with copulatory behavior in male rats when a fine-grained analysis of copulatory behavior is used as in the work of Heimer and Larsson (Paper 21). The effect is not attributable to endocrine changes (Larsson, 1969). There are species differences, with the deficits in hamsters being especially severe (Murphy and Schneider, 1970). Although some studies indicate that the effect is due to loss of olfactory input rather than to the loss of neural tissue that occurs when the olfactory bulbs are removed (for example, Devor and Murphy, 1973; Larsson, 1971), tissue loss can be important (for example, Edwards and Burge, 1973). Again, species differences may be significant. A role for the vomeronasal organ also has been uncovered (Powers and Winans, 1975). Various studies have shown that intact male rats can discriminate and prefer the odors of receptive females over those of nonreceptive females (for example, Carr et al., 1966). The possible role of olfactory stimuli in the copulatory behavior of rhesus monkeys is quite controversal (for example, Goldfoot, et al., 1976; Michael and Keverne, 1968; Rogel, 1978).

In the early studies of sensory function, Stone (Paper 3) and Brooks (Paper 22) found little effect of interference with auditory function on rodent copulatory behavior. More recently, various studies have shown that many species of rodents communicate during copulatory activity via ultrasonic calls (for example, Barfield and Geyer, 1972; Sales, 1972). It is remarkable that what appears to be one of the more important communication channels during copulation was overlooked for so many years because the signals fell outside the range of detectability of the unaided human senses.

REFERENCES

Adler, N. T., and G. Bermant, 1966, Sexual behavior of male rats; Effects of reduced sensory feedback, *J. Comp. Physiol. Psychol.* **61**:240–243.

Aronson, L. R., and M. L. Cooper, 1969, Mating behavior in sexually inexperienced cats after desensitatization of the glans penis, *Animal Behav.* **17**:208–212.

Barfield, R. J., and L. A. Geyer, 1972, Sexual behavior: Ultrasonic post-ejaculatory song of the male rats, *Science* **176**:1349–1350.

Beach, F. A., 1942, Analysis of factors involved in the arousal, maintenance and manifestation of sexual excitement in male animals, *Psychosom. Med.* **4**:173–198.

Beach, F. A., and J. R. Wilson, 1963, Mating behavior in male rats after removal of seminal vesicles, *Proc. Nat. Acad. Sci. U.S.A.* **49**:624–626.

Bermant, G., and W. H. Westbrook, 1966, Peripheral factors in the regulation of sexual contact by female rats, *J. Comp. Physiol. Psychol.* **61**: 244–250.

Carlsson, S. G., and K. Larsson, 1964, Mating in male rats after local anesthetization of the glans penis, *Z. Tierpsychol.* **21**:854–856.

Carr, W. J., L. S. Loeb, and N. R. Wylie, 1966, Responses to feminine odors in normal and castrated male rats, *J. Comp. Physiol. Psychol.* **62**:336–338.

Carter, C. S., 1974, *Hormones and Sexual Behavior*, Dowden, Hutchinson &Ross, Stroudsburg, Penn.

Dahlöf, L. G., and K. Larsson, 1976, Interactional effects of pudendal nerve section and social restriction on male rat sexual behavior, *Physiol. Behav.* **16**:757–762.

Devor, M., and M. R. Murphy, 1973, The effect of peripheral olfactory blockade on the social behavior of the male golden hamster, *Behav. Biol.* **9**:31–42.

Diakow, C., 1974, Male-female interactions and the organization of mammalian mating patterns, *Adv. Study Behav.* **5**:227–268.

Edwards, D. A., and K. G. Burge, 1973, Olfactory control of the sexual behavior of male and female mice, *Physiol. Behav,* **11**:867–872.

Goldfoot, D. A., M. A. Kravetz, R. W. Goy, and S. K. Freeman, 1976, Lack of effect of vaginal lavages and aliphatic acids on ejaculatory responses in rhesus monkeys: Behavioral and chemical analyses, *Horm. Behav.* **7**:1–27.

Hård, E., and K. Larsson, 1968, Visual stimulation and mating behavior in male rats, *J. Comp. Physiol. Psychol.* **66**:805–807.

Komisaruk, B. R., and J. Wallman, 1977, Antinociceptive effects of vaginal stimulation in rats: Neurophysiological and behavioral studies, *Brain Res.* **137**:85–107.

Larsson, K., 1969, Failure of gonadal and gonadotrophic hormones to compensate for an impaired sexual function in anosomic male rats, *Physiol Behav.* **4**:733–737.

Larsson, K., 1971, Impaired mating performances in male rats after anosmia induced peripherally or centrally, *Brain, Behav. Evol.* **4**:463–471.

Lodder, J., and G. H. Zeilmaker, 1976, Effects of pelvic nerve and pudendal nerve transection on mating behavior in the male rat, *Physiol. Behav.* **16**:745–751.

Michael, R. P., and E. B. Keverne, 1968, Pheromones in the communication of sexual status in primates, *Nature* **218**:746–749.

Murphy, M. R., and G. E. Schneider, 1970, Olfactory bulb removal eliminates mating behavior in the male golden hamster, *Science* **67**:302–304.

Powers, J. B., and S. S. Winans, 1975, Vomeronasal organ: Critical role in mediating sexual behavior of the male hamster, *Science* **187**:961–963.

Rogel, M. J., 1978, A critical evaluation of the possibility of higher primate reproductive and sexual pheromones, *Psychol. Bull.* **85**:810–830.

Sales, G. D., 1972, Ultrasound and mating behavior in rodents with some observations on other behavioral situations, *J. Zoology* **168**:149–164.

16

Reprinted from *J. Comp. Psychol.* **3**:469–473 (1923)

FURTHER STUDY OF SENSORY FUNCTIONS IN THE ACTIVATION OF SEXUAL BEHAVIOR IN THE YOUNG MALE ALBINO RAT[1]

CALVIN P. STONE

Department of Psychology, Stanford University

Researches of Sherrington ('06) and others have shown that certain complex reflexes may be evoked in the higher animals by mechanical stimuli applied to definite regions of the skin (i.e., scratch reflex, pinna reflex, extensor thrust, etc.). Assuming that the copulatory response of the male albino rat is fundamentally similar from the standpoint of its neuro-muscular mechanism to other complex patterns of congenital response, the question may well be asked as to whether it is aroused through cutaneous stimuli. This inquiry is especially pertinent in view of previous experiments which would seem to favor assigning to the cutaneous and kinaesthetic receptors the primary rôle of mediating external stimuli responsible for sexual activities.

It has been demonstrated (Stone, 1922) that, under laboratory conditions, neither visual, olfactory, nor gustatory stimuli are necessary for the arousal of the initial copulatory response in the young male rat. Elimination of these sensory functions prior to copulatory experience does not cause any observable change in the pattern of the copulatory act or the responsiveness of young males to females in heat. To the foregoing list of sensory controls the present study adds the control of cutaneous sensibility of the anterior belly wall, the inguinal region, and the scrotum— parts of the skin readily stimulated as the young male crawls over, jumps upon, or attempts to mount other rats.

[1] This study was begun at the University of Minnesota under the direction of Dr. Karl S. Lashley and completed at Stanford University. It was financed from funds granted through the Committee for Research on Sexual Problems of the National Research Council.

TECHNIQUE

1. Animals used. Seven young male rats survived all operations necessary for the control of sensory functions herein described. Three of these were reared together in a large wire cage from the date of weaning to the end of the experiment. The other four were reared in individual cages from the age of forty days to the end of the experiment, in order that all possibility of copulatory attempts upon cage mates might be prevented.

2. Operations. The technique followed for the anesthetization of the skin of the belly wall, inguinal region, and scrotum was as follows: An incision was made in the skin at the xiphoid process of the sternum (fig. 1, *A*) and continued distally in a curvilinear course (fig. 1, *B*) on the right side of the body. This incision was brought to an end at *C*, in the middle of the scrotum, approximately 2 mm. anterior to the anus. The skin of the belly wall, *D*, medial to the incision *B* was then dissected free from the muscle of the belly wall, the inguinal region, and scrotum, and laid back medialward until the linea alba was exposed. This dissection freed the penis from the integument except at the point where the prepuce is attached; this attachment was not severed until a later operation. When the foregoing dissection had been completed, the flap of skin was immediately replaced into its original position and the edges of the incision brought together and sutured by interrupted stitches as shown in figure 2. The incision was coated with a film of thin celloidin and the animal was then placed into a clean cage to await healing of the wound before undergoing further operations.

At the end of fourteen days, when the skin-wound of the foregoing operation was completely closed, an operation was performed on the left side of the body, in a manner duplicating that on the right side. An incision like that shown in figure 1, *B*, connecting proximately and distally with the incision of the first operation, was made on the left side of the body. The integument of this side was then freed from the underlying muscle, by dissecting medialward from the incision until the medial limit of the previous dissection was reached. The prepuce was cut free from the integument.   *Thus, by two operations, an island of skin of the anterior belly wall, the inguinal region, the external genitalia, and the ventral and lateral portions of the scrotum was raised from the surrounding and underlying tissues.*   In this manner *all nerve connections to this island of skin were severed, thereby rendering it completely anesthetic.*   All tests were completed before sensi-

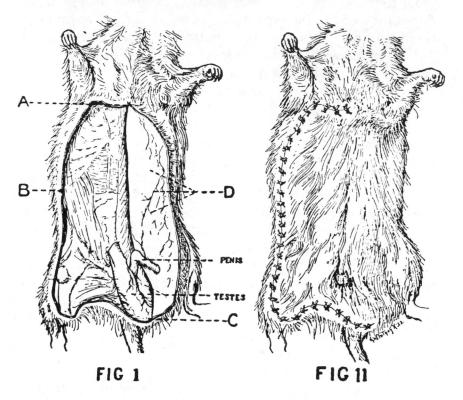

FIG 1   FIG 11

bility could be regained through the ingrowth of nerve filaments into this island of skin from the underlying or surrounding tissue.

Twenty-eight days after the first operation, and fourteen days after the second, the olfactory lobes were destroyed.   On the thirty-second day the eyes were removed and the vibrissae cut. On the thirty-fourth day the tympanic membranes of the ears were destroyed.   No control of the taste receptors was required since the anosmic males did not sniff at or lick the ano-vaginal zones while undergoing tests for copulation and consequently did not receive any gustatory stimuli whatever from an outside source. As a result of these operations, any substitute or vicarious functioning of the receptors which we wished to bring completely under control in this experiment was prevented. After the foregoing operations copulatory response could be activated only by receptors other than visual,[2] olfactory, gustatory, and the cutaneous receptors located at the roots of the vibrissae and within this island of skin rendered anesthetic by the destruction of nerves connecting it with the surrounding and underlying tissues.

RESULTS FROM TESTS FOR COPULATION

The report of tests for copulation follows in the form of case histories for individual rats or groups of rats similarly treated. Rats 1, 2 and 3 were reared together; rats 4, 5, 6 and 7, in individual cages. All tests were made between 8:00 and 11:00 p.m.

Rat 1. This rat was first used to test the effects of the skin operation alone. Twenty-eight days after the first skin operation, fourteen days after the second, a receptive female was placed into his cage. He copulated a few minutes thereafter. Two copulations were allowed. Six days later, after the elimination of eyes, olfactory bulbs, vibrissae, and tympanic membranes of the ears, he was re-tested. Copulated readily and frequently.

Rats 2 and 3. These animals were tested for copulatory ability on the thirty-fourth day after the first skin operation,—twenty days after the second skin operation, six days after the olfactory bulbs were transected and two days after the eyes were removed. Both animals were much slower in making their first copulatory attempts than normal animals, but after the initial acts copulated freely and vigorously. The physical condition of these animals was far below that of normal animals. For this reason an unequivocal cause for the delayed attempts at copulation cannot be assigned. In the light of experimental data of the earlier study (Stone, 1922), and that from rats 4, 5, 6 and 7 of

[2] Auditory control cannot be included since it was incomplete. Destruction of the tympanic membranes greatly reduces auditory sensitivity, but does not produce deafness.

the present study, I am inclined to favor "poor physical condition" rather than the loss of essential receptors as the primary cause of delay in copulatory attempts.

Rats 4, 5, 6 and 7. These animals were given their first opportunity to copulate on the thirty-sixth day after the first skin operation. At this time the animals were anosmic, blind, without vibrissae, and with reduced sensitivity of hearing. In addition, the skin of the belly wall, inguinal region, and scrotum were anesthetic. Each of the rats copulated readily and vigorously when the receptive females were introduced into their caes. Although the males experienced some difficulties in pursuing and clasping the females, their behavior indicated a well-defined sexual impulse.

SUMMARY

The results of this study may be briefly summarized as follows:

The initial copulatory response can be aroused in sexually inexperienced male rats when afferent impulses from all of the following sources have been excluded: the skin of the anterior belly wall, the inguinal region, and the ventral and lateral portions of the scrotum; from the vibrissae; and from the visual, olfactory, and gustatory receptors. To this list of sensory controls, auditory receptors should probably be added, since diminished auditory sensitivity had no observable effect.

REFERENCES

SHERRINGTON, C. S.: The Integrative Action of the Nervous System. London, 1906.

STONE, CALVIN P.: The congenital sexual behavior of the young male albino rat. Jour. Comp. Psychol., 1922, ii, 95–153.

STONE, CALVIN P.: Experimental studies of two important factors underlying masculine sexual behavior: the nervous system and the internal secretion of the testis. Jour. Exper. Psychol., 1923, iv, 85–106.

17

Reprinted from pp. 163, 202–207 of *J. Comp. Psychol.* **33**:163–207 (1942)

ANALYSIS OF THE STIMULI ADEQUATE TO ELICIT MATING BEHAVIOR IN THE SEXUALLY INEXPERIENCED MALE RAT[1]

FRANK A. BEACH

The American Museum of Natural History, New York, N. Y.

Received June 20, 1941

INTRODUCTION

One of the most important problems in the field of animal behavior is the accurate definition of the stimuli activating complex patterns of response. The need for such analysis is especially pressing in studies of innately organized behavior. No description of an instinctive response is complete without a knowledge of the sensory stimuli leading to the motor reactions.

It is really imperative that we make a serious effort to define the adequate stimulus, not only in studies of instinct but equally in studies of reflexes and of learning. Psychological theories based upon the relations of stimulus and response remain sheer nonsense so long as the stimulus is defined only as whatever the experimenter puts in front of the animal. We have gone far enough in this work to be sure that the animal rarely reacts to what the experimenter regards as the stimulus. In any complex situation the true basis of reaction can be discovered only by systematic variation of all the parts and properties of the supposed stimulus. (Lashley 1938, p. 455.)

In addition to varying the "parts and properties" of the stimulus, the experimenter may progressively and selectively reduce the sensory capacities of the reacting organism.

[*Editor's Note*: Material has been omitted at this point.]

---

[1] Supported by a grant in aid of research from the Committee for Research in Problems of Sex, National Research Council.

[*Editor's Note*: Material has been omitted at this point.]

SUMMARY

Procedure

Thirty-five male rats were raised in individual cages from 21 days of age until the completion of the experiments. Twelve males served as a normal control group. At 80 to 100 days of age the remaining males were subjected to various operations designed to eliminate certain types of sensory stimulation.

(1) Five animals were subjected to destruction of the olfactory bulbs. (2) Five males were subjected to enucleation of the eyes. (3) Five were subjected to transection of the sensory branches of the trigeminal nerve supplying the skin of the snout and lips. (4) Two rats were blinded and deprived of olfactory bulbs. (5) Two were subjected to transection of the olfactory bulbs and the sensory nerves supplying the skin of the snout and lips. (6) Two males were subjected to destruction of the eyes, olfactory bulbs, and sensory branches of the fifth nerve.

Normal and operated males were tested with the following incentive animals: (1) a sexually receptive female rat, (2) a female rat in heat but immobilized by the injection of general anaesthetic, (3) a nonreceptive, castrated female rat, (4) a young male rat smaller than themselves, (5) a young female guinea pig, and (6) a young female rabbit. These incentive animals were presented in counterbalanced order so that each sex object was presented to some male in each test. Sex tests were conducted at intervals of 5 to 7 days. Each test lasted 15 minutes from the introduction of the incentive animal; or, if a receptive female was used, 15 minutes from the time of the initial copulation.

During all sex tests motion pictures were taken to record the male's typical reactions to each incentive animal. Descriptions of the male's behavior were dictated to an assistant. The male's responsiveness to each incentive animal was rated on a 4-point scale ranging from total lack of interest to intense excitement. When copulation with the receptive female occurred the records included notes on the latent period before the beginning of mating, number of copulations, attempts, and vaginal plugs.

Normal males that failed to copulate when presented with the receptive female were retested with all incentive animals after the injection of testosterone propionate. Desensitized males that did not copulate were given three supplementary tests with the receptive female.

Fourteen sexually experienced males (all vigorous copulators) were tested with receptive females before and after the elimination of various sensory receptors. (1) Five males were tested before and after enucleation of the eyes. (2) Five animals were tested before and after complete destruction of the olfactory bulbs. (3) Four rats were tested before and after section of the sensory nerves supplying the skin of the snout and lips. (4) Six cases were tested before and after enucleation of the eyes plus section of the sensory branches of the fifth nerve. (5) Two males were tested before and after removal of the eyes and olfactory bulbs. (6) One rat was tested as a normal, retested after blinding, tested again after fifth nerve section, and tested finally after destruction of the olfactory bulbs.

Results

Inexperienced males

1. Eight of 12 unoperated males copulated with the receptive female. The mating pattern was well organized and sexual vigor was equal to that of experienced copulators.

2. One of the 12 unoperated males attempted copulation with a young male presented in the first test. With this exception no actual mating was attempted by any male prior to the achievement of normal sexual intercourse with a receptive female. There were several rats that showed marked excitement in response to different incentive animals presented before the receptive female, but no copulatory attempts occurred.

3. Males which had copulated with the receptive female, frequently attempted to mate with other incentive animals offered in subsequent tests. Not all copulators showed this behavior.

4. Some blind males, anosmic males, and males deprived of cutaneous sensitivity in the snout and lips copulated when given receptive females.

5. The relative proportion of copulators in each of the three groups deprived of a single sensory modality was lower than in the control group.

6. No animal deprived of more than one type of sensory receptor copulated with the receptive female or attempted to mate with any other incentive animal.

Experienced copulators

1. All sexually experienced males continued to copulate after the elimination of any two of the three sensory receptors investigated.

2. Elimination of all three modalities abolished mating responses in the experienced copulator.

Conclusions

1. For the majority of sexually inexperienced male rats only the receptive female rat constitutes a stimulus adequate to elicit the initial mating reactions.

2. There is a small proportion of extremely excitable males in whom the copulatory threshold is quite low. Prior to contact with a receptive female rat such males may attempt to copulate with other incentive animals (such as the young male).

3. There is a third class of inexperienced males in whom the copulatory threshold is very high. Such animals showed very little interest in the incentive animals used, and were very sluggish in their reactions to the receptive female. A few of these individuals did not copulate with the highly receptive estrous female.

4. The specificity of the adequate stimulus to initial mating behavior is positively related to the height of the copulatory threshold in each individual.

5. Neither olfaction, vision, nor cutaneous sensitivity in the snout and lips is essential to the appearance of copulatory behavior in the inexperienced male rat.

6. Elimination of any one of these modalities may result in a reduction of the excitability of the male. If the rat possesses a low copulatory threshold he will mate despite reduced excitation. If the threshold is high the excitation resulting from reduced sensory data is not sufficiently intense to produce mating behavior.

7. Combined elimination of any two of the senses studied greatly reduced the probability of copulation, and in all of our rats without sexual experience mating behavior was prevented by such operation. The absence of the motor pattern of coition

is probably a reflection of lowered excitation consequent to the reduction of sensory impulses.

8. Sexually experienced males continue to copulate after the elimination of any two of the above-mentioned senses. This survival of copulatory responses is probably related to the lowering of the copulatory threshold as a result of previous sexual intercourse.

9. Despite the possession of a low copulatory threshold, the experienced copulator deprived of vision, olfaction, and cutaneous sensitivity in the snout and lips cannot obtain from the receptive female sufficient stimulation to raise excitability to this threshold.

10. The normal male receives from the sexually receptive female visual, olfactory, and cutaneous stimuli; all of which contribute to the arousal of sexual excitement and the elicitation of the motor pattern of copulation; but none of which are essential if the copulatory threshold is low enough to permit the appearance of motor patterns in the presence of reduced excitability consequent to elimination of one or more sensory receptors.

REFERENCES CITED

ANDERSON, E. E. 1936a Consistency of tests of copulatory frequency in the male albino rat. J. Comp. Psych., 21, 447–459.

ANDERSON, E. E. 1936b Interrelationship of drives in the male albino rat. I. Intercorrelations of measures of drives. J. Comp. Psych., 24, 73–118.

BALL, J. 1937 A test for measuring sexual excitability in the female rat. Comp. Psych. Monog., 14, 1–37.

BEACH, F. A. 1938 Techniques useful in studying the sex behavior of the rat. J. Comp. Psych., 26, 355–359.

BEACH, F. A. 1939 The neural basis of innate behavior. III. Comparison of learning ability and instinctive behavior in the rat. J. Comp. Psych., 28, 225–262.

BEACH, F. A. 1940 Effects of cortical lesions upon the copulatory behavior of the male rat. J. Comp. Psych., 29, 193–245.

BEACH, F. A. 1941a Comparison of copulatory behavior of male rats raised in isolation, cohabitation, and segregation. J. Genet. Psych. (In press).

BEACH, F. A. 1941b Copulatory behavior of male rats raised in isolation and subjected to partial decortication prior to the acquisition of sexual experience. J. Comp. Psych., 31, 457–471.

BEACH, F. A. 1941c Effects of brain lesions upon running activity in the male rat. J. Comp. Psych., 31, 145–179.

Brooks, C. Mc. 1937 The rôle of the cerebral cortex and of various sense organs in the excitation and execution of mating activity in the rabbit. Am. J. Physiol., 120, 544–553.

Greene, E. C. 1935 Anatomy of the rat. Trans. Am. Phil. Soc., 27, 1–339.

Jenkins, M. 1928 The effect of segregation on the sex behavior of the white rat as measured by the obstruction method. Genet. Psych. Monog., 3, 457–471.

Lashley, K. S. 1938 Experimental analysis of instinctive behavior. Psych. Rev., 37, 1–24.

Nissen, H. W. 1929 The effects of gonadectomy, vasotomy, and injections of orchic extracts, on the sex behavior of the white rat. Genet. Psych. Monog., 5, 451–547.

Noble, G. K., and Aronson, L. R. 1942 Sexual behavior of Anura. I. The normal mating pattern of *Rana pipiens*. Ms. in preparation.

Noble, G. K., and Bradley, H. T. 1933 The mating behavior of lizards: its bearing on the theory of sexual selection. Ann. N. Y. Acad. Sci., 35, 25–100.

Richter, C. P., and Hawkes, C. D. 1939 Increased spontaneous activity and food intake produced in rats by removal of the frontal poles of the brain. J. Neurol. and Psychiat., 2, 231–242.

Spiegel, E. A., Miller, H. R., and Oppenheimer, M. J. 1940 Forebrain and rage reactions. J. Neurophysiol., 3, 538–548.

Stone, C. P. 1922 Congenital sexual behavior of young male albino rats. J. Comp. Psych., 2, 95–153.

Stone, C. P. 1923 Further study of the sensory functions in the activation of sexual behavior in the young male albino rat. J. Comp. Psych. 3, 469–473.

Stone, C. P. 1925 The effects of cerebral destruction on the sexual behavior of male rabbits. I. The olfactory bulbs. Am. J. Physiol., 71, 430–435.

Stone, C. P. 1938 Activation of impotent male rats by injections of testosterone propionate. J. Comp. Psych., 25, 445–450.

Stone, C. P. 1939 Copulatory activity in adult male rats following castration and injections of testosterone propionate. Endocrinol., 24, 165–174.

Stone, C. P., Tomlin, M. I., and Barker, R. G. 1935 A comparative study of sexual drive in male rats as measured by direct copulatory tests and by the Columbia obstruction apparatus. J. Comp. Psych., 19, 215–241.

Swann, H. G. 1934 The functions of the brain in olfaction. II. The effects of destruction of olfactory and other structures upon the discrimination of odors. J. Comp. Neur., 59, 175–201.

18

Reprinted from *J. Comp. Physiol. Psychol.* 41:252–257 (1948)

THE EFFECTS OF REMOVING SEMINAL VESICLES, PROSTATE, AND TESTES ON THE MATING BEHAVIOR OF THE GOLDEN HAMSTER *CRICETUS AURATUS*[1]

ROSLYN S. PAUKER

From the Department of Animal Behavior, American Museum of Natural History, New York City

Received January 19, 1948

The mechanisms responsible for elicitation of sexual behavior are not completely known. It is reasonably well established that testicular androgen in male vertebrates is a primary factor for the mating response in the intact animal (2, 4, 9). Castration of sexually mature male rodents however does not usually result in an immediate cessation of mating behavior. There is a gradual waning of sexual activity, but parts of the whole mating pattern may be observed long after ablation of the testes (4, 5, 6, 14, 15).

The sexual behavior that exists after castration obviously does not depend on the presence of the gonads. One explanation of post-castration behavior considers that androgen may have a long range effect on the neural mechanisms responsible for mating. However, it is possible that the interplay of other factors besides testicular androgen may be important in the mating behavior of the intact animal and must also be considered in any display of mating after castration.

Some workers have advanced the hypothesis that the major accessory reproductive glands are influential factors in mating behavior. This investigation was designed to test the validity of this hypothesis, specifically to determine the significance of the seminal vesicles and prostate in the mating behavior of the castrate male hamster.

REVIEW OF THEORIES ON THE ROLE OF THE ACCESSORY REPRODUCTIVE GLANDS IN MATING BEHAVIOR

Tarchanoff (1887) believed that the tension of the distended seminal vesicles was the mechanical stimulus to nerves and brain that aroused the "sex instincts" in frogs and possibly mammals (16). Steinach (13) removed the seminal vesicles of rats and "normal sex behavior" was still retained, " . . . manifestations not only lasted for a short period after the operation but were so far as one could tell permanent." Steinach concluded that the vesicles are not associated with the "sex instinct." However there is no evidence that he considered quantitative measures of sexual activity as a basis for comparing the pre- and post-operative performances of his animals.

More recently F. H. Allport (1) and others (14) have considered the possible function of the seminal vesicles in sex behavior. It is their contention that one component of the sexual "drive" is afferent stimuli from the distended glands of the reproductive tract. However,

[1] This investigation was supported by a grant from the Committee for Research in Problems of Sex, National Research Council, to Dr. L. R. Aronson. The author wishes to express her appreciation to Drs. Lester R. Aronson and T. C. Schneirla, Mrs. Marie Holz-Tucker, and Miss Sarah Lichtenberg for their helpful suggestions.

in continually breeding mammals there is no periodic variation in the size of the glands. To be sure, in animals that have a rutting season the seminal vesicles are known to undergo a periodic enlargement, but "sexual desire" may exist before the seminal vesicles are full (of their own secretion) (8). Thus if the seminal vesicles are responsible for initiating a sexual response it must be through a mechanism other than distension.

Steinach's results indicate that the seminal vesicles are certainly not decisive factors in mating behavior. However, that they do not play a role in the full mating pattern of the intact animal and the partial pattern of the castrate has not been conclusively shown.

The prostate besides its exocrine secretion is also considered by some workers to have an endocrine function. Serralach and Pares (1908) maintained that an internal secretion of the prostate controls testicular function (12). It has been shown more recently that the prostate as well as the seminal vesicles is kept in a functional state by the testes (10). Nissen (1929), considering the possible factors responsible for post-castration behavior, suggests that the theoretical prostatic hormone although dependent on the testes for its production may be the most immediate stimulus to the nervous mechanisms responsible for mating behavior (11).

METHODS

Six male hamsters from 2 litters were isolated in individual cages at 21 days of age. At the time of the first mating test, the animals were 71 and 76 days old.

The females used in the tests were spayed and brought into heat by subcutaneous estrogen and progesterone injections (7). [Progynon B and Proluton were furnished through the courtesy of Dr. Edward Henderson of the Schering Corporation.] Between 7.5 and 10 gamma of estradiol benzoate in sesame oil were injected 42 to 36 hours before tests were to be run. Twenty-four hours after the estrogen priming, the animals were injected with 0.75 to 1.0 mg. of progesterone. The females were in heat on the morning of the next day.

Testing technique: To determine if the injected females were receptive, they were placed individually on the testing day in a chamber with a sexually active indicator male (which was not an experimental animal in this study). A female in heat will usually assume the lordosis position when the male sniffs or licks the pudendal region.

Glass aquaria, 14 x 12 x 10½ inches, were used as testing chambers. A male was taken from his home cage and placed in the chamber. After an "orientation" period of at least five minutes, a receptive female was placed with the male. The female is presumably the only new aspect of the situation in the chamber, and the active male begins immediately to sniff and lick the female; whereas a male not given sufficient time to adjust to the chamber will not respond as quickly to the female. Such situations may eventuate in a fight if the female takes the more active role and "investigates" the male.

The observer sat quietly within close range of the chamber, a safe procedure since the nearness of an observer is apparently not disturbing to the hamster. The following features of the mating pattern were recorded in sequence on a form testing sheet:

1. *head* or *side mounting*—mounting at the head or flanks of the female
2. *rear mounting*—mounting (in copulatory position) with pelvic thrusts but without intromission

3. *copulation*—mounting with intromission. [In the hamster, so far as we
have been able to determine, copulation with an ejaculation is not correlated with any behavioral phenomenon that can be used to distinguish
it from a copulation without an ejaculation.]

TABLE 1

*Mean copulatory frequency in positive tests before and after removal of the seminal vesicles
and prostate*

MALE NO.	BEFORE OPERATION		AFTER OPERATION	
	No. of positive tests*	Mean copulatory frequency	No. of positive tests†	Mean copulatory frequency
511	10	28.8	5	37.3
519	10	32.9	5	35.2
525	8	24.3	1	39.0
527	9	36.6	4	37.0
529	10	34.6	2	31.5
531	9	33.1	2	31.0

Mean Difference in copulatory frequency = 4.3
P Value = 0.14

* Total number of tests given = 10.
† Total number of tests given = 6 (first post-operative test not counted, hence total considered as 5).

TABLE 2

*Comparison of number of positive tests before and after removal of the seminal vesicles
and prostate*

MALE NO.	NUMBER OF POSITIVE TESTS		
	Tests 1–5 (pre-operative)	Tests 6–10 (pre-operative)	Tests 12–16* (post-operative)
511	5	5	5
519	5	5	5
525	5	3	1
527	4	5	4
529	5	5	2
531	5	4	2

Mean Difference between pre-operative tests 6–10 and post-operative tests 12–16 = 1.1
P Value = 0.16

* First post-operative test (test 11) omitted.

4. *copulatory latency period*—time between introduction of the female into the
testing chamber and the first copulation.

The test was timed from the moment the female was introduced into the
chamber. If within 10 minutes no copulation had occurred the test was terminated and considered negative. If copulation occurred within this interval

(positive test) the test was continued for 10 minutes starting from the time of the first copulation.

Each male received ten weekly pre-operative tests. The seminal vesicles and prostate were then removed abdominally while the male was under ether anesthesia. Six weekly post-operative mating tests were given starting on the fourth day after removal of the accessory glands. Thirty-nine days after the first operation all the animals were castrated scrotally. Two weekly mating tests were given beginning on the seventh day after gonadectomy, and a final

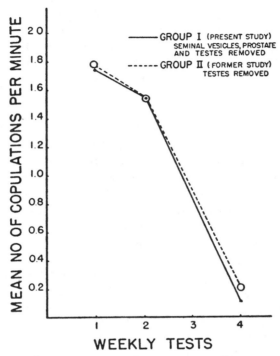

FIGURE 1. COMPARISON OF COPULATORY FREQUENCY OF CASTRATED HAMSTERS WITH HAMSTERS SUBJECTED TO CASTRATION AFTER ABLATION OF THE ACCESSORY REPRODUCTIVE GLANDS

mating test one month later. Post mortem gross inspection, done by Miss Sarah Lichtenberg, did not reveal regeneration of prostate or seminal vesicles.

RESULTS

The organization and the frequency of occurrence of the mating pattern and of its essential features are not significantly altered by the removal of the seminal vesicles and prostate glands.

The copulatory latency period does not appear to be affected. Copulatory frequency is not decreased (see table 1). Actually, in four of the six animals copulatory frequency increased after ablation of the accessory reproductive glands, an increment which is not statistically significant.

In four of the animals the number of positive tests after seminal vesicle and prostate removal appears to decrease; however this decrease is not significant (see table 2). There is an apparent tendency for negative trials to occur more frequently with repeated testing even before operation; in pre-operative tests 6–10 the number of positive trials tends to be less than the number of positive tests in the first 5 pre-operative tests. Hence the decrease in positive tests after operation probably is not due to the absence of the seminal vesicles and prostate and might have occurred even if the animals had not been operated.

Mating behavior may still be evoked in animals subjected to castration after accessory gland removal. The copulatory frequency of the males (in this study) after the seminal vesicles, prostate and testes were removed was compared with a group of four gonadectomized males used in a previous investigation (5) and tested 6, 14, 20, and 27 days after castration (fig. 1). The rate of waning in sexual activity after removal of the testes and accessory glands is similar to that of animals that are only castrated.

DISCUSSION

The accessory reproductive glands, although maintained by the testes, do not appear to function in the elicitation of mating behavior. The results of this experiment offer proof that sexual behavior is not dependent on nervous stimulation from the accessory glands. If the accessory glands, in particular the prostate, do secrete a hormone it has no noticeable effect on sexual behavior.

The results reported indicate that the extra-gonadal factors responsible for sexual behavior are not to be sought in the accessory reproductive glands. The theories still to be tested and made more specific are concerned with the role of the central nervous system, as well as with the specific actions and possible sites of activity of the endocrine secretions.

SUMMARY AND CONCLUSIONS

Six male golden hamsters were given pre- and post-operative mating tests. Two operations were performed on each animal: the first, removal of the seminal vesicles and prostate; the second, castration. There was an interval of six weeks between operations.

The results indicate that the seminal vesicles and prostate are not functional in the mating behavior of the non-castrate hamster, and the sexual behavior persisting after castration does not appear to be directly dependent on these accessory glands.

BIBLIOGRAPHY

1. ALLPORT, F. H.: *Social Psychology*. Boston: Houghton Mifflin Co., 1924.
2. BEACH, F. A.: Analysis of factors involved in the arousal, maintenance and manifestation of sexual excitment in male animals. *Psychosom. Med.*, 1942, **4**, 173–198.
3. ——: Relative effects of androgen upon mating behavior of male rats subjected to forebrain injury or castration. *J. exp. Zool.*, 1944, **97**, 249–295.
4. ——: Experimental studies of sexual behavior in male mammals. J. clin. Endocr., 1944, **4**, 126–134.

5. BEACH, F. A., AND PAUKER, R. S.: The effect of castration on the mating behavior of the male golden hamster. Unpublished data.

6a. CARPENTER, C. R.: I. The effects of complete and incomplete gonadectomy on the primary sexual activity of the male pigeon. *J. compar. Psychol.*, 1933, **16**, 25–57.

6b.——: II. The effect of complete and incomplete gonadectomy on secondary sexual activity with histological studies. *J. comp. Psych.*, 1933, **16**, 59–97.

7. FRANK, A. H., AND FRAPS, R. M.: Induction of estrus in the ovariectomized golden hamster. 1945, *Endocr.*, **37**, 357–361.

8. MARSHALL, F. H. A.: *The Physiology of Reproduction.* London: Longmans, Green and Co., 1922.

9. MOORE, C. R., AND GALLAGHER, T. F.: On the prevention of castration effects in mammals by testis extract injection. *Amer. j. Physiol.*, 1929, **89**, 388–394.

10. MOORE, C. R., HUGHES, W., AND GALLAGHER, T. F.: Rat seminal vesicle cytology as a testis hormone indicator and the prevention of castration changes by testes-extract injection. *Amer. j. Anat.*, 1930, **45**, 109–135.

11. NISSEN, H. W.: The effects of gonadectomy, vasotomy and injections of placental and orchic extracts on the sex behavior of the white rat. *Genet. psychol. Monogr.*, 1929, **5**, 451–550.

12. SERRALACH, N., ET PARES, M.: Quelques donnees sur la physiologie de la prostate et du testicule. C. R. de la Soc. Biol., 1907, **63**, 790–792.

13. STEINACH, E.: *Sex and Life.* New York: The Viking Press, 1940.

14. STONE, C. P.: The retention of copulatory ability in male rats following castration. *J. comp. Psychol.*, 1927, **7**, 369–387.

15. ——: The retention of copulatory ability in male rabbits following castration. J. genet. Psychol., 1932, **40**, 296–305.

16. TARCHANOFF, J. R.: Zur Physiologie des Geschlectsapparatus des Frosches. *Pflüg. Arch. f. d. Gesammte Physiol. des Menschen u. Tiere*, 1887, **40**, 330–351.

19

Reprinted from *J. Comp. Psychol.* **18**:419–422 (1934)

SEX BEHAVIOR OF THE RAT AFTER REMOVAL OF THE UTERUS AND VAGINA

JOSEPHINE BALL

Psychobiological Laboratory, Phipps Psychiatric Clinic, Johns Hopkins University

Since in the experiments of Allen and coworkers (1) and Hemmingsen (2) injection of oestrin into female mice and rats was followed by mating behavior, and since this hormone produces profound changes in the structure and activity of the sex tract, the question arose as to whether or not the behavior pattern might be initiated by afferent nerve impulses from these organs. The following experiment shows, however, that the changes in the sex tract cannot be the cause of "heat" behavior, a result in consonance with Wang, working in Richter's laboratory (3), who found that removal of the uterus was without effect upon the general spontaneous activity of the female rat.

PROCEDURE

The sex tract exclusive of ovaries and Fallopian tubes was removed from six rats when they were twenty-six to thirty-six days old. The vaginal membrane was intact at this time except in the case of one rat.

The operation was performed in the following manner. The branches of the uterine artery which supply the uterus were first tied off. The distal portion of this artery which anastomoses with the ovarian artery was allowed to remain unrestricted. The tubes were sectioned at the point where they join the uterus. The uterine horns were then lifted, stripped of their blood vessels and the entire uterus as far as the cervix was removed. After the cervix had been carefully freed from the ureters it was held with forceps while the upper vagina was separated from surrounding tissues. Blood vessels in this region were not tied off but bleeding was not persistent. The skin around the external opening was then cut at about the hair line and the lower vaginal wall was dissected away from surrounding tissue until the whole of the vagina and cervix could be removed by pulling it through the pelvic floor. The edges of the circular wound thus made were then brought together and sutured, as was also the abdominal incision.

The animals made a prompt, uneventful recovery.

At autopsy, two months later, the site of the operation was found to be perfectly healed with no evidence of infection. In one animal the bladder was greatly distended; the fluid was easily expressed through the urethra, showing that it was not due to mechanical restriction but the cause of the retention was not discovered. This animal has not been included in the following report. In every case the periovarian sac on one or both sides was abnormally distended with fluid, often to the size of a large pea, but the ovaries were found upon histological examination to be normal.

The rats were put into revolving drum cages (Wang (4), Slonaker (5)) at the ages of thirty-eight to fifty-five days, in order to determine through their activity cycles when they were in heat, since with vaginal mucosa removed it was, of course, not possible to follow ovarian cycles by means of vaginal smears.

The animals were tested for sex behavior at irregular intervals during the six to eight weeks they were in these cages before autopsy.

RESULTS

Four of the five healthy rats exhibited normal, active sex behavior with typical darting, crouching and arching. The early age at operation together with the fact that the vaginal membrane was intact at that time precludes the possibility that this was due to learning.

There was some tendency to a lower degree of sexual excitability in the later tests as if failure to complete the mating pattern had resulted in lowered responsiveness to sexual stimuli with the initial parts of the pattern. This interpretation, however, cannot be accepted without further experimentation, since it is quite possible that the endocrine balance was disturbed either by the absence of the sex tract or by the fluid which had not escaped from the periovarian sac. Although upon microscopic examination no abnormalities were found in the ovaries, nevertheless, cycles of activity had been somewhat irregular in one of the four rats·that showed normal sex behavior and were quite irregular in the rat that failed to exhibit any sex activity. It is possible that these irregularities of behavior were early indications of endocrine disturbance that might have shown itself in the ovaries only after several months.

The animal which failed to respond to males with any of the typical heat behavior pattern showed an activity curve which was high but so irregular that cycles can scarcely be discerned. This curve is similar to the activity curves of rats whose ovaries have been made cystic by traumatisation (Wang and Guttmacher (3)) and the sexual excitability of such rats is always very low (unpublished data of the writer).

DISCUSSION

The fact that 4 of the animals exhibited typical sex activity upon being placed with males after the removal of vagina and uterus proves conclusively that these organs, although markedly altered at the time of oestrus, are nevertheless, not essential in the production of the condition called heat, in which the animal responds to certain stimuli with behavior that cannot be elicited during dioestrum. This does not exclude the possibility that these organs may play some part in the psychobiological mating reaction of the intact animal. It does, however, make it apparent that the hormone responsible for the heat behavior pattern affects some other part of the organism in producing this changed reactive condition.

SUMMARY

Normal sex behavior was demonstrated in rats after removal of the uterus and vagina, showing that afferent nerve impulses from these organs are not essential for mating activity in these animals.

REFERENCES

(1) Allen, E., Francis, B. F., Robertson, L. L., Colgate, C. E., Johnson, C. G., Doisy, E. A., Kountz, W. B., and Gibson, H. V.: The hormone of the ovarian follicle; its localization and action in test animals, and additional points bearing upon the internal secretions of the ovary. Am. Jour. Anat., 1924, xxxiv, 133.

(2) Hemmingsen, A. M.: Studies on the oestrus-producing hormone (oestrin). Skand. Arch. f. Physiologie, 1933, lxv, 97.

(3) Richter, Curt P.: Animal behavior and internal drives. Quart. Rev. Biol., 1927, ii, 307–343

(4) Wang, G. H.: The relation between "spontaneous" activity and oestrous cycle in the white rat. Comp. Psychol. Monog., 1923, Series 2, No. 6; also Am. Nat., 1924, lviii, 36–42.

(5) Slonaker, J. R.: The effect of pubescence, oestruation and menopause on the voluntary activity in the albino rat. Amer. Jour. Physiol., 1924, lxviii, 294–315.

(6) Wang, G. H., and Guttmacher, A. F.: The effect of ovarian traumatization in the spontaneous activity and genital tract of the albino rat, correlated with a histological study of the ovaries. Amer. Jour. Physiol., 1927, lxxxii, 335.

20

Reprinted from *Science* **152**:226–230 (1966)

SEASONAL VARIATION IN MATING BEHAVIOR IN CATS AFTER DESENSITIZATION OF THE GLANS PENIS

L. R. Aronson and M. L. Cooper
Department of Animal Behavior, American Museum of Natural History, New York

Behavior in Cats after Desensitization of Glans Penis

Abstract. *The glans penis in 14 sexually experienced cats was desensitized by section of the nerves dorsalis penis. These males mounted the estrous female readily but they were so disoriented that they could not achieve intromission. Reduced sensory feedback resulting from the operation and from lack of intromissions caused a decided drop in sexual activity in the fall with recovery in early winter. A latent sexual cycle in male cats is revealed, which corresponds in time to the established female cycle.*

In the comprehensive theory of the regulation of mammalian sexual behavior developed by Frank Beach (1), the sensory input into the system is considered nonspecific and additive. This conclusion, which was derived from his own research on rodents and from a survey of the literature (2), stems from a variety of observations and experiments that show that sensory deprivations (visual, auditory, tactile from snout and genitalia), regardless of modality or area, cause a decline in sexual activity, but do not cause qualitative changes in sexual behavior. Conversely, increasing the stimuli derived from the sexual partner or from the environment increases sexual activity, while the gonadal hormones and neural activity, particularly of the neocortex, adjust the threshold for appearance of the various behavioral acts.

In 1962 we reported (3) preliminary observations on a sensory deprivation in male cats, observations that seemed, at first thought, to be at variance with this part of Beach's theory. By surgical procedure the nerves dorsalis penis of several males were severed bilaterally. This operation desensitizes the glans penis but does not interfere with erection. These males showed no observable decrease in sexual activity. They mounted the female as readily as before operation but were so disoriented that they were unable to insert the penis into the vagina. Thus, by a small circumscribed sensory deprivation we produced major qualitative changes in behavior with no immediate loss in sexual arousal. As testing continued, however, decrements in sexual behavior appeared; not as a continuous decline, which would be predicted from Beach's theory, but as a pronounced seasonal decline in the fall with a return to

higher levels of sexual activity in the winter.

We are now presenting an interim report of this experiment based on the behavior of 14 male cats that have been observed for 2 to 26 months after operation (average 18 months). Seven of these males are still being observed. The subjects were domestic short-hair males of unknown ancestry. Seven were obtained as adults and were presumed to have had sexual experience. The other seven were obtained as kittens and raised in laboratory cages, and all of the mating activities of these animals were controlled and observed. Since our analysis is not concerned with the effects of experience, and since there were no apparent differences in sexual behavior between the two groups, the data of the two groups were pooled.

Sex tests and methods of observation were similar to those used by Rosenblatt and Aronson (4). The major items of the normal mating pattern observed are (i) the male grips the back of the female's neck with teeth; (ii) mounts the back of the female; (iii) makes stepping movements with hind limbs; (iv) exhibits pelvic thrusting which is followed by (v) a single brief intromission with ejaculation after which (vi) the male dismounts. The females used as test animals were spayed and brought into heat by weekly injections of 0.15 mg of estradiol benzoate in 0.15 ml of sesame oil. All males were given a 20-minute sex test in a specific test room once a week except for occasional gaps resulting from fortuitous circumstances. They were given 5 to 86 tests prior to operation (Table 1) and had from 8 to 58 intromissions. The preoperative records thus served as control data.

Under nembutal anesthesia, sections of the nervus dorsalis penis, about 3 mm in length, were removed along with equivalent portions of the accompanying dorsal arteries (Fig. 1). To be sure that removal of the dorsal blood vessels or other surgical procedures did not interfere with penile function, we performed a sham operation on an additional male, removing the dorsal arteries but leaving the dorsal nerves intact. Intromission occurred in six successive postoperative tests, and no decrements or changes in behavior were observed.

None of the animals achieved intromission during the first 1½ months after operation, and in ten subjects this loss continued as long as the animals were tested. In two males, HA and RO, intromissions reappeared after 6 and 18

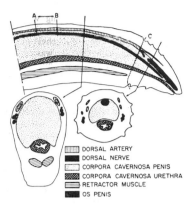

Fig. 1. Diagram of cat penis showing relation of dorsal artery and nerve to the corpora cavernosa. Lines *A–B* show segment of nerve removed. Attachment of prepuce is indicated by line *C*.

weeks, respectively. Two animals still being tested (PH and OD) show occasional intromissions.

Failure to achieve intromission was correlated completely with improper orientation of the male on the back of the female during the mount. At most times the pelvic region of the male was so far forward and highly elevated that the penis seldom came near the genital area of the female. In four animals disorientation was much more profound. By 13 weeks after operation, TM developed the habit of lying on his side while holding the neck-grip. In this position he would thrust vigorously with his penis several centimeters from the female. Male RO (during the first 14 postoperative weeks) and male MV would circle the female in a frenzied manner while holding the neck-grip. These males were frequently observed thrusting vigorously while standing at right angles to the female. These abnormal patterns were never observed prior to operation. With PH and OD, the cats that intromit occasionally, disorientation is less pronounced.

After operation, with the loss of intromission, the duration of mounting increased markedly. This usually took the form of long, protracted mounts, many lasting the full 20 minutes of the test period. The behavior of a few males was characterized by a large number of shorter mounts from momentary to a few minutes.

Stepping behavior did not change in any animals. Pelvic thrusting remained vigorous in most subjects, but declined in forcefulness in two. At 6 weeks we noticed in PG that thrusting was less vigorous. By 4 months thrusting was reduced to occasional twitches of the thigh muscles. This change was readily demonstrated by comparing motion pictures taken at 3 and 22 months after operation.

Before operation most of the males would bite the skin on the back of the neck and hold this neck-grip as long as they were mounting. After operation a small number of males would occasionally release the neck-grip during the mount and then retake the neck-grip after a few seconds or minutes, particularly when the female started to move forward. The amount of neck-licking did increase postoperatively but this was related to the fact that the duration of mounting was so much greater and was not attributable to any qualitative change in behavior. No other change in neck-grip behavior was observed in any of the 14 subjects.

Since the dorsal nerves of the penis carry some sympathetic nerve fibers, an important question raised in this study is whether our surgical procedures not only anesthetized the penis but also interfered to some extent with erection. This has been answered in three ways:

1) Examination of the neuroanatomical and neurophysiological literature (5) shows that erection is controlled by the pelvic nerve (nervus erigens) and not by the pudendal nerve from which the dorsal nerve of the penis arises (6).

2) Occasional observations of erection were made in some of the experimental animals when they lay on their sides during a mount. In all instances erection seemed complete.

3) In five additional males, K. K. Cooper (7) performed sacral laminectomies and stimulated the ventral roots of S_2 bilaterally. With the use of a square-wave stimulator at a pulse frequency of 20 per second and of 0.4-msec duration, the voltage was adjusted (from 0.25 to 20 volts) to produce protrusion and full erection within 1 minute of stimulation. After erections had been elicited several times in this manner, the dorsal nerves were sectioned. With the same stimulus parameters as above, the ventral roots of S_2 were again stimulated. Full erection occurred as rapidly as before nerve section in all five subjects. Completeness of nerve section was verified histologically in each case.

Completeness of the denervation of

Table 1. Summary of tests.

Animal	Preoperative			Postoperative		Date of operation (week of year)
	No. of tests*	Intromissions		No. of tests	Intromissions (total No.)	
		Total No.	Average in last 5 tests			
BG†	15	21	1.8	52	0	39
CG	47	28	1.0	105	0	10
HA†	86	73	2.6	11	7	13
LI†	12	23	2.6	84	0	11
MV	6	17	3.4	95	0	15
OD	9	20	2.6	86	6	27
PG†	24	28	2.0	85	0	9
PH	6	8	1.4	98	9	13
RO†	43	36	2.0	24	6	26
RS	39	40	1.4	70	0	41
ST	5	12	2.8	64	0	39
SY†	9	12	2.1	14	0	24
TM†	24	58	2.4	80	0	27
WD	11	23	2.2	87	0	29

* Starting from first test in which mounting occurred. † Postoperative observations on these animals have been terminated.

the glans was determined in two ways: (i) The penises of the seven males in which observations had been terminated were sectioned serially and were stained by the Bodian silver-proteinate–gold-chloride technique. In three, frozen sections were also made near the tip of the glans and were stained by the method of Winklemann (8). Bundles of functional nerves were seen in RO and HA, the two males in which normal sexual behavior had returned, indicating incomplete surgery or regeneration, or both. In three cats, SY, TM, and PG who never intromitted after operation, the nerves were absent distal to the site of operation except for a few scattered fibers that may represent anastomoses with the nerve cavernosus. In the two others animals (LI and BG), who also did not intromit after desensitization, some nerve bundles were seen distal to the operation and in LI definite signs of regeneration through the scar tissue were evident. Apparently desensitization of the glans was sufficient in these two animals to preclude the return of normal sexual behavior. (ii) Terminal operations were performed on three animals by K. K. Cooper. The dorsal nerve of the penis was exposed deeply, just distal to its origin from the pudendal nerve. Action potentials were displayed on a cathode ray oscilloscope with bipolar electrodes. In PG and BG, increased firing could not be obtained when the glans penis was stimulated with a camel's hair brush. When the electrodes were shifted to the scrotal branch of the pudendal nerve, stimulation of the scrotum gave a marked increase in firing. In RO, a male in which regular intro-

missions returned, increased action potentials were readily elicited by stimulation of the tip of the glans.

From the foregoing evidence we cannot claim complete desensitization in any of our 14 subjects, but the evidence clearly demonstrates an extensive and persisting decrement in glans sensitivity in all but two of our subjects examined so far.

Rosenblatt and Aronson (4) devised an index figure (sex score) based on the highest level of sexual activity reached in a given test. The median sex score of all the animals for the last five preoperative tests was 10, which indicates two intromissions. After operation, the scores were 8 or less (no intromissions) in all tests except for the four males where intromissions returned.

A striking feature of the level of sexual activity when followed over many months is its cyclical fluctuation. Regardless of the week of year that the operation was performed (Table 1), the sex scores were high in winter and spring (Figs. 2 and 3). They dropped precipitously in late summer (around the 35th week of the year) and rose again in the early winter (around week 50). All except one animal showed this cycle, and eight of the males that were tested for 2 years showed two similar cycles. It is also of interest that in PH and OD, the two animals that have occasional intromissions, these only appeared during the months of high sexual activity.

Analysis of the total duration of mounting per test reveals the same phenomenon, namely, a decided decline in mounting from late summer to early winter (Fig. 2). This period of minima

196

sexual activity corresponds to the cycles obtained by the sex score, but not precisely in most cases.

Mount latency, the time from the beginning of the test to the first mount, may be used as an additional measure of level of sexual activity, where high latencies reflect low levels of arousal. Here again we found a clear-cut cycle in 'all subjects except BG and PG. While the peaks in latency correspond to the troughs in mount duration and sex score, the overlap is not exact in most cases (Figs. 2 and 3).

Five different sexual functions can be attributed to the penis of the cat, namely: (i) transfer of semen; (ii) stimulation of the female; (iii) provide sexual stimulation leading to ejaculation; (iv) orientation to the female while mounting; and (v) sensory feedback for the maintenance of sexual arousal.

· The first function is obvious and well established. The second is equally clear since the female issues a loud, characteristic cry at the moment of intromission, and then becomes highly resistant to the male. Also, the act of intromission will cause an intact female to ovulate (9). The third function, well established for some mammals is presumably true for cats also. While ejaculation has been obtained in physiological experiments, it has not been observed during copulation in the cat as a discreet act separable from the brief intromission. Smears taken from the penis of the tractable nerve-cut males during a mount or immediately thereafter usually had some sperm, but we cannot be sure that their presence represented a normal ejaculation. The fourth, namely, orientation to the female while mounting has not been recognized heretofore as a function of the penis. The only other indication of this function that we could find in the literature is a Swedish veterinary report describing the failure of a bull to achieve intromission following pathological degeneration of the pudendal nerve (10).

The fifth, namely, sensory feedback for the maintenance of sexual arousal, is equally new. A test of this hypothesis was actually the major reason for the experiment. We had predicted, after denervation, a gradual decrease in sexual activity similar to that in castrated cats rather than cyclical fluctuations. It is commonly believed that intact male cats are ready to mate throughout the year (11). There is limited evidence, however, that some may be slightly less active in the fall

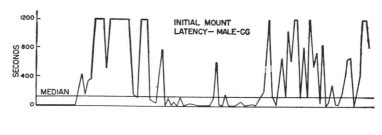

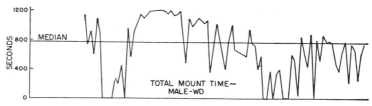

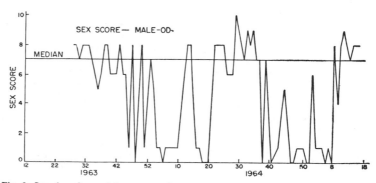

Fig. 2. Samples of complete postoperative records of three quantitative measures for three different males.

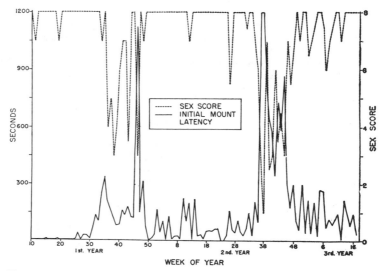

Fig. 3. Median sex scores and initial mount latencies for all tests of all males showing cyclical decline in sexual activity in the fall of each year.

197

(12). In the present experiment, data of two intact control males tested for 1 year and 1½ years, respectively, are available. The sex scores of one declined slightly during the fall of two successive years, but in the other no cyclical changes are evident. We have, in addition, evidence from previous experiments (4) on four postpuberally castrated males that persisted in high levels of sexual behavior for one to two full years after operation. PW had mount latencies of zero or nearly zero for 2 years; AP and MK for 1 year. The latencies of SV started high after operation and then declined to zero or nearly so and stayed at this level for over a year. The sex score for PW declined from 10 to 8 over a 2-year period. In 1 year, AP went from 13 to 8; MK went from 7 to 4; and SV stayed between 8 and 10 in most of the tests. In none of these data were there any indications of a cycle.

Wild felines, like most mammals, have a definite breeding season (13). The wildcat of Scotland, for example, breeds only in the spring. Under domestication, seasonal breeding in many mammals largely disappears (14). The domestic cat certainly breeds all year around, especially in captivity (11), but a vague cycle attributed to the female does seem to persist, showing heightened activity in the late winter and spring with a lull in the fall. Based on histological and physiological evidence, Foster and Hisaw (15) identified, in the case of the female cat, an anoestrus period, extending from September to January, and this has been related to variations in day length. This period coincides fairly well with the time of low activity found in our males. Our experimental desensitization of the penis has apparently brought out a comparable latent cycle in the male. As a working hypothesis we recognize two processes necessary for the maintenance of high levels of mating behavior in the male, namely, (i) environmental influence, especially variations in day length impinging on the neuroendocrine mechanism controlling sexual behavior, and (ii) sensory feedback from the penis (especially during mounting and intromission) acting on the central mechanisms for sexual arousal. When either one or both of these processes are fully operative, high levels of sexual behavior are evident, but when both processes are minimal, a definite lull in sexual activity occurs.

We had expected that our desensitized animals would decline in sexual behavior in a manner somewhat similar to those of castrated males, particularly those that continued to mate for a great many months after operation. This expectation was based on the fact that castration results in structural changes of the skin of the glans (16), which may mean lowered sensory input. This prediction was realized in part for our desensitized males. One of them, BG, ceased all sex behavior after 8 months; in three, pelvic thrusting became minimal; after the low period, six males never quite reached the level of total mounting time shown just after operation. A qualitative decrease in sexual excitability was clearly evident in PE, LN, and MV. If there is a real loss in penile sensitivity after castration, it probably occurs gradually in the long-persisting animals. Eventually it is accompanied by loss of erection, which further decreases penile stimulation, so that in the long run loss of feedback may be greater in castrated animals than in our operated animals. This accounts, in part, for the greater loss in arousal which eventually occurs in most castrates.

After the publication of our original abstract, two reports have appeared on the use of topical anesthetics on the glans penis of rats, resulting in a loss of intromission and apparently incomplete erection. One report (17) indicated no loss of sexual arousal; the other (18) indicated a decline in arousal as the test proceeded. We have tried a topical anesthetic (5 percent lidocaine ointment, 19) on one additional intact male and produced, in three tests, disorientation in mounting lasting 26 to 30 minutes, after which the male achieved intromission. Full erection was observed during the period of disorientation. In three control tests with blank ointment, intromission occurred after 5 to 8 minutes. In two additional tests, when a solution of 2 percent tetracaine hydrochloride (20) was sprayed on the penis, disorientation and failure to achieve intromission persisted for 37 minutes when the observations were terminated. Experiments such as these are limited by the fact that anesthesia wears off during the course of the test, so that sensory feedback is delayed, not necessarily reduced. Also, the tests in rats were not continued, and a feedback process of the kind found in our cats was not detected.

In summary, long-lasting desensitization of the glans penis causes disorientation in mounting behavior which precludes intromission. This, in turn, causes further decrements in sensory feedback which leads to a pronounced seasonal decline in sexual arousal. In this last aspect our results are in agreement with Beach's theory as stated in the introduction, although the specificity of the sensory mechanism remains unanswered.

References and Notes

1. F. A. Beach, J. Comp. Psychol. 33, 1 (1942).
2. ———, Physiol. Rev. 27, 240 (1947).
3. M. Cooper and L. R. Aronson, Amer. Zool. 2, 514 (1962).
4. J. Rosenblatt and L. R. Aronson, Behaviour 12, 285 (1958).
5. J. Langley and H. K. Anderson, J. Physiol. 19, 85 (1895).
6. A. Kuntz, The Autonomic Nervous System (Lea & Febiger, Philadelphia, 1953), p. 29; P. Bessou and Y. Laporte, Arch. Sci. Biol. Ital. 101, 90 (1963); W. S. Root and P. Bard, Amer. J. Physiol. 151, 80 (1947).
7. K. Cooper, M. Cooper, L. R. Aronson, Am. Zool. 4, 301 (1964).
8. R. K. Winkelmann and R. W. Schmit, Proc. Staff Meetings Mayo Clinic 32, 217 (1957).
9. W. W. Greulich, Anat. Rec. 58, 217 (1934).
10. S. Rubarth, Skand. Vet. Tidskr. 36, 7 (1946).
11. J. Rosenblatt and T. C. Schneirla, in The Behaviour of Domestic Animals, E. Hafez, Ed. (Bailliere, Tindall and Cox, London, 1962), p. 453.
12. P. Leyhausen, Z. Tierpsychol. Suppl. 2 (195); R. Michael, personal communication.
13. L. H. Mathews, Proc. Zool. Soc. London 111, 59 (1941).
14. E. C. Amoroso and F. H. A. Marshall, Marshall's Physiology of Reproduction, A. Parkes, Ed. (Longmans, London, 1960), 707.
15. H. A. Foster and F. L. Hisaw, Anat. Rec. 62, 72 (1935).
16. M. Cooper and L. R. Aronson, in preparation.
17. N. Adler and G. Bermant, J. Comp. Phys. Psychol., in press.
18. S. G. Carlson and R. Larsson, Z. Tierpsychol. 21, 854 (1964) .
19. Xylocaine, courtesy of Astra.
20. Cetocaine, Cetylite Industries.
21. Supported by grant HD-00348 and grant M 08600 from NIH. Several students of Undergraduate Research Participation program supported in part by NSF grant C 6387 assisted in the research. We thank Mr. Alba D. Plescia for technical assistance and Dr. Ethel Tobach and K. K. Cooper for reading the manuscript and for helpful suggestions.

21

Reprinted from *Physiol. Behav.* **2**:207–209 (1967), by permission of Microforms International Marketing Corp. as exclusive licensee of Pergamon Press journal back files

Mating Behavior of Male Rats after Olfactory Bulb Lesions[1]

LENNART HEIMER AND KNUT LARSSON[2]

Departments of Anatomy and Psychology, University of Göteborg

(Received 14 November 1966)

LARSSON, K. AND HEIMER, L. *Mating behavior of male rats after olfactory bulb lesions.* PHYSIOL. BEHAV. **2** (2) 207–209, 1967.—Interruption of the afferent olfactory connections by lesions in the olfactory bulbs caused a marked impairment of the sexual behavior of male rats. The animals showed a prolonged latency in their responses to estrous females, a reduction in the number of ejaculations and an increased tendency to remain totally unresponsive to the sexual stimulation. It is suggested that olfactory impulses exert a strong influence upon the preoptic-anterior hypothalamic region, which is known to be of crucial importance for mating behavior.

Reproduction   Behavior   Olfactory lobes   Lesion   Rat

It is generally assumed that the sense of smell plays a major role in mammalian reproduction. Experimental evidence in support of this view was recently provided by Whitten [16], Bruce [3] and Parker and Bruce [10], who showed that in the female mouse, olfactory stimulation is a prerequisite for normal estrous cyclicity and is even involved in maintaining the pregnancy. With respect to the male the experimental data are less affirmative. Removal of the olfactory bulbs does not abolish copulatory activity of male rabbits [Brooks, 2] or rats [Beach, 1] indicating that in neither species is olfaction indispensable for the elicitation of mating. However, Beach found that sexually inexperienced rats were less likely to engage in mating after lesions in the olfactory bulbs than were normal animals. Some reduction in the sexual activity was also observed in experienced rats. In the present study the problem was reinvestigated, using sexually highly experienced and active rats which were subjected to detailed behavior analysis before and after destruction of the olfactory bulbs.

METHOD

Animals

The subjects were male albino rats bred and maintained in this laboratory. The animals were selected on the basis of their sexual performances in previous tests. Only animals showing a consistently high level of mating activity were used. At the time of the operation, the subjects were approximately 8 months old.

The subjects were divided into two groups according to a randomization table. One group (A, N = 12) was subjected to olfactory bulb lesions and the other group (B, N = 9) served as a control.

The experimental animals had at operation an average

weight of 396 g (362–400) and at sacrifice weighed 423 g (355–500). Corresponding figures in the control group were 432 g (385–475) and 455 g (395–550) respectively.

Surgical Technique and Histological Procedure

Before the operation the subjects were anesthetized with pentobarbital sodium. The bulbs on both sides were exposed by the removal of a small part of the frontal region of the skull roof and incision of the dura mater. Removal of the rostral portion of the bulbs was done by suction under a dissecting microscope. Attempts were made to destroy all olfactory filaments at their emergence through the cribriform plate. Animals used as controls were similarly treated except that, after opening the skull, no incision was made in the brain.

After the completion of the experiments, at four months after the operation, the animals were again anesthetized and perfused through the heart with physiological saline followed by neutralized 10% formalin. The brain, including the remaining olfactory bulb tissue, was removed under the microscope and fixed in formalin. It was embedded in paraffin and serially sectioned at 10 μ. Depending on the proximity to the lesion, every second to tenth section was stained with the Klüver–Barrera stain.

Behavior Testing Procedure

The subjects were observed in 30-min tests given at 10-day intervals. Three preoperative and eleven postoperative tests were performed. The operation was performed within a week after the last preoperative test. The first postoperative test was made 10 days after the operation.

[1]This study was supported by the National Institute of Child Health and Human Development, US Public Health Service (HD 00344) and by a grant from the Swedish Council for Social Research.
[2]The assistance of Mrs. Birgit Notbeck and Mrs. Lydia Aigartz is thankfully acknowledged.

The sexual behavior was tested according to a standard procedure described in detail elsewhere [Larsson and Essberg, 7]. Prior to testing, stimulus females were brought into estrous by hormone treatment, consisting of an intramuscular injection of 0.15 mg of estradiol benzoate in oil approximately 30 hr before the test, followed 24 hr later by 1.0 mg of progesterone.

The mating behavior was observed during the dark period of the artificially reversed light-dark cycle (14 hr light, 10 hr dark). The following behavior variables were measured:

1. *Mounting frequency.* Number of mounts without intromission.

2. *Intromission frequency.*

3. *Intromission latency.* Time from the entrance of the female into the observation cage to the first intromission.

4. *Ejaculation latency.* Time from the first intromission until ejaculation.

5. *Average intercopulatory interval.* Average delay between each intromission computed by dividing the ejaculation latency by the intromission frequency.

6. *Postejaculation interval.* Time from ejaculation to the next intromission.

By "series of copulations" is meant the sequence of mounts and intromissions culminating in ejaculation.

RESULTS

The Lesions

Some bulbtissue remained in all animals. The lesions involved approximately the rostral half of the olfactory bulbs. In three animals bleeding had occurred into one of the olfac-

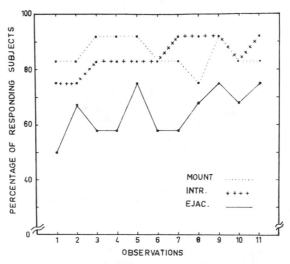

FIG. 1. Percentage of experimental animals showing at least one mounting, intromission or ejaculation during each of the post-operative tests.

TABLE 1

MATING BEHAVIOR DISPLAYED BY THE EXPERIMENTAL (A) AND CONTROL (B) ANIMALS. THE FIGURES REFER TO PERFORMANCES IN THE LAST PREOPERATIVE TEST (1) AND IN THE FIRST (2) AND THE LAST (3) POST-OPERATIVE TESTS IN WHICH EJACULATION OCCURRED

| | Pre-op. | | Post-op. | | | | | | | |
	A_1 Md	B_1 Md	A_2 Md	B_2 Md	A_3 Md	B_3 Md	A_2/B_2 p	A_3/B_3 p	A_2/A_3 p	B_2/B_3 p
Intr. lat.	22.5	0.15	0.53	0.20	0.69	0.18	>0.05	<0.05	>0.05	>0.05
Ejac. lat.	4.49	4.85	15.43	4.07	11.77	4.28	<0.002	<0.02	>0.05	>0.05
ICI	0.69	0.57	0.86	0.68	2.17	0.55	>0.05	<0.02	>0.05	>0.05
PEI	5.76	5.21	7.58	5.33	9.33	5.43	<0.02	<0.02	>0.05	>0.05
Intr.	6.5	8.0	7.0	7.0	4.0	8.0	>0.05	>0.05	<0.02	<0.05
Mounts	3.0	3.0	4.0	2.0	2.0	2.0	>0.05	>0.05	>0.05	>0.05
Ejac/30 m	3.0	3.0	2.0	3.0	1.5	3.0	<0.02	<0.002	<0.05	>0.05
N	12	9	10	9	10	9				

Intergroup differences were, for independent samples evaluated by the Mann–Whitney U-test and for related samples by the Wilcoxon *t*-test. Two tailed probabilities were used.

tory ventricles, causing some degeneration of the deep rostral part of the anterior olfactory nucleus on the affected side.

The Behavior

Complete sexual behavior, including ejaculation, occurred in all animals in each of the preoperative tests. Following the operation no deficits were observed in the mating behavior of the control animals. However, in the experimental group, a marked decrease in sexual activity took place after the operation. When presented with the receptive female, the male usually mounted her and made some intromissions but in each of the tests a number of males failed to ejaculate (Fig. 1). Great individual differences occurred in the reaction to the operation. Thus, some operated animals ejaculated in all of the tests performed, while others showed a marked inconsistency in their behavior. No animal, however, was wholly unable to ejaculate.

Table 1 shows the performance in the last preoperative test (A_1B_1) and in the first and the last postoperative tests in which ejaculation occurred. Two males which ejaculated in only one of the postoperative tests were excluded from this analysis. As shown by the table, the operation caused a prolongation of all response latencies, and a reduction of the total number of ejaculations achieved during the test. The tendency was apparent in the first postoperative (A_2B_2) test and was more marked as the animal's age increased during the postoperative period (A_3B_3). Old animals are known to be more dependent upon a maximum of sensory stimulation for sexual activity than are younger ones [Larsson, 5, 6] and this may explain why the last postoperative test, compared with the first one, showed a greater impairment of this behavior. The number of mounts and intromissions preceding ejaculation changed only slightly.

DISCUSSION

Confirming and extending previous findings of Beach [1], the present results provide conclusive evidence of the importance of olfactory stimuli in the male rat mating behavior. After operation the animals were less likely to engage in mating, the response latencies were much prolonged and the ejaculatory reflex was delayed or absent.

The present results are particularly interesting in view of the findings, in this and other laboratories, that destruction of the medial preoptic-anterior hypothalamic continuum abolishes mating in the male rat [Soulairac, 12; Larsson and Heimer, 8; Heimer and Larsson, 4]. Neuroanatomical studies by Lohman [9], Powell, Cowan and Raissman [11], as well as those performed in our laboratory [Heimer, unpublished], suggest that olfactory impulses have a relatively oligosynaptic access to this area. The functional state of the medial part of the preoptic-anterior hypothalamic continuum is largely dependent upon impulses arriving by way of the medial forebrain bundle. The rostral part of the bundle, bordering the medially located region involved in mating behavior, receives a large number of fibers originating in secondary olfactory structures such as the olfactory tubercle and the prepiriform cortex. The stria terminalis, in part originating from the cortico-medial subdivision of the amygdala, may also be an important connexion by which olfactory stimuli can modify the activity taking place in the medial preoptic-anterior hypothalamic continuum.

REFERENCES

1. Beach, F. A. Analysis of the stimuli adequate to elicit mating behavior in the sexually inexperienced male rat. *J. Comp. Psychol.* **33**: 163–207, 1942.
2. Brooks, C. M. The role of the cerebral cortex and of various sense organs in the excitation and execution of mating activity in the rabbit. *Am. J. Physiol.* **120**: 544–553, 1937.
3. Bruce, H. M. Olfactory block to pregnancy among grouped mice. *J. Reprod. Fert.* **6**: 451–460, 1963.
4. Heimer, L. and K. Larsson. Impairment of mating behavior in male rats following lesions in the preoptic-anterior hypothalamic continuum. *Brain Res.* **3**: 248–263, 1966/1967.
5. Larsson, K. Age differences in the diurnal periodicity of the sexual behavior. *Gerontologia* **2**: 64–72, 1958.
6. Larsson, K. Non-specific stimulation and sexual behavior in the male rat. *Behavior* **20**: 110–114, 1963.
7. Larsson, K. and L. Essberg. Effect of age on the sexual behavior of the male rat. *Gerontologia* **6**: 113–143, 1962.
8. Larsson, K. and L. Heimer. Mating behavior of male rats after lesions in the preoptic area. *Nature, Lond.* **202**: 413–414, 1964.

9. Lohman, A. H. M. The anterior olfactory lobe of the guinea pig. *Acta Anat. Suppl.* **53**: 1–109, 1963.
10. Parker, A. S. and H. M. Bruce. Olfactory stimuli in mammalian reproduction. *Science* **134**: 1049–1054, 1961.
11. Powell, T. P. S., W. M. Cowan and G. Raissman. The central olfactory connexions. *J. Anat.* **99**: 791–813, 1965.
12. Soulairac, M. L. Etude expérimentale des régulations hormono-nerveuses du comportement sexuel du rat mal. *Annls. Endocr., Suppl.* **24**: 1–98, 1963.
13. Stone, C. P. The effects of cerebral destruction on the sexual behavior of rabbits. The olfactory bulbs. *Am. J. Physiol.* **71**: 430–435, 1925.
14. White, L. E. Olfactory Bulb projections of the rat. *Anat. Rec.* **152**: 465–479, 1965.
15. Whitten, W. K. Modification of the oestrus cycle of the mouse by external stimuli associated with the male. Changes in the oestrus cycle determined by vaginal smears. *J. Endocr.* **17**: 307–313, 1958.
16. Whitten, W. K. Occurrence of anestrus in mice caged in groups. *J. Endocr.* **18**: 102–107, 1959.

Editor's Comments
on Papers 22 Through 28

NEURAL CORRELATES

The transition from material on the sensory bases of copulatory behavior to its neural and pharmacological correlates, especially neocortical function, is an easy one. In the early literature on sensory factors in copulatory behavior, the nonspecific, mutually facilitative interaction of stimulation from different modalities was emphasized. Beach (1942) noted that this pattern of sensory interaction appeared phylogenetically "with the appearance of forebrain association areas receiving impulses from several independent sensory systems and providing the anatomical substratum for multi-sensory interaction, summation, and facilitation" (p. 219). Thus the cerebral cortex, the location of primary receptive areas for the various sensory modalities, was viewed as the locus of multisensory interaction and facilitation. The early data were generally consistent with such a view.

In 1937 Brooks (Paper 22) published a study of the role of the cerebral cortex and various sensory modalities in regulating copulatory behavior in male and female rabbits. A variety of lesions and ablations failed to eliminate sexual behavior in Brook's rabbits, although the possibility of quantitative changes in behavior was not addressed. Not even complete removal of the neocortex in four males and six females eliminated normal mating. When the olfactory bulbs were also removed, males, but not females, finally ceased to mate. Apparently because of the more precise orientation and finer integration required for male sexual behavior, males appeared to be somewhat more affected by ablations than were females. These results were generally consistent with the earlier studies of Stone (for example, 1926). Davis (1939) found copulatory behavior to survive decortication in male and female rats.

An extensive series of studies of the role of the neocortex in male and female rats and cats was conducted by Beach and his associates (for example, Beach, 1940, 1944; Zitrin et al., 1956). The extent of the deficits observed in rats was a function of the amount of cortical tissue removed. No male stopped copulating after removal of less than 50 percent of the neocortex. Location of the lesion appeared less important than the extent of tissue removed. This result was consistent both with the data on sensory influences and with the views of "mass action" and "equipotentiality" of cortical action popularized by Lashley (1933). Female rats were less affected by cortical lesions than were males. Removal of 20 to 50 percent of the neocortex had little effect; destruction of 97 to 100 percent of the cortex increased the variability of lordotic behavior but little else. Two of the seven females in the latter group displayed an exaggerated lordosis, in which they held the receptive posture longer than normal after the male dismounted (Beach, 1944).

Larsson (Paper 23; 1964) studied the effects of lesions in specific areas of the cerebral cortices of male rats. Like Beach, Larsson found little change in the motor patterns of males with cortical damage. However, Larsson looked for and found clearcut evidence that the effect of cortical lesions depended on the location of the insult. Thus the more recent data indicate that lesions of the cortex appear to affect sensory and motor function in more specific location-dependent ways than had been implied in the earlier data (see also Zitrin et al., 1956).

Several investigators have studied the role of the limbic system in regulating copulatory behavior. Most attention has been directed at the amygdala, the destruction of which has been reported to produce a "hypersexuality" in several mammalian species (for example, Kluver and Bucy, 1939). The adequacy of the behavioral measures and controls in these early studies has been questioned. More recent evidence indicates that amygdaloid lesions interfere with discriminative function rather than increase "sexuality" (for example, Aronson and Cooper, 1979). In rats, amygdaloid lesions produce deficits in copulatory behavior, although hippocampal lesions produce but very small effects (Bermant et al., 1968; Dewsbury et al., 1968; Harris and Sachs, 1975).

The region of the hypothalamus has received more attention than any other from researchers on sexual behavior, largely because of its overwhelming importance for sexual activity. A representative example of the early work on hypothalamic regulation of sexual behavior is Paper 24 by Brookhart, et al. They demonstrated not only that destruction of portions of the hypothalamus disrupt sexual behavior but that such disruption is not reversed by the administration of exogenous hormones. Thus the hypothalamus must act rather directly on behavior instead of indirectly via the regulation of endocrine function. More recent studies of hypothalamic function in copulatory behavior have emphasized not only such effects but also neurohormonal integration, as by using techniques such as the implantation of small amounts of hormone into specific hypothalamic loci (for example, Davidson, 1966; Lisk, 1967; Michael, 1962) and autoradiographic techniques (for example, Pfaff and Keiner, 1973). One way in which the hypothalamus affects sexual behavior is via relatively direct action of the hormone-releasing factors that it produces (for example, Moss and McCann, 1973). Because it permits the study of particular connecting pathways, many investigators have turned from the lesion method to the use of specific knife cuts in the brain (for example, Szechtman et al., 1978). Remarkably, lesions in some sites in the region of the hypothalamus produce facilitation rather than inhibition of sexual behavior (for example, Law and Meagher, 1958; Powers and Valenstein, 1972). Several important papers dealing with the role of the hypothalamus in copulatory behavior were reprinted by Carter (1974), and Hart (1974a, b) reviewed much relevant literature.

Both the midbrain and the region of the diencephalic-mesencephalic border have been shown to be important in the regulation of sexual behavior in male rats. Heimer and Larsson (Paper 25) found a dramatic facilitation of the copulatory behavior of some male rats following lesions in the region of the diencephalic-mesencephalic border. The appearance of such a facilitation following the destruction of a part of the brain implies the existence of some inhibitory region in the normal brain. Other demonstrations of facilitative effects of lesions in the same region or in various regions of the midbrain have been reported by Barfield et al., (1975), Clark et al. (1975), and Lisk (1966).

According to Beach (1942), "Most of the evidence suggests that in all classes of vertebrates the discrete acts involved in patterns of courtship, mating, and parental care are mediated by nervous mechanisms located in the brain stem and spinal cord" (p. 219). This theory was elaborated by Beach (1967), when he reviewed many data consistent with the view that the reflexes underlying integrated copulatory behavior are organized in the hindmost portions of the brain and the cord and normally inhibited by higher structures. Hormones were viewed as acting primarily on the higher centers to effect the removal of this tonic inhibition. The data showing increased sexual behavior after hypothalamic lesions, cited above, appear consistent with this view. The inhibition hypothesis has been reviewed critically by Clemens (1978).

The study by Maes (Paper 26) provided one of the first indications of the extent to which the reflexes comprising mating behavior are organized in the spinal cord. Whereas intact female cats showed no evidence of sexual reflexes under Maes's testing conditions, such responses were observed after the spinal cord was transected, thus removing the connections between the cord and the brain.

Hart and Kitchell (Paper 27) provided a further indication of the complexity of organization within the cord. Reflexes that could be identified as constituents of the complete, and rather complex, mating pattern of male dogs could be elicited with tactile stimulation in spinal, as well as intact, male dogs. Hart went on to study spinal function in the copulatory patterns of dogs and rats, and the endocrine control thereof in an extended research program

(Hart, 1978). This research provides the most comprehensive evidence available of the complexity of the neural organization of behavior-related reflexes in the spinal cord.

The decade of the 1970s may be remembered as the time in which the study of the pharmacological control of sexual behavior matured. Whereas at the beginning of the decade very few researchers were conducting pharmacological research, a large number were active in this area at its close.

Pioneers in pharmacological work on sexual behavior were Soulairac and Soulairac. Working in France and publishing in French, the Soulairacs were somewhat separated from much of the mainstream of research on sexual behavior. Soulairac and Soulairac (Paper 28) studied the effect of reserpine on male copulatory behavior. They found reserpine, a depletor of brain monoamines, to facilitate sexual behavior in relatively specific ways; by increasing the number of ejaculations and decreasing the number of intromissions required to reach ejaculation. The effect of reserpine described by Soulairac and Soulairac was replicated by Dewsbury and Davis (1970). Paper 28 is an original translation of the Soulairac's paper by P. J. Salis, [Ms. Salis also translated M.-L. Soulairac's (1963) review paper of the effects of lesions, hormones, and drug treatments on sexual behavior; copies are available from the editor at the cost of duplicating the materials.]

Much of the pharmacological research of the 1970s was stimulated by the proposal that in both males and females high levels of certain brain monoamines, especially serotonin, inhibit sexual activity; lowering the levels of these chemicals facilitates sexual behavior. The hypotheses were first popularized for females by Meyerson (1964) and for males by Tagliamonte et al. (1969). Although the mechanisms through which these drugs actually produce their effects remain controversial (see Clemens, 1978) and the measures used in some early studies were inadequate, there has been much interest in the effects. For reviews, see Sandler and Gessa (1975) and Carter and Davis (1977).

Other research was stimulated by an interest in the behavioral effects of a variety of drugs widely used by humans. For example, alcohol has been found to have generally depressive effects of sexual behavior and reflexes (Dewsbury 1967; Hart 1969).

REFERENCES

Aronson, L. R., and M. L. Cooper, 1979, Amygdaloid hypersexuality in male cats re-examined, *Physiol Behav.* **22**:257–265.

Barfield, R. J., C. Wilson, and P. G. McDonald, 1975, Sexual behaivor: Extreme reduction of postejaculatory refractory period by midbrain lesions in male rats, *Science* **189**:147–149.

Beach, F. A., 1940, Effects of cortical lesions upon the copulatory behavior of male rats, *J. Comp. Psychol.* **29**:193–239.

Beach, F. A., 1942, Central nervous mechanisms involved in the reproductive behavior of vertebrates, *Psychol. Bull.* **39**:200–226.

Beach, F. A., 1944, Effects of injury to the cerebral cortex upon sexually-receptive behavior in the female rat, *Psychosomat. Med.* **6**:40–55.

Beach, F. A., 1967, Cerebral and hormonal control of reflexive mechanisms involved in copulatory behavior of male rats, *Physiol. Rev.* **47**:289–316.

Bermant, G., S. E. Glickman, and J. M. Davidson, 1968, Effects of limbic lesions on copulatory behavior of male rats, *J. Comp. Physiol. Psychol.* **65**:118–125.

Carter, C. S., 1974, *Hormones and Sexual Behavior*, Dowden, Hutchinson & Ross, Stroudsburg, Penn.

Carter, C. S., and C. Davis, 1977, Biogenic amines, reproductive hormones and female sexual behavior: A review, *Biobehav. Rev.* **1**:213–224.

Clark, T. K., A. R. Caggiula, R. A. McConnell, and S. M. Antelman, 1975, Sexual inhibition is reduced by rostral midbrain lesions in the male rat, *Science* **190**:169–171.

Clemens, L. G., 1978, Neural plasticity and feminine sexual behavior in the rat, in *Sex and Behavior: Status and Prospectus*, T. E. Mcgill, D. A. Dewsbury, and B. D. Sachs, eds., Plenum, New York, pp. 243–266.

Davidson, J. M., 1966, Activation of the male rat's sexual behavior by intracerebral implantation of androgen, *Endocrinology* **79**:783–794.

Davis, C., 1939, The effect of ablation of neocortex on mating, maternal behavior and the production of pseudopregnancy in the female rat and copulatory activity in the male, *Am. J. Physiol.* **127**:347–380.

Dewsbury, D. A., 1967, Effects of alcohol ingestion on copulatory behavior of male rats. *Psychopharmacol.* **11**:276–281.

Dewsbury, D. A., and H. N. Davis, 1970, Effects of reserpine on the copulatory behavior of male rats, *Physiol. Behav.* **5**:1331–1333.

Dewsbury, D. A., E. D. Goodman, P. J. Salis, and B. N. Bunnell, 1968, Effects of hippocampal lesions on the copulatory behavior of male rats, *Physiol. Behav.* **3**:651–656.

Harris, V. S., and B. D. Sachs, 1975, Copulatory behavior in male rats following amygdaloid lesions, *Brain Res.* **86**:514–518.

Hart, B. L., 1969, Effects of alcohol on sexual reflexes and mating behavior in the male rat, *Psychopharmacol.* **14**:377–382.

Hart, B. L., 1974a, Medial preoptic-anterior hypothalamic area and sociosexual behavior of male dogs: a comparative neuropsychological analysis, *J. Comp. Physiol. Psychol.* **86**:328–349.

Hart, B. L., 1974b, Gonadal androgen and sociosexual behavior of male mammals: A comparative analysis, *Psychol. Bull.* **81**:383–400.

Hart, B. L., 1978, Reflexive mechanisms in copulatory behavior, in *Sex and Behavior: Status and Prospectus*, T. E. McGill, D. A. Dewsbury, and B. D. Sachs, eds., Plenum, New York, pp. 205–242.

Kluver, H., and P. D. Bucy, 1939, Preliminary analysis of functions of the temporal lobes in monkeys, *Arch. Neurol. Psychiatr.* **42**:979–1000.

Larsson, K., 1964, Mating behavior in male rats after cerebral cortex ablation. II. Effects of lesions in the frontal lobes compared to lesions in the posterior half of the hemisphere, *J. Exp. Zoology* **155**:203–214.

Lashley, K. S., 1933, Integrative functions of the cerebral cortex, *Physiol. Rev.* **13**:1–42.

Law, T., and W. Meagher, 1958, Hypothalmic lesions and sexual behavior in the female rat, *Science* **128**:1626–1627.

Lisk, R. D., 1966, Inhibitory centers in sexual behavior in the male rat, *Science* **152**:669–670.

Lisk, R. D., 1967, Neural localization for androgen activation of copulatory behavior in the male rat, *Endocrinology* 80:754–761.

Meyerson, B. J., 1964, Central nervous monoamines and hormone induced estrus behavior in the spayed rat, *Acta Physiol. Scand.* 63(Supplement 241):1–32.

Michael R. P., 1962, Estrogen-sensitive neurons and sexual behavior in female cats, *Science* **136**:322–323.

Moss, R. L., and S. M. McCann, 1973, Induction of mating behavior in rats by luteinizing hormone-releasing factor, *Science* **181**:177–179.

Pfaff, D., and M. Keiner, 1973, Atlas of estradiol-concentrating cells in the central nervous system of the female rat, *J. Comp. Neurol.* **151**: 121–158.

Powers, B., and E. S. Valenstein, 1972, Sexual receptivity: Facilitation by medial preoptic lesions in female rats, *Science* **174**:1003–1005.

Sandler, M., and G. L. Gessa, eds., *Sexual Behavior: Pharmacology and Biochemistry*, Raven, New York.

Soulairac, M.-L., 1963, Experimental study of the neurohormonal regulation of male rat sexual behavior, *Ann. Endocrinol.* 24(Supplement): 1–98.

Stone, C. P., 1926, The effects of cerebral destruction on the sexual behavior of male rabbits. III. The frontal, parietal, and occipetal regions, *J. Comp. Psychol.* **6**:435–448.

Szechtman, H., A. R. Caggiula, and D. Wulkan, 1978, Preoptic knife cuts and sexual behavior in male rats, *Brain Res.* **150**:569–591.

Tagliamonte, A., P. Tagliamonte, G. L. Gessa, and B. B. Brodie, 1969, Compulsive sexual activity induced by *p*-Chlorophenylalanine in normal and pinealectomized male rats, *Science* **166**:1433–1435.

Zitrin, A., J. Jaynes, and F. A. Beach, 1956, Neural mediation of mating cats. III. Contributions of occipetal, parietal and temporal cortex, *J. Comp. Neurol.* **105**:111–125.

22

Reprinted by permission from *Am. J. Physiol.* **120**:544–553 (1937)

THE RÔLE OF THE CEREBRAL CORTEX AND OF VARIOUS SENSE ORGANS IN THE EXCITATION AND EXECUTION OF MATING ACTIVITY IN THE RABBIT[1]

CHANDLER McCUSKEY BROOKS

From the Department of Physiology, The Johns Hopkins University School of Medicine

Received for publication May 24, 1937

Copulation and the associated mating behavior, although very primitive and essential in character, require a delicate coördination of muscular activity. It is obvious that these activities constitute a reaction pattern which is essentially a form of emotional behavior. Elicitation of such behavior under specific conditions must depend upon stimuli received by olfactory, auditory, visual, tactile or some other group of afferent endings. It is conceivable that the relative importance of each of these various sensory stimuli might vary from time to time or might be changed by the removal of other organs of special sense. Loss of certain types of sensation due to cortical ablation might likewise increase the importance of other kinds of stimuli. Stone in 1925 demonstrated that removal of the olfactory bulbs and portions of the frontal half of the neocortex of male rabbits in no way interfered with their sexual activity. Bard (1934) observed treading, rolling, and other signs of sexual excitement in a female cat from which the neocortex had been completely removed. This animal also mated and exhibited all the indications of heat. Except for this work and other studies by Bard (Bard, 1936; Bard and Rioch, 1937) very little has been done to ascertain to what extent sexual behavior of mammals is dependent upon the integrity of various parts of the cerebral cortex and subcortical structures.

In the work here described rabbits were studied. This species is of peculiar interest in that ovulation normally occurs only on coitus or as a result of strong sexual excitation. Thus the occurrence of ovulation furnishes an additional indication of the normality of behavior and a means of judging the effectiveness or intensity of emotional excitement. A minor advantage is that the female rabbit is relatively constantly receptive and males are invariably active sexually.

[1] Preliminary notes published in This Journal **113**: 18, 1935, and Proc. XVth Internat. Physiol. Congress, 251, 1935.

This work was aided by a grant (to Dr. Philip Bard) from the Committee for Research in Problems of Sex, National Research Council.

The sexual behavior of male and female rabbits has been described by numerous workers (Hammond, 1925). The male mounts almost immediately when the female is placed with him. If the female is receptive intromission occurs very quickly and the male falls off, presumably when he ejaculates, usually uttering a characteristic cry as he does so. After falling off the male hops about for a few minutes stamping with his hind feet upon the floor of the cage and then mounts again. The female does not exhibit such a complex type of response. If receptive, she responds when mounted, chiefly by raising her tail and hind parts. Failure of the female to react in this way prevents intromission. These are the reac‑ tions which will be spoken of as normal sexual behavior. Normality of mating behavior does not indicate that a female is capable of caring for and rearing young. This paper does not include a consideration of all the reproductive activities of the female.

METHODS. In testing the receptivity of females and the activity of males the female was always placed in the buck's cage. Mating generally occurs more promptly if that procedure is followed rather than the reverse. In cases in which there was any doubt as to the occurrence of intromission and ejaculation the presence or absence of sperm in a vaginal smear settled the question. Occasionally a female rabbit will receive a male and not ovulate. Young females will mate before ovulation is possible (Hammond, 1925) and adult females, though permitting intromission, frequently fail to ovulate when in poor physical condition. Much experience has shown, however that if a mature healthy female receives a male ovulation follows. In all the cases reported here sexual activity was not considered normal unless ovulation did occur subsequent to coitus. The occurrence of ovulation was determined by laparotomy and examination of the ovaries for corpora lutea four days after mating. During this interval the corpora lutea attain a degree of development which enables easy and unmistakable identification. Occasionally a female will mount another rabbit of the same sex, make copulatory movements, eventually fall off uttering a cry like that of a male and as a result ovulate. To prevent such activity all animals were kept in individual cages.

Under the rather artificial conditions existing in the laboratory, training and conditioning of the rabbits probably play some part in the elicitation of sexual activity. Females to be tested were invariably placed in the male's cage. Males accustomed to this practice mount cats, kittens, guinea pigs and even dead animals or inanimate objects which are placed with them. This nonspecificity of behavior may be due in part to their training. At any rate, all the males and females used were accustomed to this procedure before operation and the same method was employed in testing reactions to the opposite sex after operation .

Removal of certain cortical areas produces sensory deficiencies. Abolition of sexual activity as a result of a cortical ablation might therefore be due to a sensory loss, a motor deficiency, a loss of ability to translate a particular type of stimulation into an appropriate response or any combination of the three. An effort was made to ascertain the importance of pure sensory deficiencies uncomplicated by cortical insult. Vision was eliminated by enucleation of the eyes. Complete anosmia was easily produced by exposing and removing the olfactory bulbs through a small opening made in the rostro-dorsal aspect of the skull. An attempt was made to inactivate the auditory apparatus by destroying the labyrinths through openings in the bullae. For this an occipital approach (Camis, 1930) was used. Although hearing was apparently destroyed these operations failed to produce marked signs of labyrinthine deficiency. Absence of such deficiencies indicated that the labyrinths were only partially destroyed and that the ability to hear might have been impaired but not abolished. To make sure that no auditory sensibility remained these operations were supplemented by destruction of the middle ear. This complex was destroyed with a probe passed through the external auditory meatus. Since ovulation normally occurs only on coitus it has been suggested that stimulation of the genitalia and genital tract might be essential to this reaction. An estimation of the importance of this sensory factor was attempted by deafferentation of the pelvic region. This was done by removal of the lower segments of the lumbar and the entire sacral cord or by transection of the cord in the lumbo-sacral region. In addition the abdominal sympathetic chains were removed.

Brain operations were performed following exposure of the cortex by removal of the skull over the region to be ablated. Cortical tissue was carefully dissected away from the underlying structures with a small sharp tonsil dissector and an ear spoon. The moderate bleeding encountered was easily controlled by means of small cotton pledgets. The incision was closed by subcutaneous as well as skin sutures. Pentobarbital sodium (0.7 grain per kgm. of body weight) injected intraperitoneally was the anesthetic used; it was supplemented with ether when necessary.

The chief difficulty encountered was in maintaining the proper nutrition of those animals which were subjected to extensive cerebral ablations. Unless a rabbit begins to eat spontaneously within a week or two it is almost impossible to keep it alive and maintain its weight by feeding either with a stomach tube or by hand. This difficulty has thus far prevented the satisfactory study of chronic preparations which have undergone complete decortication plus extensive ablation of subcortical tissue. All animals ate normally if the cortical removal had not involved the olfactory tract. A few completely decorticate animals survived removal of the olfactory bulbs or severance of the olfactory tracts but they apparently could not find their food with ease and they did not eat enough spontaneously to preserve their normal body weights.

Following each operation the animals were studied and their ability to react tested before further operations were tried.   Animals which failed to respond were kept for periods of a year to eighteen months and tested frequently before it was concluded that they would no longer mate.

RESULTS.   In three males and two females the olfactory bulbs were removed without rendering these animals at all abnormal in their sexual activities.   Enucleation of the eyes of three males and two females likewise failed to produce any deficiency.   In two males and one female an essential part of the auditory apparatus was destroyed with the same lack of effect. Finally two males and two females were deprived of olfactory, visual and auditory end organs.   These animals were as active sexually as before operation.   It can be concluded that these three special senses are not essential for the elicitation of emotional excitement and sexual activity if the animals are otherwise normal.   It follows, therefore, that any sexual inactivity resulting from cortical ablations cannot be referred solely to a loss of the cortical response to afferent impulses originating in nose, eye, or ear.

Six males and eight females were completely hemidecorticated with no resulting deficiencies in their ability to become sexually excited and sexually active.   Two females were completely hemidecerebrated on the left side by unilateral removal of all central tissue lying ahead of the mesencephalon (fig. 1-D).   The pituitary gland and some of its hypothalamic connections were left intact.   The left optic tract was ablated but the chiasm and left optic nerve were not damaged.   Two months were required for complete recovery but after that time they received males and ovulated normally. One male was prepared with a similar unilateral lesion and two others sustained operations in which not only the cortex but also the striatum and hippocampus of one side were removed.   These three males showed normal sexual activity in spite of the unilateral motor deficiencies which rendered them more awkward than normal rabbits.   It is safe to conclude 1, that both in the female and the male one-half of the forebrain can be removed without abolishing sexual activity, and 2, that the mere removal of a large amount of cerebral tissue does not in itself put an end to this form of behavior.

In several rabbits in which complete hemidecortication had been carried out various cortical areas of the remaining hemisphere were extirpated. Two males and one female sustained complete removal of the cortex of the left hemisphere and ablation of the occipital half of the right neocortex (fig. 1-B).   These animals mated normally with one another as well as with normal rabbits of the opposite sex.   Subsequent removal of the olfactory bulbs and enucleation of the eyes did not abolish their activity.   One male survived a still more extensive cortical removal, namely, ablation of the entire left cortex, the occipital half of the right neocortex and most of

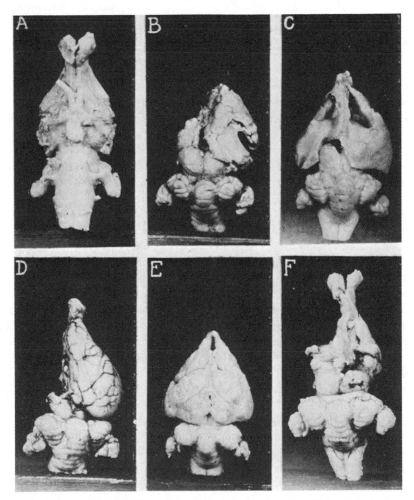

Fig. 1. A. Bilateral pyriform lobe removal in a female rabbit. Ventral aspect of brain. This animal mated and ovulated.

B. Brain of male rabbit which mated after ablation of olfactory bulbs and all cortex except the pyriform lobe and a fragment of the frontal portion of the neocortex of one hemisphere.

C. Left neocortex removal supplemented by ablation of the olfactory bulbs and of the frontal half of the neocortex of the right hemisphere. This male was able to mate even after enucleation of the eyes.

D. The brain of hemidecerebrate rabbit. This animal showed normal mating behavior.

E. Bilateral neocortex removal. Olfactory bulbs ablated. This male never mated after inactivation of the olfactory system.

F. Dorsal aspect of brain of a completely decorticate female. Olfactory bulbs present but the olfactory tracts severed. On left side striatum almost completely destroyed. Superior colliculus of right side injured superficially. This animal mated and ovulated.

the right pyriform lobe. This animal was also without olfactory bulbs. The small remaining fragment of sensori-motor cortex was sufficient to enable the animal to mate. Removal of the frontal third of one cortex, even when supplemented by ablation of the neocortex of the opposite hemisphere, rendered two females and two males in no wise abnormal in respect to their sexual reactions. Here again destruction of the olfactory bulbs did not interfere with the activity. In one male and one female of this group the eyes were also enucleated. The female reacted normally but the male was usually unsuccessful in his mating attempts. He was very active and when disturbed in any way made copulatory movements. He experienced difficulty in locating the female and making proper orientations but on two occasions he intromitted successfully (fig. 1-C).

In two males and two females the pyriform lobes alone were extirpated (fig. 1-A). These rabbits mated normally after a few days.

Complete removal of the neocortex in four males and six females did not prevent normal mating. These animals still possessed intact olfactory bulbs, intact pyriform lobes and should have possessed some visual ability (van Herk and Ten Cate, 1933) and auditory sensibility if they are comparable to cats (Bard, 1934; Dusser de Barenne, 1934; Bard and Rioch, 1937). They showed marked motor deficiencies and some indication of slight diminution of tactile sensitivity due to the removal of sensori-motor areas (Brooks and Woolsey, 1938). Ablation of the olfactory bulbs of the four males abolished their mating reactions and when other rabbits of either sex were placed in their cages they either ignored them or dashed about wildly as if frightened. They never coöperated in any sexual activity although the females frequently tried to mount them. Figure 1-E shows the brain of one of these rabbits after removal of the olfactory bulbs. In one male both olfactory bulbs and the cortex of one hemisphere were removed. The animal mated after those operations but following removal of the remaining neocortex mating never occurred although the animal was kept for eighteen months. Apparently in the normal male the olfactory system is not essential to mating but after complete ablation of the neocortex olfaction is indispensable. In these anosmic males without neocortex non-olfactory sources of sensation were incapable of initiating any specific sexual activity.

The six females from which all neocortex had been ablated continued to accept males after removal of the olfactory bulbs. In one of them all possibility of visual or auditory responses was also destroyed by extirpation of the end organs, but sexual activity continued and ovulation followed mating. Such operations abolish mating behavior in the male. There are at least two possible explanations of this difference between the two sexes. First, the male's mating behavior requires a more exact orientation and involves a more complex and delicately coördinated set of activities than is required of the female.   This greater complexity of behavior would conceivably necessitate a more exact sensory discrimination and would be dependent upon higher levels of integration.   Secondly, since the female is the more passive partner in mating, special senses are apparently of lesser importance to this sex than they are to the male.   It seems reasonable to assume that somatic stimuli (cutaneous, deep, proprioceptive) connected with the experience of being mounted and clasped by the male constitute an effective stimulus to the female.   These sources of stimuli are not abolished by the operations described above.

Three completely decorticate females (rhinencephalic cortex as well as neocortex removed) mated but only one of these ovulated as a result of coitus.   These animals did not eat spontaneously and slowly lost weight until killed.   They were kept for periods of from four to six weeks.   One which would not mate spontaneously was given estrogenic material in the form of Progynon-B (injected intramuscularly).   As a result of the injection she mated frequently but ovulation did not follow.   The other two mated a few times shortly after the operation.   They showed less excitement than normal females but did perform the characteristic mating responses.   One of the animals (fig. 1-F) ovulated but the other failed to do so, apparently because of her poor nutritive condition.   The removals were grossly identical in extent.

Three female rabbits from which the sacral cord and the lowest lumbar segment had been removed ovulated after coitus.   In another it was found that transection of the cord through the lowest lumbar segment did not stop the ovulatory response to mating.   Even when these procedures were supplemented by complete abdominal sympathectomy, hysterectomy and extirpation of the proximal half of the vagina, coitus was still followed by ovulation.   All these preparations exhibited signs of full anesthesia and analgesia of the vagina, vulva and of the skin surrounding the vulva.   This was tested by the application of strong electrical and mechanical stimuli.   The animal did not respond in any way to these traumatizing stimuli.   The existing paralysis of the sphincters and bladder tended to confirm the completeness of the denervation.   Some intrinsic bladder and sphincter tone developed after several weeks.   These experiments prove that under normal conditions specific genital stimulation is not an indispensable factor in the induction of mating behavior and ovulation, but its importance may become greater when other sources of sensation have been removed.   In one of these four females removal of the olfactory bulbs and destruction of vision did not abolish mating.   Other sensory clues were evidently sufficient to produce excitement.   This animal on one occasion mounted another female and reached a peak of excitement which resulted in ovulation.

Females whose hind legs had been paralyzed by lumbar (4 rabbits) or thoracic (2 rabbits) cord section failed to ovulate, though males mounted them, intromitted, ejaculated and fell off when the paralyzed animals were placed in such an attitude that this was mechanically possible. In three females both hind legs were completely denervated. For a few days the paralyzed limbs were completely flaccid but as muscular atrophy developed the legs became rigidly extended, possibly because of the greater strength of the extensor muscle group, and the joints became ankylosed. This extension of the hind legs made general movement difficult for the animals; they failed to ovulate after coitus. Two animals with almost completely denervated limbs did mate and ovulate. These animals were capable of executing some hip movement and did respond when mounted. Such observations indicate that the female must coöperate if the sexual activity and associated excitement is to be effective in inducing ovulation.

DISCUSSION. These experiments suggest that several factors coöperate in the production of sexual excitement and the resulting activity. If the neocortex or any considerable portion of it is intact the animals are not dependent on olfactory stimuli or vision. After removal of the neocortex of male rabbits the olfactory apparatus is essential to the initiation of mating behavior. The somewhat impaired tactile sensibility and the remaining sources of stimuli are not adequate to arouse a sufficient degree of sexual excitement to induce a male with neocortex ablated and the olfactory bulbs or tracts destroyed to mount and copulate. These experiments likewise show that the female rabbit becomes sexually excited when stimulated in any one of a variety of ways. Many types of sensation probably normally play a part in the initiation of excitement but practically every one of them is dispensable provided some other source of stimuli remains. In the normal animal olfactory, visual, auditory and genital stimuli must be important but in their absence the female can be excited through other channels. Removal of the cortex creates numerous sensory and motor deficiencies but these are not sufficient to abolish excitement, mating behavior and ovulation when the animal is mounted by a male. The afferent, central and motor components of the mating response are sufficiently intact to permit a practically normal reaction.

The results obtained lead to the conclusion that ovulation occurs as a result of intense sexual or emotional excitement rather than as a result of a reflex initiated by stimulation of any specific group of sensory endings. A considerable mass of evidence can be marshalled in support of this contention. Since ovulation normally follows coitus it has been thought that it might be dependent upon genital stimulation. Marshall and Verney (1935) have produced ovulation by strong repeated stimulation of the lumbo-sacral cord but no one has been able to produce ovulation merely by mechanical stimulation of the vulva or vagina. Mating frequently fails to cause ovulation when the female does not become intensely excited.

216

Parks and Fee (1930) found that anesthetization of the vaginal and vulval region with "percain" did not prevent mating and ovulation. The experiments here described show that females with deafferented genital regions ovulate when mounted and when excited sexually. Sexual excitement is hard to arouse if the female cannot coöperate in the sexual activity. Animals with hind legs and hips completely paralyzed fail to show signs of sexual excitement and fail to ovulate when forced to receive males. An additional piece of evidence in support of this theory of the importance of excitement is the observation that when one female mounts another ovulation follows if after executing male-like copulatory movements she falls off as does the male on ejaculation. The afferent stimulation pattern in these cases must be quite different from that of normal mating. There is no intromission or genital stimulation; the animal is not clasped or mounted; olfactory stimuli are probably quite dissimilar. Despite this dissimilitude of stimulation intense sexual excitement unquestionably develops and ovulation ensues. Similar inferences can be drawn from the occasionally-reported examples of isolated females which become excited by proximity to males or other rabbits and ovulate without any mating behavior whatsoever. In these cases the ovulatory stimulus must be chiefly psychic or emotional.

Excitement may normally cause an orgasmal contraction of uterus, vagina or other genital tissue. This in turn may reflexly initiate ovulation in some way. However, hysterectomized animals with truncated vaginas continue to ovulate. The manner in which ovulation is induced is not fully known. Apparently in the rabbit there is some nervous mechanism involved in the excitation of the endocrine activity (increased secretion of gonadotropic substance from the anterior pituitary, Free and Parkes, 1929; Smith and White, 1931; Brooks, 1937) known to occur. It is felt that this work supports the thesis that ovulation in the rabbit is the direct result of strong sexual excitement and that this excitement can be initiated by stimuli reaching the higher subcortical centers from a variety of sensory endings. In the female the cerebral cortex is not essential to the initiation of the excitement nor to the normality of the response.

SUMMARY

Bilateral destruction of the labyrinths and auditory apparatus, enucleation of the eyes and removal of the olfactory bulbs does not abolish sexual activity in either the male or female rabbit.

Ovulation occurs normally following coitus in rabbits whose sacral cords have been removed. Even when this denervation of the genital region is supplemented by complete abdominal sympathectomy, hysterectomy, and extirpation of the proximal half of the vagina coitus accompanied by signs of emotional excitement results in ovulation.

Males which have undergone bilateral removal of all neocortex mate. After ablation of the olfactory bulbs such animals do not mate.

Females continue to mate and ovulate after removal of the neocortex and destruction of the olfactory bulbs. Three completely decorticate females exhibited typical mating behavior and one ovulated following coitus.

Evidence in behalf of the interpretation that ovulation in the rabbit is dependent upon the development of sexual excitement is discussed. It is felt that in the female this excitement can be induced by a variety of stimuli.

REFERENCES

BARD, P.   Psychol. Rev. **41**: 424, 1934.

   Proc. Am. Physiol. Soc., This Journal **116**: 4, 1936.

BARD, P. AND D. McK. RIOCH.   Bull. Johns Hopkins Hosp. **60**: 73, 1937.

BROOKS, C. McC.   Proc. Am. Physiol. Soc., This Journal **119**: 280, 1937.

BROOKS, C. McC. AND C. N. WOOLSEY.   To be published. 1938.

CAMIS, M.   The physiology of the vestibular apparatus.   Transl. by R. S. Creed. Oxford, 1930.

FEE, A R. AND A. S. PARKES.   J. Physiol. **67**: 383, 1929.

DUSSER DE BARENNE, J. G.   Localization of function in the cerebral cortex.   Chap. V.   Baltimore, 1934.

HAMMOND, J.   Reproduction in the rabbit.   London, 1925.

MARSHALL, F. H. A. AND E. B. VERNEY.   J. Physiol. **85**: 12P, 1935.

PARKES, A. S. AND A. R. FEE.   J. Physiol. **70**: 385, 1930.

SMITH, P. E. AND W. E. WHITE.   J. A. M. A. **97**: 1861, 1931.

STONE, C. P.   This Journal **72**: 372, 1925.

VAN HERK, A. W. H. AND J. TenCATE.   Acta brev. neerl. **3**: 96, 1933.

23

Reprinted from J. Exp. Zool. **151**:167–176 (1962)

Mating Behavior in Male Rats after Cerebral Cortex Ablation

I. EFFECTS OF LESIONS IN THE DORSOLATERAL AND THE MEDIAN CORTEX

KNUT LARSSON[1]

Department of Psychology, University of Göteborg, Sweden

Detailed studies of the influence of the cerebral cortex upon the male rat mating behavior have been performed by Beach and collaborators (Beach, '40, '44; Beach, Zitrin and Jaynes, '55, '56; Zitrin, Jaynes and Beach, '56). They found that complete removal of the cerebral cortex totally abolished the copulatory behavior of rats and cats. Partial decortication did not give the same results in the two species. In the male rat, small lesions, not including more than 20% of the cortical surface, did not affect the behavior, while lesions including 60% of the cortex or more, independent of localization, destroyed the mating behavior. In the cat, the effect varied with the particular area removed. In this species lesions localized to the motor cortex modified the copulatory reactions causing prolonged latency before mounting, and an increased proportion of mounts without intromissions. Removal of the temporal and parietal areas in the cerebral hemispheres left the mating behavior unchanged and occipital injury did not cause any disturbances unless it was extensive enough to produce absence of vision. On the basis of these results, Beach et al., suggested that in the smooth brained rodent, the main function of the cerebral cortex in the organization of the mating behavior is to facilitate its arousal. In the more complex brain of the cat, the cortex is specifically involved in the coordination and execution of the mating behavior.

A detailed examination of the investigations described does not exclude the further possibility that the different effects of partial decortication in the two orders of animals could be due to differences in the operational and behavioral techniques

applied. While the studies of the cat included a detailed analysis of the operational effect upon the latency and frequency of the various responses composing the mating pattern, the studies of the rat were performed by the less adequate behavioral technique then available.

The present paper which is the first in a series, describes changes in the mating of male rats after bilateral lesions in the dorsalateral and median cortical surfaces.

SUBJECTS AND METHODS

Subjects

The experimental subjects were 40 sexually experienced, male rats approximately 150 days old at the beginning of experimentation. Both hooded and albino types were included. Sexually active males were selected on the basis of their mating performances in preliminary tests, and randomly assigned to one of two groups.

One group included animals which were to receive lesions of the dorsolateral cortical surface, the dorsolateral group, and the other group included those to receive lesions of the median cortical surface, the median group. The subjects in the dorsolateral group had an average weight of 530 gm before the operation, and 549 gm in the first postoperative testing. The subjects in the median group weighed 509 gm before the operation, and 524 gm at the beginning of the postoperative testing. The subjects were housed in individual cages with free access to food.

[1] The study was performed while the author held a Senior Postgraduate Fellowship in physiological psychology from the National Academy of Sciences at the University of California, Berkeley, under the sponsorship of F. A. Beach.

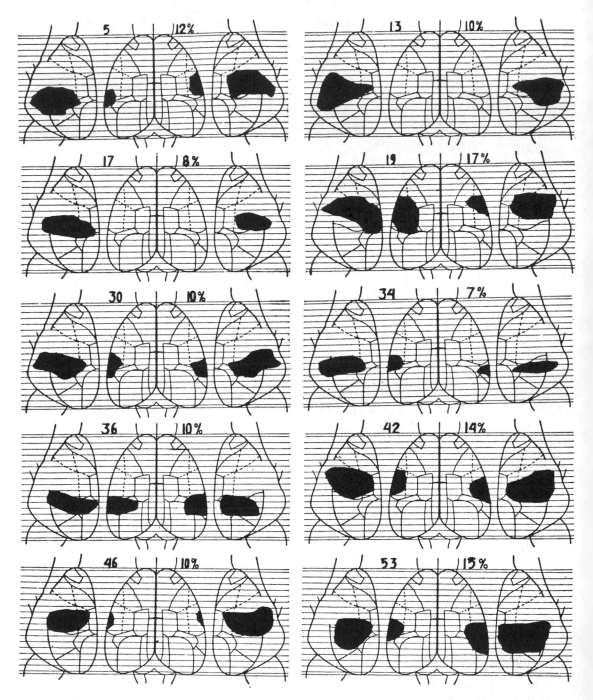

Fig. 1 Reconstruction of the group of dorsolateral cortical lesions. For each diagram the left hand number indicates the identification of the animal, the right hand number the percentage of neocortex removed.

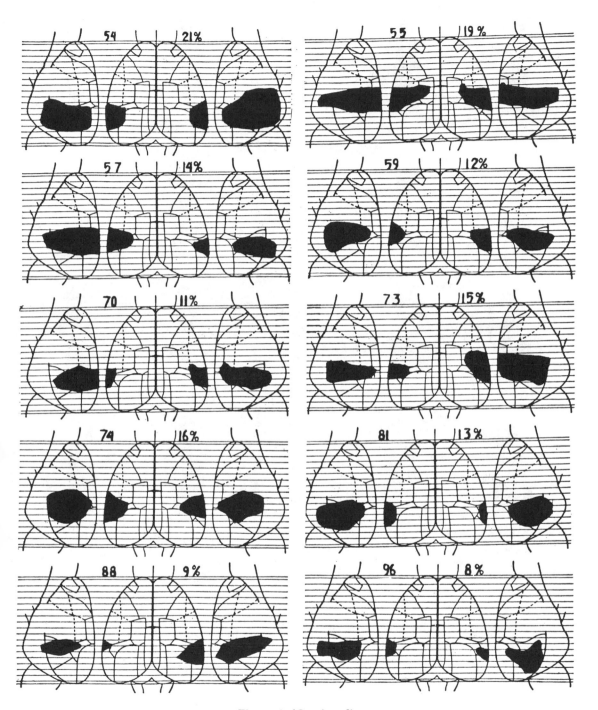

Figure 1 (*Continued*)

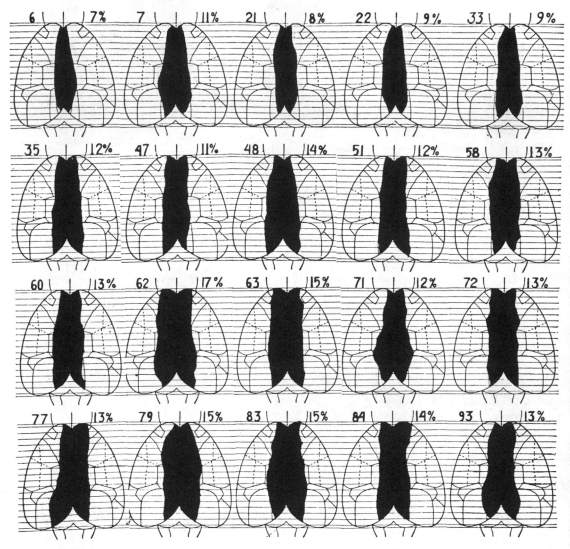

Fig. 2 Reconstruction of the group of median cortical lesions. The solidly marked areas indicate complete removal of the underlying cortical tissue. For each diagram the number on the left indicates the identification of the animal, the number on the right the percentage of cortex removed.

Apparatus

The apparatus included special observation cages and a recording device. The cages were semicircular, having a straight glassfront and sheet-metal sides. The cages were 27 in wide at the front, 16 in deep and 16 in high. The rear half of the top was surmounted by a release can in which a receptive female was placed just before the test. The bottom of the can was hinged, and could be dropped by raising a counter-weight. Four observation cages were placed upon a rack in such a position that they could be watched simultaneously. The experimenter recorded all sexual responses on a constant-speed kymograph.

Procedure

Prior to testing, stimulus' females were brought into estrous by hormone treatment. Thirty-six hours before the test .06 mg of estradiol benzoate was injected subcutaneously and this was followed six hours before the test by the injection of 1.0 mg of progesterone.

All tests were given during the dark phase of the 24–hr light and dark cycle. The subject was placed in the observation cage five minutes before a test. At the end of this period a female was presented to the male. Each subject was observed during a 60 minutes' testing period.

During each test the following items were measured: (1) *mounting frequency:* number of mounts without intromission, (2) *intromission frequency:* number of mounts with intromission, (3) *intromission latency:* time from entrance of the female into the cage till the first intromission, (4) *ejaculation latency:* time from the first intromission until ejaculation, (5) *intercopulatory interval (ICI):* average delay between the intromissions, (6) *postejaculatory interval (PEI):* time from ejaculation to the next intromission.

By Series is meant the sequence of mounting and intromission, culminating in ejaculation. Thus Series I indicates the sexual responses preceding and including the first ejaculation; Series II refers to those leading up to, and including the second ejaculation etc.

Subjects were observed in three preoperative, and three postoperative tests. The postoperative testing was begun three weeks after the operation, at which time all subjects had regained and slightly surpassed their preoperative weight. Between each one of the pre- and postoperative tests, the animals rested for 12 – 14 days.

Subjects that did not ejaculate during any of the postoperative tests were treated with hormones. Daily injections were made of 250 µg testosterone propionate in oil per 100 gm bodyweight. Two one-hour tests of the sexual behavior in these animals were made, the first eight days after the beginning of the hormonal treatment and a second after an additional eight days.

METHOD

Lesions were inflicted by the suction method under Nembutal anesthesia. Depending upon the group, ablation was made either of the dorsolateral or of the median cortex.

After the completion of the experiment the animals were anesthetized and perfused with physiological saline, followed by 10% formalin. The brains were embedded in paraffin and cut in sections of 10 µ. Every twenty-fifth section was preserved and stained with gallocynanide. The testes were removed and fixed in Bouin solution.

Before the hormonal treatment small pieces of testicular tissue had been removed and preserved for histological analysis.

The cortical lesions were reconstructed by help of Lashley brain diagrams and the sections examined for subcortical damage. Figures 1 and 2 show the extent of cortical damage and figure 3 presents two representative sections.

The dorsolateral group. The lesions in the dorsolateral cortex comprised between 7 and 21% of the surface. The corpus

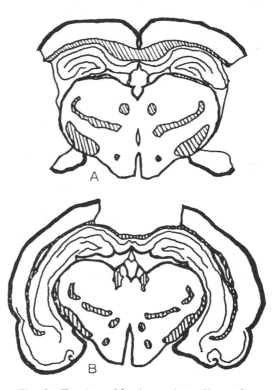

Fig. 3   Tracing of brain sections. Upper figure represents a section of rat number 81 belonging to the dorsolateral group; lower figure represents a section in rat number 79 belonging to the median group.

callosum was injured in all of these cases. Eleven animals showed some damage to the hippocampus which was mostly confined to the superficial layers. The nucleus caudatus was slightly injured in eight cases, and the capsula interna in five cases. The ventricular system and thalamus were left intact.

The median group. The injured median area comprised between 7 and 17% of the cortical surface. All of these subjects showed damage to the corpus callosum, and in all cases the ventricular system was opened at the frontal pole. In nine animals the superficial layers of the hippocampus were injured. The thalamus was intact in all cases.

RESULTS

Effects of dorsolateral lesions of the cerebral cortex

After the operation there was a complete suppression of the sexual behavior in four subjects, and a lowering of mating activity in the other 16 animals of the group. In the inactive animals no mating response took place. When presented with the receptive female, the operated animal followed her for some minutes but never mounted her and then quietly went into a corner of the cage and went to sleep for the rest of the hour. Intact animals that become sleepy during the observation period can often be aroused by non-specific handling. No such stimulation, however, could arouse the operated animals to copulate.

The other animals, although sexually active, were slower in mounting the female after the operation. Several subjects copulated and ejaculated in some tests, and not in others. Either the male did not mount the female at all, or he did mount her but failed to ejaculate, showing an unproportionally large number of mounts without intromission. An inspection of table 1 reveals this type of behavior in six animals. This behavior is in marked contrast to the preoperative performances where all subjects ejaculated in all tests.

Even when aroused to exhibit a complete sexual mating pattern including ejaculation, the animals were less active showing a lower ejaculation frequency. A

TABLE 1

Individual performances of those subjects in the dorsolateral group, which ejaculated in one or several of the post-operative sessions

No.	Postoperative tests		
	1	2	3
5	+	+	+
19	+	+	+
36	+	+	−
42	+	+	+
46	+	−	−
53	+	+	+
54	+	−	−
55	+	+	(−)[1]
57	+	+	+
59	+	+	+
70	+	+	+
73	+	(−)[2]	(−)[3]
74	+	+	+
81	+	+	+
88	(−)[4]	+	−
96	+	+	+

+: Sexual activity with ejaculation.
−: No sexual activity.
(−): Intromission but no ejaculation.
Notes:
[1] 11 intr. 47 mounts.
[2] 4 intr. 28 mounts.
[3] 20 intr. 14 mounts.
[4] 1 intr. 26 mounts.

comparison between the performances in these postoperative tests where the Ss attained one or more ejaculations with the performances in the corresponding preoperative tests revealed a difference statistically significant on the 5% level (Wilcoxon two-tail test, Walker and Lev, '53).

Besides the decrease in ejaculation frequency, the most notable change occurring in the mating behavior was a prolongation of the intromission latency. In the preoperative tests, the subjects responded to the female with a median intromission latency of 17 seconds and only occasionally showed a longer latency than one minute. After the operation the median intromission latency was prolonged to 155 seconds and several males failed to mount the female until ten or fifteen minutes after her presentation (table 3). As shown by tables 2 and 3 no other significant changes appeared in the mating behavior.

The postoperative performances were determined to a large extent by the preoperative activity level of the individual animal (rho = .48 p < .05). Among the four less active animals in the preopera-

TABLE 2

Number of intromissions, mounts and ejaculations before and after operation. The median numbers represent performances in three preoperative and three postoperative tests respectively. The statistical tests were made by help of Wilcoxon and Mann-Whitney non-parametric two tail tests

| Behavioral component | Series | Median | | Dorsolateral | | Median preop.– postop. p | Dorso-lateral preop.– postop. p | Postop. median– dorso-lateral p |
		Preop. Md	Postop. Md	Preop. Md	Postop. Md			
Intromissions	I	7.4	6.8	7.8	7.5	< 0.05	NS	NS
to ejaculate	II	4.3	3.9	4.1	3.7	< 0.01	NS	NS
	III	4.3	3.8	4.2	4.0	< 0.02	NS	NS
Mounts before	I	2.9	3.3	3.1	5.1	NS	NS	NS
ejaculations	II	1.4	2.3	1.6	2.1	NS	NS	NS
	III	0.5	0.7	2.0	3.8	NS	NS	< 0.002
Ejaculations per hour		4.1[1]	4.3[1]	4.3[2]	3.3[2]	NS	< 0.01	< 0.05

[1] N = 18.
[2] N = 16.

TABLE 3

Response latencies before and after the operation

| Behavioral component | Series | Median | | Dorsolateral | | Median preop.– postop. p | Dorso-lateral preop.– postop. p | Postop. median– dorso-lateral p |
		Preop. Md	Postop. Md	Preop. Md	Postop. Md			
Intr. latency in seconds		25	45	17	155	< 0.05	< 0.01	< 0.05
Ejaculatory	I	470	358	373	395	NS	NS	NS
latency in	II	235	198	169	173	NS	NS	NS
seconds	III	208	183	160	168	< 0.01	NS	NS
Average inter-	I	59	56	49	63	NS	NS	NS
copulatory	II	44	48	40	48	NS	NS	NS
interval in seconds	III	50	46	30	43	NS	NS	NS
Postejaculatory	I	425	375	398	358	NS	NS	NS
interval in	II	500	467	477	450	NS	NS	NS
seconds	III	578	528	496	507	NS	NS	NS

tive tests, two became sexually inactive following the operation. The other two copulated but showed an activity level which was much below the average of the group (Md = 1.8 ejaculations). The four most active animals in the preoperative tests, on the contrary, were all sexually active following the operation and showed a relatively high ejaculation frequency (Md = 4.1).

Within the dorsolateral group there is no significant correlation between the extent of the cortical injury and the decrease in ejaculation frequency (rho = .27). Numbers 17 and 34 with lesions which com-prised less than 10% of the cortical surface became inactive sexually, while numbers 54 and 55 having lesions which comprised about 20% of the surface remained active. Possibly, the animals that received relatively large lesions showed a low ejaculation frequency also before the operation. No statistically significant correlation was found, however (rho = .12).

Since the lesions inevitably damaged subcortical tissue, the possibility exists that the deficiency in mating was dependent upon the subcortical injury. A careful examination of the brain section, however, failed to reveal any relationship be-

TABLE 4

Effect of treatment with testosterone propionate in nonactive subjects eight and sixteen days after the first injections

Group	No.	Tests	
		1	2
Dorsolateral	13	No activity	No activity
	17	No activity	No activity
	30	No activity	1 ejac.
	34	No activity	11 mounts
Median	48	1 ejac.	14 mounts
	93	2 intr. 51 mounts	2 intr. 55 mounts

tween the extent or localization of the subcortical injury and the behavioral deficit.

In an attempt to determine whether the loss of sexual activity could be due to interference with the pituitary-gonad system, the testes of the four inactive males were studied histologically. No abnormality of the interstitial tissue could be detected, and the tubules showed normal spermatogenesis.

A second check on the possibility of endocrine disturbance involved treating the inactive males with testosterone propionate. As shown by table 4 the hormonal treatment only restored the complete mating pattern in one of the four inactive males, namely number 30. In the first test performed eight days after the beginning of the hormonal treatment, the male did not mount the female but in the second test when the animals had received 18 mg testosterone propionate, the male showed a normal mating pattern including three ejaculations.

Effects of lesions in the median cerebral cortex

Following the operation two animals numbers 48 and 93 failed to ejaculate. The remaining 18 subjects ejaculated in all of the postoperative tests, and attained approximately the same number of ejaculations as before the operation. As was the case in the dorsolateral group, the intromission latency was slightly prolonged. Once the male had started to copulate, however, he ejaculated after fewer intromissions than he did before the operation and with a shorter latency.

Those animals that showed the highest ejaculation frequency before the operation remained the most active after (rho = .37 p < .05). In contrast to the dorsolateral group there was a positive correlation between the size of the lesions and the postoperative performance (rho = .46 p < .05). No significant correlation could be found between the preoperative performance level and the size of the brain lesion (rho = .17).

A histological examination of the testes of males 48 and 93 did not reveal any abnormality. Treatment with testosterone propionate restored sexual behavior in male 48. After having received 10 mg of this hormone for eight consecutive days, this animal ejaculated once in the first test. Compared with the preoperative performances, the ejaculation latency and postejaculatory intervals were greatly prolonged. In the second test performed 16 days after the beginning of the hormonal treatment the same animal mounted the female but he did not penetrate or ejaculate. The other animal, number 93, did not ejaculate in any of the tests after the hormonal treatment. In the first test he responded by 51 mounts and two intromissions and in the second by 55 mounts and two intromissions.

Before being submitted to hormonal treatment both numbers 48 and 93 were observed in a fourth test performed two weeks after the end of the experimental series. While number 48 was not sexually active, number 93 started to copulate after a delay of 233 seconds. Since the animal apparently was quite capable of mating although having a heightened threshold for arousal, it is a notable fact that even very big doses of testosterone failed to induce the sexual activity.

DISCUSSION

The chief findings of the present study are, that contrary to what was earlier believed, the effect of a lesion of the cerebral cortex upon the pattern of mating in the male rat is dependent upon the particular area destroyed. Thus in the median group two animals failed to ejaculate in the tests immediately following the operation. In no case, however, was the sexual behavior permanently destroyed. One of these males mated in a test performed after the end of the regular series, and the other after hormonal treatment. Dorsolateral lesions, on the contrary, permanently destroyed the sexual behavior in three subjects and lowered the sexual activity in all the other animals of the group. Although lesions both in the median and the dorsolateral regions may lower the male's responsiveness to sexual stimulation, the most severe disturbances occur after destruction of the parietal-temporal areas.

The results indicate the importance of the sensorimotor cortex to mating. Recently similar findings have been reported by Rasmussen, Kaada and Bruland ('60). Using a modified obstruction box, where the male rat had to cross an electrically charged grill to reach the female, they found a reduction of the number of crossings after lesions in the fronto-parietal cortex and in the anterior temporal lobes. No changes were found after lesions in other cortical areas, including the cingulate gyrus, the hippocampus, and the entorhinal area.

The changes in mating following removal of the dorsolateral regions might be interpreted as indicating a heightened threshold for elicitation of the mounting response. Out of 16 sexually active animals in this group, six subjects, although perfectly capable of executing a normal mating pattern either did not ejaculate or failed to start copulating in one or two of the postoperative tests. The complete lack of sexual activity seen in four animals may be due to heightened threshold for arousal of the mounting response in these animals.

In the sexually active animals there did not seem to be any impairment of the animal's ability to integrate and execute the motor pattern. Every male that attempted to copulate was able to carry out a complete coital pattern.

Following the operations there was a decrease in the intromission frequency. This change was statistically significant only in the median group, but a tendency in the same direction was also observed in the dorsolateral group. It is presently not possible to give any explanation of this effect. Experiments performed have shown that a lowering of the number of intromissions preceding ejaculation occurs when the intervals between the copulations become prolonged above their normal lengths, and when the animals are exposed to traumatic experiences or when the rat is ageing (Beach and Fowler, '59; Larsson, '56, '62).

Besides its localization and extent, the effect of a brain lesion upon the mating seems to be highly dependent upon qualities inherent in the individual animal. The results described, clearly indicate that those animals which are most active before the operation remain most active after. Unfortunately the groups were too small and the factors involved too many to admit a statistical evaluation of the relative importance of preoperative mating activity level and extent of injury inflicted upon the brain.

SUMMARY

The sexual behavior of 40 adult male rats was observed before and after removal of the dorsolateral and median areas of the cerebral cortex. The area removed comprised approximately 12% of the cortical surface. To control for possible hormonal deficiencies attendant upon operative intervention, postoperative noncopulators were given extensive androgen treatment and observed in a final series of tests. The following main results were obtained:

1. After ablation of the dorsolateral surface there was a permanent suppression of the sexual behavior in three Ss. A fourth inactive male copulated after hormonal treatment.

2. After removal of the median surface two Ss failed to ejaculate. In no case, however, was the sexual performances permanently destroyed. One of the inactive

males ejaculated in an additional test performed two weeks after the regular tests, and the other after hormonal treatment.

3. In the dorsolateral group, the sexually active Ss attained fewer ejaculations per hour after the operation than they did before. Six animals mated and ejaculated in only some of the postoperative tests. No corresponding decrease in the sexual activity was observed in the median group.

4. In both groups following the operation a slight prolongation was observed of the interval between the reception of the female and the first mounting.

5. A histological examination of the testes did not reveal any abnormality in the sexually inactive Ss. This suggests that the suppression of mating behavior was not due to a hormonal deficiency. The interpretation is supported by the fact that androgen treatment failed to restore sexual behavior in all but two of the hormone treated Ss.

Comparison of present findings with earlier work shows that changes in mating can be produced by relatively small lesions in the cerebral cortex. They also show that the effect of a lesion is dependent upon the particular area destroyed. Although removal of the median cortex may cause some impairment in the sexual activity, by far the most severe disturbances occur after lesions in the sensori-motor cortex.

LITERATURE CITED

Beach, F. A. 1940 Effects of Cortical Lesions upon the Copulatory Behavior of Male Rats. J. Comp. Physiol., 29: 193–244.
——— 1944 Relative Effects of Androgen upon the Mating Behavior of Male Rats Subjected to Forebrain Injury and Castration. J. Exp. Zool., 97: 249–295.
Beach, F. A., and H. Fowler 1959 Effects of "Situational Anxiety" on Sexual Behavior in Male Rats. J. Comp. physiol. Psychol., 52: 245–248.
Beach, F. A., A. Zitrin and J. Jaynes 1956 Neural Mediation of Mating in Male Cats: I. Effects of Unilateral and Bilateral Removal of the Neocortex. Ibid., 49: 321–327.
——— 1955 Neural Mediation of Mating in Male Cats: II. Contribution of the Frontal Cortex. J. Exp. Zool., 130: 381–401.
Larsson, K. 1956 Conditioning and Sexual Behavior. Stockholm pp. 209.
——— 1962 Spreading Cortical Depression and the Mating Behavior in Male and Female Rats. Zeitschr. Tierpsychologie, 19: 321–331.
Rasmussen, E. W., B. R. Kaada and H. Bruland 1960 Effects of Neocortical and Limbic Lesions on the Sex Drive in Rats. (Abstract) Acta Physiol. Scand., 50: suppl. 175, 126–127.
Walker, H. M., and J. Lev 1953 Statistical Inference. New York, pp. 510.
Zitrin, A., J. Jaynes and F. A. Beach 1956 Neural Mediation of Mating in Male Cats: III. Contribution of the Occipital, Parietal and Temporal Cortex. J. Comp. Neur., 105: 111-125.

24

Reprinted from *Proc. Soc. Exp. Biol. Med.* **44**:61–64 (1940)

Failure of Ovarian Hormones to Cause Mating Reactions in Spayed Guinea Pigs with Hypothalamic Lesions.*

J. M. Brookhart, F. L. Dey and S. W. Ranson.

From the Institute of Neurology, Northwestern University Medical School.

It has been reported that female cats with small lesions in the hypothalamus in such a position as to interrupt the supraoptico-hypophysial tract were not observed to come into heat and were never bred in the laboratory.[1] It has more recently been found that the production of small lesions in a comparable part of the hypothalamus of the female guinea pig is followed by a complete lack of the mating response and in some cases also by disturbances in the ovarian cycle.[2] Although the disturbances in the sexual cycles may be attributable to a secondary disruption of hypophysial function, the majority of the animals showed regular sexual cycles which were normal so far as the physical changes in ovaries, uteri and vaginae were concerned, suggesting that their hypophyses were functioning normally. The present investigation was undertaken in order to determine whether the lack of the mating response in guinea pigs following hypothalamic lesions of the type described is due to an hormonal insufficiency or to the destruction of neural elements indispensable to the estrous or mating reflex.

Marrian and Parkes[3] have shown that vaginal estrus may be brought about by an amount of estrogen which is insufficient to induce uterine changes or copulatory behavior. On the other hand, Dempsey and Rioch[4] were unable to induce behavioral estrus in a guinea pig following removal of the brain rostral to a plane extending between the anterior limits of the superior colliculus and the posterior edge of the mammillary bodies, although the reflex arc remained intact when the ventral limit of the section was in front of the mammillary bodies. On the basis of this evidence they have postulated a sexual center located in the ventral hypothalamus at the

---

* Aided by a grant from the Committee for Research in Problems of Sex of the National Research Council.

1 Fisher, C., Magoun, H. W., and Ranson, S. W., *Am. J. Obstet. and Gynec.*, 1938, **36**, 1.

2 Dey, F. L., Fisher, C., Berry, C. M., and Ranson, S. W., *Am. J. Physiol.*, 1940, **129**, 39.

3 Marrian, G. F., and Parkes, A. S., *J. Physiol.*, 1930, **69**, 372.

4 Dempsey, E. W., and Rioch, D. M., *J. Neurophysiol.*, 1939, **2**, 9.

level of the mammillary bodies.   Bard,[5] however, has reported that estrous responses may be elicited in cats following massive lesions in the posterior hypothalamus which destroy all known descending paths from that part of the brain, and believes that the integration of the reflex is a mesencephalic function.

A series of 27 young, adult female guinea pigs, weighing between 400 and 600 g, were ovariectomized.   Following a recovery period they were each brought into full behavioral estrus several times by the subcutaneous injection of 12.5-15.0 IU of estrogen[†] on hours 0, 24, 48, and 60, followed by 0.2 IU of progesterone[‡] on hour 72, after the method described by Collins, Boling, Dempsey and Young.[6] After the constancy of the response to ovarian hormones had been established in each animal, lesions were placed in the hypothalamus at the level of the posterior border of the optic chiasma with the aid of a Horsley-Clarke instrument bearing a unipolar electrode.   Three lesions were placed in each animal, one in the midline and one on each side of the midline at a distance of one millimeter, by passing a direct current of 3.0 ma for 30 seconds.   In 22 animals the lesions were placed 1 mm above the ventral surface of the brain, and in 5 animals the lesions were placed 6 mm above the ventral surface.   Five of the animals with the low lesions failed to survive the operation. Gross inspection of the brains from these animals indicates that the lesion occurs just posterior to the optic chiasma.   The remaining 22 animals recovered completely, grew normally, and remained in excellent condition for the duration of the experimental period.   Aside from the diabetes insipidus which developed in some animals, and a transitory period of depression which lasted for approximately 12 hours after the operation, there were no criteria by which the operated animals could be differentiated from normal anestrous female guinea pigs.

At least 2 attempts have been made to induce estrus in 17 of these animals with lesions near the ventral surface of the brain, using the dose of ovarian hormones which was sufficient to alter the behavior of the animals before the lesion.   None of the animals so treated showed either proestrous or estrous behavior.   It was impossible to elicit the estrous reflex by manual stimulation of the vulva or the lumbo-sacral region of the back, and none of these animals would

[5] Bard, P., *Res. Publ. Assn. Res. Nerv. Ment. Dis.*, 1940, **20**, 551.

[†] Theelin, through the courtesy of Dr. Oliver Kamm, Parke, Davis and Co.

[‡] Proluton, through the courtesy of Dr. Erwin Schwenk, Schering Corp.

[6] Collins, V. J., Boling, J. L., Dempsey, E. W., and Young, W. C., *Endocrinology*, 1938, **23**, 188.

accept the male. All gave good avoiding responses to such stimulation, after the manner of a normal anestrous female. In subsequent trials, 8 of the animals were injected with double the usual dose of hormones and failed to show estrus, while 4 of the animals were injected with quadruple the usual dose of hormones and also failed to come into heat.

That the failure of the ovarian hormones to induce estrus in these animals is not due to a non-specific effect of destruction in the central nervous system is shown by the experiments of Bard,[7] Bard and Rioch,[8] Brooks,[9] Dempsey,[10] and Davis.[11] In addition, the 5 animals which have had lesions placed 6 mm instead of 1 mm above the ventral surface of the brain have been brought into heat with the same dose of hormone which induced estrus before the lesion was made.

These results differ from those of Dempsey and Rioch[4] and Bard.[5] Dempsey and Rioch's localization of the sexual center is based primarily upon the results of acute experiments on one guinea pig and one cat. In their chronic experiments failure to induce estrus following the removal of the anterior hypothalamic region is attributed to the debilitating effect of the operation on the animal. In their acute experiments successive transections were made in the same animal at various levels of the brain stem either with a blunt spatula or with a small sucker. In such experiments the accuracy of the localization of a "center" depends entirely upon the accuracy with which the location of the destruction to the central nervous system can be determined. We believe it may be significant that out of the 5 cats reported upon by Bard, the lesion in the one animal which failed to come into heat extended farther forward than in the other 4 animals. Although the main body of the lesion in the cats involved all known descending tracts from the hypothalamus, the possibility of the conduction of descending impulses by other paths has not yet been ruled out.

Summary. Following appropriately placed lesions at the level of the posterior border of the optic chiasma, ovariectomized guinea pigs failed to respond to previously effective dosages of estrogen and progesterone. The results reported here indicate that the failure of these animals to show estrous behavior is not due to a lack of ovarian hor-

---

[7] Bard, P., *Am. J. Physiol.*, 1936, **116**, 4.

[8] Bard, P., and Rioch, D. M., *Bull. Johns Hopkins Hosp.*, 1937, **60**, 73.

[9] Brooks, C. M., *Am. J. Physiol.*, 1937, **120**, 544.

[10] Dempsey, E. W., *Am. J. Physiol.*, 1939, **126**, 758.

[11] Davis, C. D., *Am. J. Physiol.*, 1939, **127**, 374.

mones. It is possible that the lack of response to the hormones is a result of the destruction of a portion of the central nervous system which is indispensable to the integration of a complex behavior pattern. If further control experiments prove this to be the case, then the possibility must be considered that the integrating mechanism involved is located in the midventral portion of the anterior hypothalamus instead of the region of the mammillary bodies or the mesencephalic tegmentum.

25

Drastic Changes in the Mating Behaviour of Male Rats Following Lesions in the Junction of Diencephalon and Mesencephalon

L. Heimer and K. Larsson

Departments of Anatomy and Psychology, University of Göteborg (Sweden), March 31, 1964.

Sexual abnormalities including mounting of diverse objects, intense masturbation and homosexual behaviour have been reported to follow lesions in various parts of the limbic system[1-5]. The following describes how a striking increase in the normal sexual activity of male rats was provoked by extensive lesions at the junction of diencephalon and mesencephalon.

Bilateral electrolytic lesions were made in 16 rats with the Horsley-Clarke technique by a d.c. of 1 mA for 30 sec. The lesions included the posterior part of the hypothalamus, the posterior parts of the medial thalamic nuclei, and the rostral part of the central grey matter in the mesencephalon. Reconstruction of a representative lesion is shown in Figure 1.

Prior to the experiment the animals were given 1–2 h of sexual experience. The male was presented with a female brought into oestrus by hormone treatment, and the mating behaviour observed during 30- or 60-min sessions. The sexual behaviour was tested by a standard procedure described in detail elsewhere[6]. The animals were given two preoperative sessions separated by a ten-day period of rest. The operations were performed one to three days after the last preoperative test.

Postoperative testing was begun ten days after the operation and was continued for 2–6 months. The following behavioural variables were measured: (1) mounting frequency: number of mounts without intromission, (2) intromission frequency, (3) intromission latency: time from the entrance of the female into the cage to the first intromission, (4) ejaculatory latency: time from the first intromission until ejaculation, (5) average intercopulatory interval: average delay between each intromission, computed by dividing the ejaculatory latency by the intromission frequency, (6) postejaculatory interval: time from ejaculation to the next intromission.

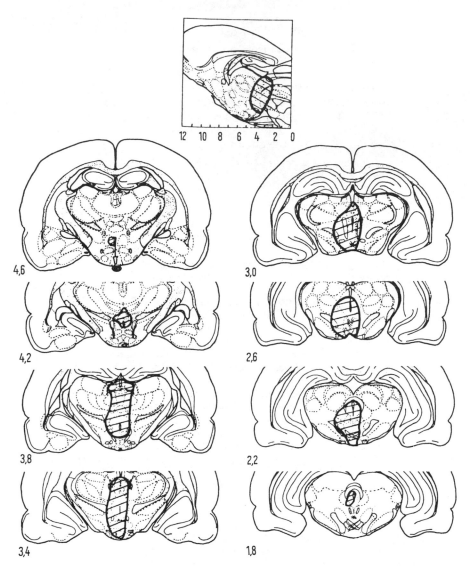

Fig. 1. Lesion at the junction of dien cephalon and mesencephalon resulting in an increase in sexual activity. (Reconstructions based upon DE GROOT[7]. The rat forebrain in stereotaxic coordinates.)

Nine rats did not show any deviations in their post-operative sexual behaviour. In seven animals, however, conspicuous changes occurred after the operation. These males tended to ejaculate after relatively few intromissions and with a short latency. The intervals of sexual inactivity following ejaculation were shortened from 5–6 min preoperatively to 1–3 min after the operation. Owing to the shortened latency periods there was an increase in the number of ejaculations achieved in a 30-min test.

The behavioural deviations appeared in the first post-operative tests and no recovery was ever seen. Exceptions to this rule were two males that did not achieve intromission and ejaculation before one or two months after the operation. When presented with the female these males approached her, followed her around the cage, and put the forepaws upon her back without making the mount with a pelvic thrust that characterizes the normal mounting responses. Frequent spontaneous penis erections accompanied by licking of the penis were displayed on these occasions. These two males also showed other abnormal features in their sexual behaviour. During the weeks following the operation they had frequent spontaneous penis erections. The erections occurred not only when these males were presented with a receptive female but also on other occasions; for instance, when being fed and weighed. The penile erections successively diminished

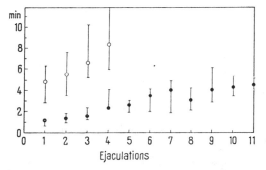

Fig. 2. The figure demonstrates the reduction of the postejaculatory intervals and the accompanying increase in ejaculation frequency in one of the operated animals observed in three 60-min tests approximately four months after the operation. Filled circles represent the median values of the postejaculatory intervals recorded during these tests. Vertical lines represent maximal and minimal values. Open circles show the corresponding performances of a group of 31 intact males.

in frequency and completely disappeared after about a month. When able to display complete copulations including ejaculation, both animals showed the same striking deviations in their sexual behaviour as described above. A demonstration of the reduction of the postejaculatory intervals and the accompanying increase in the ejaculation frequency is given in Figure 2. Histological examination of the testes did not reveal any abnormalities in any of the animals showing sexual deviations.

Comparison between the lesions in animals remaining sexually normal and in those showing sexual disturbances did not give any explanation for the behavioural differences observed. Smaller lesions within different parts of the critical region at the junction of diencephalon and mesencephalon, including complete destruction of the habenular complex, have so far not produced any sexual abnormalities [8].

Zusammenfassung. Umfangreiche Läsionen im Grenzgebiet zwischen Mittelhirn und Zwischenhirn bei Ratten führten zu stark erhöhter sexueller Aktivität. Die Ruheperioden nach der Ejakulation waren abnorm kurz, was die Zahl der Ejakulationen während des Versuches stark erhöhte. Auch 6 Monate nach Operation konnte kein Rückgang des Zustandes festgestellt werden.

[1] H. Klüver and P. D. Bucy, Arch. Neurol. Psychiat. (Chicago) *42*, 979 (1939).

[2] J. D. Green, D. D. Clemente, and J. de Groot, J. comp. Neurol. *108*, 505 (1957).

[3] L. Schreiner and A. Kling, J. Neurophysiol. *16*, 643 (1953).

[4] N.-Å. Hillarp, H. Olivecrona, and W. Silfverskiöld, Exper. *10*, 224 (1954).

[5] W. D. Hagamen and E. M. Lance, Anat. Rec. *130*, 414 (1958).

[6] K. Larsson and L. Essberg, Gerontologia *6*, 133 (1962).

[7] J. de Groot, *The Rat Forebrain in Stereotaxic Coordinates* (Verhandelingen der Koninklijke Nederlandsche Akademie van Wetenschappen, Afd. Natuurkunde, Tweede Reeks, Del L II, 1959).

[8] *Acknowledgment.* This study was supported by Public Health Service Research Grant HD 00344-03 National Institute of Child Health and Human Development. Hormones were generously supplied by Pharmacia Inc., Uppsala.

NEURAL MECHANISMS OF SEXUAL BEHAVIOUR IN THE FEMALE CAT

J. P. Maes

Laboratoire de Pathologie Générale
Université, Bruxelles

FEMALE cats in spontaneous or induced œstrus display a characteristic behaviour which may be divided into : (1) courtship activities (playful rolling, rubbing, calling, crouching with pelvis raised, and 'treading'), and (2) the after-reaction (vigorous or frantic rolling, rubbing, squirming, and licking).[1] Bard[1] has shown that extirpation of the neocortex does not change the specific pattern of the sexual response. It was my aim to determine if this peculiar performance of the female cat during 'heat' is conditioned by the activity of an encephalic 'sexual centre' or if some typical components of the complex behaviour could still be elicited from the spinal cord after transection of the brain stem below the medulla.

It is self-evident that, among the reactions mentioned above, only those the accomplishment of which is not dependent on the integrating mechanism of standing and equilibrium are suitable for this kind of experiment. Two reflexes fulfil these conditions : raising of the pelvis, and 'treading'. Treading consists in a rhythmic slow or rapid marching movement of the hind legs with flexion of the thigh and knee and dorsal flexion of the foot, as soldiers mark time. This is obtained by gentle tapping of the perineum. During the course of the investigation another reflex apparently characteristic of 'heat' was observed : when the perineum is tapped on one side the tail is swept towards the other and maintained there until the stimulus ceases, uncovering the whole perineum. This is the natural attitude of the female cat in œstrus when approached by a male. These three reflexes were used as tests for 'heat' in the spinal cats.

The investigation was carried out on thirteen cats, eight of which were normal, in the beginning or towards the middle of their breeding season. Five were ovariectomized during the breeding season and studied towards the end or after the same. In all the spinal cord was transected at the level of the first cervical segment, artificial respiration applied, and the cat kept warm on a heating pad. From three to six hours after the transection, when the spinal shock had subsided, the sexual reflexes were tested on the animal held in crouching position, hindlegs flexed, chest touching the table. Of the eight normal cats, all but one showed a complete positive response to tapping of the perineum. Vaginal smears and the condition of the ovaries indicated œstrus, pro-œstrus or metœstrus. The one negative result was obtained on an old female with anœstrus smear and smooth ovaries. She showed as the only response an elevation of the pelvis. Of the five animals ovariectomized three months previously, three were not injected with œstradiol. Their response was negative as regards treading and tail movements. Two of them showed some raising of the pelvis. When tested before the transection no response whatever could be elicited. The other two ovariectomized animals were injected twice during six days with 1 mgm. œstradiol (Schering Progynon *B* 1939) ; when they exhibited typical sexual behaviour the transection was made. Reflexes similar to those obtained from the normal cats on heat were obtained from them.

These experiments demonstrate that some components at least of the sexual behaviour are short arc reflexes, comparable to the scratch-reflex, which can be elicited independently of the higher centres, but the occurrence of which depends strictly on hormonal conditions. It seems, therefore, that the existence of a hypothetical 'sexual centre'[2] should be accepted with caution, and then not so much as a pace-maker under the influence of which unspecific activities of the spinal cord are transformed into specific sexual reactions, than as a mechanism co-ordinating certain independent activities pre-existing at different levels of the brain stem. Dempsey and Rioch's[2] failure to obtain sexual reflexes from decerebrate cats may be explained by the extensor rigidity which follows the mesencephalic transection, rhythmic reactions (as treading) and reflex raising of the pelvis requiring for their performance a normal balance of muscle tone.

[1] Bard, Ph., *Psych. Rev.*, **41**, 424 (1934) ; *Amer. J. Physiol.*, **116**, 4 (1936).

[2] Dempsey, E. W., and Rioch, D. McK., *J. Neurophysiol.*, **2**, 9 (1939).

27

Reprinted by permission from *Am. J. Physiol.* **210**:257–262 (1966)

Penile erection and contraction of penile muscles in the spinal and intact dog[1]

BENJAMIN L. HART[2] AND RALPH L. KITCHELL[3]

*Department of Veterinary Anatomy, University of Minnesota,
St. Paul, Minnesota*

HART, BENJAMIN L., AND RALPH L. KITCHELL. *Penile erection and contraction of penile muscles in the spinal and intact dog.* Am. J. Physiol. 210(2): 257–262. 1966.—Electromyographic recordings from penile muscles and observations of erection were made while stimulating various areas of the penis of dogs. Three distinct patterns of responses were observed in both the chronic spinal and intact dogs depending on the area of penis stimulated. Rubbing behind the bulbus glandis elicited tonic contraction of the ischiourethral muscle, rhythmic contraction of the bulbocavernosus and ischiocavernosus muscles, and rapid penile tumescence. During a refractory period which followed this reaction, a second similar reflex was elicited by applying pressure behind the bulbus glandis and rubbing the urethral process. The responses to this stimulus were tonic contraction of the ischiourethral muscle, rhythmic contraction of the bulbocavernosus and ischiocavernosus muscles, and slow penile tumescence. A third reflex, elicited by stimulating the corona glandis, resulted in tonic contraction of the bulbocavernosus and ischiocavernosus muscles and rapid detumescence. The possible role of these reflexes in penile erection and copulation in the dog are discussed.

electromyography; spinal cord; reflexes; sexual behavior; reproductive physiology

---

THERE ARE SEVERAL STUDIES which show that much of the activity usually labeled as sexual behavior is mediated by neural elements in lower thoracic, lumbar, and sacral areas of the spinal cord. For example, it is known that in a dog in which the spinal cord has been transected in the lower thoracic or upper lumbar region, appropriate stimulation of the penis may elicit erection and ejaculation as well as some skeletal movements suggestive of copulatory activity (1, 9). Presently con-

Received for publication 8 February 1965.

[1] This investigation was supported, in part, by Public Health Service Training Grant 5T1-GM-386-04, and was based, in part, on a Ph.D. thesis submitted by the senior author to the University of Minnesota.

[2] Present address: Depts. of Anatomy and Psychology, University of California, Davis, Calif.

[3] Present address: College of Veterinary Medicine, Kansas State University, Manhattan, Kan.

siderable effort is being directed toward understanding the role of the brain in sexual function. It is clear that studies of the sexual functions of the brain and other parts of the nervous system could be more meaningful with a detailed understanding of what reflex functions the isolated spinal cord is capable of mediating and to what extent these resemble the responses of the intact animal.

Since the extrinsic penile muscles play an important role in sexual functions such as ejaculation and erection, the purpose of this investigation was to compare the chronic spinal male dog with the intact dog in regard to the elicitation of erection and contraction of the extrinsic penile muscles in response to various types of genital stimulation.

Electromyography was used as a tool to carefully monitor and describe the patterns of contraction of the penile muscles.

METHODS

Twenty adult mongrel male dogs weighing between 8 and 12 kg were used. Ten of these animals were used for the study involving transection of the spinal cord.

Surgical procedures and postoperative care. All surgical procedures were performed under pentobarbital sodium anesthesia administered at the rate of 26 mg/kg body wt. During surgery the dogs were maintained in deep surgical anesthesia with additional doses of anesthetic if necessary.

It was necessary to dock the tails of the dogs in both the intact and spinal groups in order to eliminate interference with the recording electrode leads. This procedure also facilitated caring for the spinal animals. The surgical procedure for tail docking was that described by Lacroix (6).

For the spinal dogs a laminectomy was performed at the midthoracic (T_6–T_8) level. The dura was incised and a section of the spinal cord about .5 cm long was removed using a pair of fine scissors, after first passing a piece of suture tape under the cord to hold it away from the dura while it was being cut. There appeared to be no significant spinal cord hemorrhage. The back

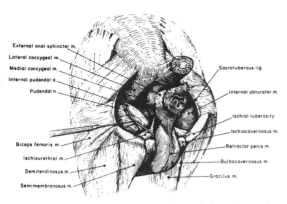

External anal sphincter m.
Lateral coccygeal m.
Medial coccygeal m.
Internal pudendal a.
Pudendal n.
Sacrotuberous lig.
Internal obturator m.
Ischial tuberosity
Ischiocavernosus m.
Biceps femoris m.
Retractor penis m.
Ischiourethral m.
Bulbocavernosus m.
Semitendinosus m.
Gracilus m.
Semimembranosus m.

FIG. 1. Drawing of deep dissection of the extrinsic penile muscles and related structures. The dissection is positioned as it would be for insertion of needle electrodes into the ischiourethral (IU), ischiocavernosus (IC), and bulbocavernosus (BC) muscles.

muscles were sutured and the skin was closed with horizontal mattress sutures.

It was not possible to section the spinal cord at precisely the same level in all animals. At necropsy it was found that the level of the segment of cord removed ranged from T_6 through T_8. Work on dogs used in preliminary studies as well as studies by others dealing with sexual functions (7–9) indicate that this level is well above spinal segments involved in erection and ejaculation.

The spinal dogs were kept in padded cages and given special care twice a day which consisted of expressing the urinary bladder, exercising the rear limbs for a few minutes by giving them full range of movement and washing off any urine or feces that may have accumulated on the animals. All animals survived the surgery and postsurgical care very well. The length of time the animals were kept as chronic spinal experimental subjects ranged from 64 to 185 days.

Stimulation of the penis. The penis was stimulated by touching, rubbing, or compressing various parts of the penis after the sheath had been pushed backward.

The glans penis of the dog is traditionally divided into a pars longa glandis and bulbus glandis. It has been reported elsewhere (3) that the classical anatomical nomenclature applied to the glans penis of the dog does not adequately describe the anatomy of this organ. Structures which have been recently described and labeled are the urethral process, corona glandis, and collum glandis (Fig. 5). For convenience these structures will be referred to in this paper in indicating where a particular stimulus was applied.

Recording apparatus and procedures. A special table was used during the recording procedures with which it was possible to suspend the chest of the spinal dogs in a canvas sling and to support the rear limbs with two metal bars placed beneath the groin region. Recordings from intact dogs were performed using the same table except that the rear limbs were left free.

The electromyographic (EMG) activity was recorded on a Grass model 5 pen-writing oscillograph with two Grass 5P3 preamplifiers and one Grass 5P5 preamplifier. Recordings were taken with the paper running at 15 mm/sec. The purpose in using electromyography was to determine when the penile muscles were contracting and to monitor the pattern of contraction. There is a very considerable attenuation of the high-frequency EMG signal with the pen-writing recorder we used so it was impossible to accurately measure the amplitude of the EMG signal. Therefore there are no vertical calibration indicators on the EMG tracings presented in Figs. 2, 3, and 4.

The EMG electrodes used were special bipolar, concentric needle electrodes developed by Hart and Kitchell (2). With these electrodes it was possible to inject a biologically inert marking substance (colored petrolatum) into the EMG recording site at the end of an experiment so that the recording site could be identified several weeks later on postmortem examination.

The muscles from which recordings were taken were the bulbocavernosus, ischiocavernosus, and ischiourethral (Fig. 1). The bulbocavernosus muscle is unpaired and arises from the external anal sphincter. The fibers of this muscle run mostly in a transverse direction covering the surface of the urethral bulbs. The paired ischiocavernosus muscle originates from the ischial tuberosity and inserts on the corpus cavernosum penis. The ischiourethral muscle (paired) also originates on the ischial tuberosity adjacent to and craniodorsal to the origin of the ischiocavernosus muscle. The insertion of the ischiourethral muscle is onto a fibrous ring encircling the common trunk of the left and right dorsal veins of the penis.

Electrode placement was guided by palpation of the muscles or their attachments. This placement was confirmed for many of the recordings by identification of the injected colored petrolatum at necropsy. Later in the studies it was possible to identify the muscles by noting the pattern of the EMG activity evoked following stimulation of various regions of the penis.

During all of the recording sessions a tape recorder was used to vocally record the degree of erection and to indicate the type and onset of genital stimulation employed. This information was correlated with the recorded EMG activity by an event marker.

A number of recordings were taken from each animal. The recordings from the spinal dogs are from after the 40th postoperative day. Judging from the strength of the return of reflexes such as the extensor-thrust, crossed-extensor and standing reflexes, it was evident that general reflex activity had recovered by this time.

RESULTS

Spinal subjects. With the preputial sheath retracted, touching the corona generally evoked a strong tonic contraction of the ischiocavernosus muscle (IC) and the bulbocavernosus muscle (BC). This response was consistent regardless of the time of stimulation (Fig. 2A). There was also rapid detumescence if the penis was erect

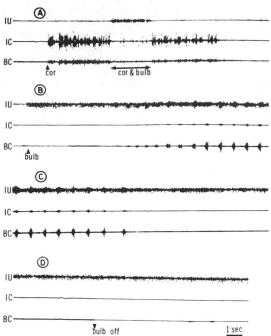

FIG. 2. Example of EMG tracings from spinal dogs. *A*: response to stimulation of the corona (cor) with the resultant inhibitory effect of applying pressure behind the bulbus (cor and bulb). *B–D* (continuous tracing): response to stimulation behind the bulbus (bulb).

when the corona was stimulated. By applying pressure to the penis just behind the bulbus glandis this response could be inhibited even though coronal stimulation was continued. The inhibition lasted as long as the area behind the bulbus was compressed and was reversed immediately on releasing the pressure.

Applying pressure behind the bulbus glandis and slowly rubbing this region of the penis resulted in immediate excitation of the ischiourethral (IU) to tonic contraction. With a few seconds' latency the IC and BC began to contract in a series of rhythmic contractions (Fig. 2, *B–D*). When the rhythmic BC and IC contractions began, there was generally some increased IU activity superimposed on the tonic IU contraction which was phasic with the rhythmic IC and BC contractions. If the pressure was maintained behind the bulbus, the rhythmic contractions would subside generally within 30 sec after the occurrence of the first contraction. The IU activity generally continued even after the release of the stimulus pressure. With the onset of rhythmic contractions the penis became rapidly engorged and complete erection was reached before the cessation of the rhythmic contractions. In some of the dogs (especially during the first few recording sessions) there was an expulsion of semen during the series of rhythmic contractions.

It was found that the above complex reaction could

not be elicited again immediately after this reaction had occurred. There was, apparently, a refractory period during which this reaction could not be again elicited. The length of the refractory period was not tested other than to ascertain that it was longer than 5 min and less than 24 hr.

During the apparent refractory period just described, rubbing the urethral process, in addition to applying pressure on the body of the penis behind the bulbus, would again initiate rhythmic BC and IC contractions and tonic IU contraction (Fig. 3). Again there were bursts of IU activity superimposed on the tonic contraction which were phasic with the BC and IC contractions. This response did not cease after a few seconds but instead lasted as long as stimulation continued and usually after all stimulation had been removed. When the rhythmic contractions began, there was a slow progression of erection. This response appeared to be a different reaction than the one which could be elicited by stimulating behind the bulbus glandis only.

Rubbing the urethral process alone usually evoked only sporadic activity in the BC and IC (at the frequency of the rubbing stimulation) and occasionally sporadic activity in the IU. When, however, while rubbing the process, pressure was applied to the area behind the bulbus, rhythmic BC and IC contractions were again initiated along with tonic IU contraction.

Touching or rubbing the collum resulted in a response very similar to that evoked by stimulation of the corona, especially when the more distal area near the corona was

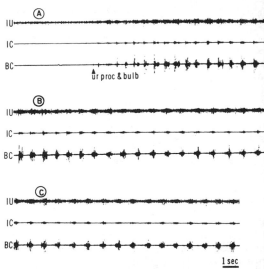

FIG. 3. EMG tracings taken a few minutes later from the same spinal dog as that used in Fig. 2, *B–C*. The reaction represented in *A–C* above (continuous tracing) occurs in the refractory period following the reaction illustrated in Fig. 2, *B–C*. Stimulation behind the bulbus was started 20 sec before the record begins in *A*. Rhythmic IC and BC contractions were initiated only when the urethral process was stimulated together with the area behind the bulbus (ur proc and bulb).

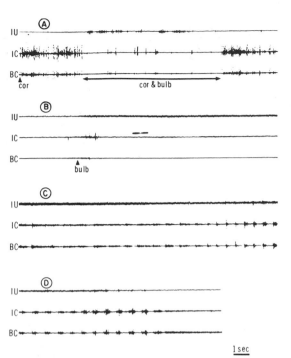

FIG. 4. Example of EMG tracings from intact dogs. *A:* response to stimulation of the corona (cor) and the resultant inhibitory effect of applying pressure behind the bulbus (cor and bulb). *B–D:* response to stimulation behind the bulbus (bulb). *C* starts 36 sec after *B, D* starts 24 sec after *C.*

stimulated. When the preputial sheath was lightly rubbed over the collum, a partial erection was usually evoked. However, in one of the dogs this kind of stimulation almost always resulted in the elicitation of complete erection.

Intact subjects. The same restraining apparatus, recording procedures, electrode placements, and stimulation techniques were used with intact dogs to determine if responses similar to those exhibited by the spinal dogs could be evoked in the intact animals.

The pattern of contraction in the penile muscles and the erectile responses in the intact animals were, as a rule, similar to those observed in the spinal subjects. There was, however, a greater variation in the intensity of recorded EMG activity. Stimulation of the corona glandis usually evoked a strong IC and BC contraction which could be inhibited by applying pressure behind the bulbus glandis (Fig. 4A). Sometimes this inhibition did not last as long as the inhibitory stimulus and, at other times, the inhibition was not complete. If the penis was erect, there was usually rapid detumescence when the corona was stimulated. Besides variation in intensity, the responses from the intact subjects were less consistent than the responses from the spinal subjects. For example, sometimes a series of rhythmic BC and IC contractions, tonic IU contraction, rapid progression of erection and seminal expulsion could be elicited by stimulation

behind the bulbus glandis (Fig. 4, *B–D*). With other dogs no such response could be evoked.

In the animals in which rubbing behind the bulbus could not elicit a reaction, rubbing the urethral process in addition to holding pressure behind the bulbus would evoke rhythmic BC and IC contractions lasting as long or longer than the period of stimulation. When the rhythmic contractions began, there was a slow progression of erection. With some subjects a small amount of semen was expelled during these rhythmic IC and BC contractions.

DISCUSSION

In spinal and intact dogs certain regions of the pars longa glandis, namely the urethral process, corona glandis, and collum glandis are not only morphologically distinct in the erect organ (3), but, as shown by the present study, are functionally distinct in the erect and the nonerect state. Previously it was thought that the glans of the penis held the receptors most important to sexual function. However, this study indicates that in the dog the distal part of the body of the penis has the most significant role in the initiation of both erectile and ejaculatory responses.

In light of the findings correlating contraction of the penile muscles with tumescence or detumescence in the penis, the role of these muscles in erection should be reviewed. It is generally felt that the contraction of the cavernosus muscles aids the erectile process by forcing a partial occlusion of venous return from the penis. It is true that the BC and IC were contracting in a characteristic rhythmic manner during erection—in fact, tumescence to complete erection was never observed to occur without this—but significant impedance of venous drainage by such a pattern of contraction seems unlikely. In fact, tonic and intense contraction of the IC and BC (with coronal stimulation) was correlated with rapid detumescence of the penis. Recordings by Kollberg, Petersén, and Stener (5) taken during erection in man show that in the human species erection can occur without participation of the bulbocavernosus, membranous urethral sphincter, and deep transverse perineal muscles. The observations reported here support the contention of Henderson and Roepke (4) and Watson (10) that the IC and BC facilitate erection by pumping blood into the distal venous sinuses.

Tonic contraction of the IU was frequently related to tumescence. It seems logical that the IU, with its insertion onto a fibrous ring surrounding the dorsal vein of the penis, would facilitate erection when tonically contracting. However, complete erection has been observed to occur without any apparent contraction of this muscle. For example, in the early stages of recovery from spinal shock, contraction of the IU could not be evoked, yet a complete, slowly progressing erection was elicited (unpublished observation). We have noted in one spinal animal that complete erection was still possible even though the IU was surgically severed from its ischial origin (unpublished observation). It is, therefore, sug-

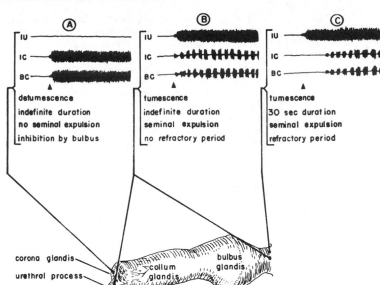

FIG. 5. Diagrammatic representation of three reactions elicited by stimulation of the penis. *A:* stimulation of the corona. *B:* stimulation of the urethral process together with pressure behind the bulbus. *C:* stimulation of the area behind the bulbus. Solid arrows indicate onset of stimulus. IU, ischiourethral muscle; IC, ischiocavernosus muscle; BC, bulbocavernosus muscle.

gested that the IU plays a role in increasing the rate of penile tumescence although it is not necessary for slow tumescence.

Expulsion of semen was observed in both the intact and the spinal subjects and was associated with rhythmic contraction of the BC and IC. However, there were many instances in intact dogs when no obvious expulsion of semen occurred even though these rhythmic contractions were observed. The amount of semen collected from the spinal dogs was always small and most of the time the rhythmic contractions were not accompanied by seminal expulsion.

Although the patterns of contraction of the penile muscles were similar in spinal and intact dogs, the reliability with which particular patterns of contraction of the penile muscles could be evoked by stimulation of specific areas of the penis was greater in the spinal animal than it was in the intact animal. Responses of spinal animals were so reproducible that at least three were easily identified. An attempt to distinguish these responses in the intact animal revealed that the reflexes were present, but that they occurred less consistently and less intensely than in the spinal animal. The most logical explanation of this is that under the testing conditions— i.e., restraint, needle electrodes, artificial stimulation— there was suprasegmental inhibition of the reflexes, whereas in the spinal preparation the reflexes were free from suprasegmental inhibition.

The role of the three reflexes reported here in mating behavior should be discussed. The two reflexes, which are characterized by rhythmic contractions in the IC and BC, are undoubtedly involved in ejaculation. When seminal expulsion did occur, it was always correlated with rhythmic IC and BC contractions.

The observation that the erectile process is much more rapid with IU contraction suggests that an important function of the IU is to induce a lock or tie of the sexual partners. Without rapid tumescence of the bulbus, twisting or jerking by the female dog during copulation could lead to withdrawal of the penis and prevention of a tie.

It has also been shown that if pressure is maintained behind the bulbus, erection does not readily subside. When the male and female are locked and standing tail-to-tail in the typical dismount fashion, it seems quite apparent that there is a good deal of pressure exerted on the body of the penis behind the bulbus. This, then would tend to prevent detumescence and continue the lock. It is generally assumed that contraction of the constrictor muscles in the bitch with a resultant compression of the dorsal vein of the penis is the chief mechanism initiating and maintaining the copulatory lock. However, this study indicates that there are adequate reflex mechanisms in the male to initiate and maintain the lock even if the role of the female genital muscles is solely that of application of pressure on the sides of the penis behind the bulbus (without compression of the dorsal vein).

The reflex elicited by touching the corona and characterized by tonic contraction of the IC and BC and rapid detumescence appears to have no obvious functional role. The data presented show that pressure behind the bulbus will inhibit the response to coronal stimulation so it does not seem likely that this reflex is responsible for the eventual break of the lock even if the corona is stimulated within the vagina. One possible function of this reflex might be to insure that the penis does not engorge so much during precopulatory thrusting that complete intromission is impossible. Rubbing of the corona against the female's perineum would serve as an appropriate stimulus to activate this reflex

REFERENCES

1. BARD, P. The hypothalamus and sexual behavior. *Res. Publ. Assoc. Res. Nervous Mental Disease* 19: 190–218, 1940.
2. HART, B. L., AND R. L. KITCHELL. An electromyographic electrode designed for use in identification of recording sites. *J. Appl. Physiol.* 20: 1094–1095, 1965.
3. HART, B. L., AND R. L. KITCHELL. External morphology of the erect glans penis of the dog. *Anat. Record* 152: 193–198, 1965.
4. HENDERSON, V. E., AND M. H. ROEPKE. On the mechanism of erection. *Am. J. Physiol.* 160: 441–448, 1933.
5. KOLLBERG, S., I. PETERSÉN AND I. STENER. Preliminary results of an electromyographic study of ejaculation. *Acta Chir. Scand.* 123: 478–483, 1962.
6. LACROIX, J. V. Docking. In: *Canine Surgery* (4th ed.), edited by K. Mayer, J. V. Lacroix, and H. P. Hoskins. Evanston, Ill.: Am. Vet. Med. Publ., 1957, p. 784–788.
7. ROOT, W. S., AND P. BARD. The mediation of feline erection through sympathetic pathways with some remarks on sexual behavior after deafferentation of the genitalia. *Am. J. Physiol.* 151: 80–90, 1947.
8. SEMANS, J. H., AND O. R. LANGWORTHY. Observations on the neurophysiology of sexual function in the male cat. *J. Urol.* 40: 836–946, 1938.
9. SHERRINGTON, C. S. The spinal cord. In: *Text-Book of Physiology*, edited by E. A. Schafer. New York: Macmillan, 1900, vol. 2, p. 782–883.
10. WATSON, J. W. Mechanism of erection and ejaculation in the bull and ram. *Nature* 204: 95–96, 1964.

28

EFFECT OF RESERPINE ON MALE RAT SEXUAL BEHAVIOR

A. Soulairac and M.-L. Soulairac

This excerpt was translated expressly for this Benchmark volume by Patricia J. Salis, Sleep Disorders and Research Center, Baylor College of Medicine, Texas Medical Center, from "Action de la réserpine sur le comportement sexuel du rat mâle" in Compt. Rend. Soc. Biol. **65**:1010–1013 (1961)

We have previously shown that experimental lesions of anterior and posterior hypothalamic structures abolish sexual behavior in the male rat. It is known, moreover, that a certain number of "psychotropic" substances act on the central nervous system through effects at the hypothalamic level. In addition, it seemed of interest in our general program of psychopharmacological research to study the effects of some of these drugs on different types of instinctive behavior.

This report describes the changes in male rat sexual behavior obtained with administration of different doses of reserpine. The study was performed on male albino rats weighing about 300 g. Sexual behavior was evaluated according to our usual procedure (Soulairac, 1957), in which we record during one-hour tests the number of ejaculations, the number of intromissions preceding each ejaculation, and the time during which these intromissions occur (ejaculation latency). We also measure the duration of the different refractory periods (Soulairac, 1952).

In addition to this quantitative estimation of the various elements of sexual behavior, it seemed of interest in this psychopharmacological study to consider variations in some of these elements with respect to each other. We therefore studied two interesting indexes:

1. The ratio of number of intromissions to ejaculation latency (in minutes), which in a sense reflects the density of neuromotor activity per minute. With this value (ANM index: neuromotor activity) one can easily determine whether a substance actually changes neuromotor activity or simply lowers the central-nervous-system sensitivity threshold.

2. The number of intromissions required for each successive ejaculation. This value I (intromissions for one ejaculation) also permits determination of changes in central-nervous-system thresholds because it may be assumed that the greater the receptivity of the nervous system, the smaller the number of intromissions required to release an ejaculation. The value is an indirect means of determining the response threshold of genital reflexes.

We point out that normally some facilitation occurs during the course of sexual behavior. In effect, the value I decreases and the ANM index

Table 1 Effect of low and high doses of reserpine

Animal Groups	Ejaculation 1				Ejaculation 2			Ejaculation 3		
	No. Ejac.	No. Intro.	Ejac. Latency (min)	Refrac. Period (min)	No. Intro.	Ejac. Latency (min)	Refrac. Period (min)	No. Intro.	Ejac. Latency (min)	Refrac. Period (min)
A. Reserpine 25 µg/100 g										
Normal (N=24)	2.50	18.1	17.11	7.99	10.95	9.92	9.48	12.72	8.17	8.71
Reserpine (N=19)	3.83	12.94	11.98	6.88	6.57	5.58	8.08	6.33	4.80	9.59
Difference	+1.33 ±0.11	-5.16 ±1.62	-5.13 ±1.52	-1.11 ±0.98	-4.38 ±0.99	-4.34 ±0.93	-1.40 ±0.51	-6.39 ±2.01	-3.37 ±0.98	+0.88 ±1.21
Significance	p=0.01	p=0.05	p=0.05	NS	p=0.02	p=0.01	p=0.05	p=0.05	p=0.05	NS
B. Reserpine 60 µg/100 g										
Normal (N=18)	2.15	22.9	17.8	7.12	14.60	10.7	8.56	12.57	6.66	9.50
Reserpine (N=6)	3.83	17.8	14.3	6.17	7.0	5.2	7.20	6.0	5.33	14.33
Difference	+1.68 ±0.46	-5.1 ±3.72	-3.5 ±3.1	-0.95 ±0.28	-7.6 ±1.99	-5.50 ±1.87	-1.36 ±1.24	-6.57 ±2.33	-1.33 ±1.3	+4.83 ±4.0
Significance	p=0.02	NS	NS	p=0.02	p=0.02	p=0.05	NS	p=0.10	NS	NS

increases. These changes demonstrate that there is a simultaneous lowering of genital reflexes and elevation of the density of neuromotor activity (Table II).

We studied two experimental groups: (1) administration of high doses of reserpine for a rather short period—60 μg/100 g of weight, daily for four days, with behavioral tests on the second and fourth days: (2) administration of low doses of reserpine for a longer period— 25 μg/100 g of weight, daily for fifteen days, with tests on the second, fourth, eighth, and tenth days. The numerical results are summarized in Table I.

These studies showed that:

1. Low doses of reserpine changed male rat sexual behavior by increasing very significantly the number of ejaculations, decreasing significantly the number of intromissions preceding each ejaculation, and decreasing significantly the different ejaculation latencies. Refractory periods changed only slightly. Examination of the different ANM indexes showed that they underwent no appreciable change except for a slight decrease for the third ejaculation. By contrast, the value I decreased significantly for every ejaculation (Table II).

Table II Variations in the value I and in the ANM index with administration of low and high doses of reserpine

Animal Groups	Ejaculation 1		Ejaculation 2		Ejaculation 3	
	I	ANM	I	ANM	I	ANM
A. Reserpine 25 μg/100 g						
Normal	18.10	1.058	10.95	1.104	12.72	1.557
Reserpine	12.94*	1.117	6.57**	1.180	6.33*	1.319
B. Reserpine 60 μg/100 g						
Normal	22.9	1.287	14.6	1.319	12.6	1.885
Reserpine	17.8	1.112	7.0**	1.265	8.7	1.507

*Significant at p = 0.05
**Significant at p = 0.02

2. Forty-eight hours after administration of high doses of reserpine, sexual behavior was again changed in a manner analogous to that observed with low doses. There was an increase in the number of ejaculations, but little or no change in intromissions or ejaculation latencies. The ANM indexes varied exactly as in the first study, but the value I did not undergo an equivalent significant decrease. Only for the second ejaculation was there a significant decrease.

The most noteworthy result was that the changes in sexual behavior produced by high doses of reserpine were very transient, for tests performed on the fourth day were essentially all negative: the animals showed no inclination toward sexual activity, and, in most cases, remained in a stuporous state throughout the behavioral test. Normal behavior in these animals had returned five days after discontinuation of reserpine treatment. This suggests that the disappearance of sexual behavior with high doses of reserpine was a direct effect of the drug.

In conclusion, low doses of reserpine changed male rat sexual behavior not by increasing specific neuromotor activity but by lowering the neural receptivity threshold and facilitating genital reflexes. Excessively high doses of reserpine rapidly inhibited all sexual activity.

REFERENCES

Soulairac, A., 1952, La signification de la période réfractaire dan le comportement sexuel du rat mâle, *J. Physiol.* **44**:99–113.

Soulairac, A., 1957, Données expérimentales sur le comportement sexuel du rat mâle, *Psychol. Fr.* **2**:1–9.

Editor's Comments
on Papers 29 Through 33

MOTIVATIONAL AND SOCIAL FACTORS

Copulatory behavior is affected by many factors operating at the time of its occurrence. Although the study of many of these factors involves manipulation of characteristics inside the animal, others can be studied with no such manipulation. Students of copulatory behavior have considered the effects of a broad range of independent variables on copulatory behavior. The rubric of "motivation" is convenient as a label for bringing together the results from these diverse studies.

One of the enduring questions appropriate for consideration under the rubic of motivation is that of the reward or reinforcing value of sexual activity. There appear to have been three waves of research on the reward value of sexual activity. The first wave occurred in the 1920s and 1930s and was characterized by the use of the Columbia Obstruction Box for the comparison of the relative strengths of a different "drives." An animal was placed in one end of a test apparatus separated from some incentive by an electrified grid. The relative frequency with which the animal crossed the grid was taken as a measure of the strength of the drive. The study of Warner (Paper 29) is an excellent example of the use of this method. Warner studied males running to females and females running to males, and compared both to the tendency for animals to cross when the goal box was empty. Warner found that males would indeed cross the grid to reach females, although they did eventually become satiated (see also Paper 4). The tendency of females to cross the grid to reach males varied with their estrous cycles, as inferred from vaginal cytology. Researchers have been particularly interested in the reward value of sex of females, probably for cultural reasons. Other applications of the obstruction-box technique were made by Moss (1924), Jenkins (1928), Nissen (1929), and Stone, et al., (1935a, 1935b). The study of Seward and Seward (1940) is also relevant.

A second wave of research on the reward value of copulation occurred in the 1950s and centered on the question of the necessity of drive reduction for reinforcement as proposed in certain learning theories of the day. It was generally found that sex was reinforcing for males even when ejaculation was prohibited, although the opportunity to ejaculate was even more reinforcing (Kagan, 1955; Sheffield et al., 1951; Whalen, 1961).

The third wave of research occurred primarily in the 1960s and 1970s and was less theoretically stimulated but nevertheless focused on the reward value of copulation. Numerous investigators studied the factors affecting the learning of arbitrarily chosen tasks by male rats when the opportunity to engage in sexual activity was the reward (for examples, Sachs et al., 1974; Schwartz, 1956). It was found that females can play a role in the pacing of sexual activity, performing a response leading to sexual activity sooner after a mount than after an intromission or an ejaculation (Bermant, 1961; Peirce and Nuttall, 1961). Although there was some question as to whether sexual activity is indeed reinforcing for female rats apart from the reward value of social interaction per se (Bolles et al., 1968), the matter was eventually settled in the affirmative (Drewett, 1973).

Much interest has been generated in the effects of aversive stimulation, especially electrical shock, on sexual activity. The effects of the delivery of shock depend on the delivery schedule used. If shocks are arranged so as to be contingent on a male's copulatory activity, inhibition of sexual behavior occurs (Beach et al., 1956). However, if the shocks are delivered in a noncontingent manner, they can facilitate the copulatory behavior of both virgin and experienced males (Barfield and Sachs, 1968; Caggiula and Eibergen, 1969). In Paper 30, Beach and Fowler demonstrate the lasting effects of experience with shock in a particular situation. Males previously shocked in the test chamber ejaculated more readily when later tested in that same chamber than did the males of two control groups.

Much of the research on motivational factors in rodent copulatory behavior has been performed in relation to a hypothetical model of the control of copulation proposed by Beach and modified by various workers. Some fundamentals of the theory were outlined by Beach (1942). The theory was further developed by Beach (1956) and is summarized in the last section of Paper 4 by Beach and Jordan. An excellent review of the various modifications proposed for this model and of much research relevant to it has been prepared by Sachs and Barfield (1976). In essence, the model was designed on the premise that at least two hypothetical mechanisms are involved in the control of copulation in male rats. One, the Sexual Arousal Mechanisms (SAM) functions in generating sexual arousal to the point where copulatory activity is initiated. Once copulation starts, a different mechanism, the Copulatory Mechanism (CM) controls the sequence. Ejaculation was treated as occurring when the level of excitation in the CM reaches a certain threshold value. Some form of inhibition, varying with the particular version of the model, then was thought to operate to prevent further copulation until the close of the postejaculatory refractory period. The SAM and CM were treated as relatively independent of each other. The independence of measures of copulation considered to be a function of these two mechanisms has been confirmed by factor-analytic techniques (Dewsbury, 1979a; Sachs, 1978). More recently, forms of these models have been computer based (for example, Toates and O'Rourke, 1978).

One of the sources of controversy in the development of theoretical models of male rat copulatory behavior has been the nature of the hypothetical processes acting to trigger ejaculation (Sachs and Barfield, 1976). According to a quantal hypothesis, each intromission added one unit of excitation to a summator that eventually would trigger ejaculation. According to a temporal hypothesis, intromissions functioned only to maintain activity in the CM for a period of time, that time being critical for the triggering of ejaculation. On a nonquantal hypothesis, the increment in the CM resulting from each intromission was believed to increase with time after the intromission. According to this view, males normally copulate at rates so fast that the full incremental effects on the CM of each intromission are not achieved unless they are somehow delayed.

This model both stimulated and was affected by research on the "enforced interval effect," the first demonstration of which is attributed to an unpublished manuscript by E. W. Rasmussen. Larsson (1956) completed the first published study. The basic manipulation necessary for demonstrating the enforced interval effect simply entails separating the partners for predetermined periods of time within the ejaculatory series. In Paper 31, Larsson demonstrates that as one increases the interval between intromissions within the range tested, the number of intromissions required to attain ejaculation decreases. These results are consistent with, though they do not necessitate, the nonquantal hypothesis. Various investigators have refined the enforced interval effect (Bermant, 1964; Hård and Larsson, 1970). One reason for the effect may be that individual intromissions are longer after enforced intervals than when copulation occurs ad lib (Carlsson and Larsson, 1962). The enforced-interval effect also has been found in guinea pigs (Gerall, 1958).

Just as prolongation of interintromission intervals can produce a facilitation of copulatory behavior, so can prolongation of the length of the postejaculatory refractory periods (Larsson, 1958, 1959; Beach and Whalen, 1959; Dewsbury and Bolce, 1970).

Different species vary with respect to the number of ejaculations typically attained in a single episode of copulatory activity. Whereas male rats typically attain seven ejaculations before reaching a criterion of satiation (Paper 4), guinea pigs often satiate after just one. In many species sexual activity by a male will be reinitiated with a change of partners. In the study by Grunt and Young (Paper 31), replacement of the female with which a male guinea pig had earlier ejaculated with another estrous female resulted in a reinitiation of sexual activity in comparison with the effects of other control treatments. This demonstrates the effectiveness of a change of mating partner for reawakening sexual activity in satiated males, a phenomenon that has become known as the "Coolidge effect" (Wilson et al., 1963). A similar effect has been proposed in male rats (for example, Hsiao, 1965; Fowler and Whalen 1961; Wilson et al. 1963), although the effect appears less dramatic than suggested by the earlier results (Fisher 1962). The Coolidge effect has been studied in a variety of species (Wilson et al., 1963), may not occur in pair-bonding species (Wilson et al., 1963; Dewsbury, 1971), but has been demonstrated in sheep (Pepelko and Clegg,

251

1965). A review of studies of the Coolidge effect is provided by Dewsbury (1981).

The phenomena surrounding the Coolidge effect provide a natural transition to a consideration of social influences on sexual activity. As discussed in the introduction, it must always be remembered that in most species, sexual activity occurs within a social context. Most of the work discussed thus far was conducted with just a single male-female pair present under controlled conditions. This has been an effective strategy for the study of many of the phenomena of copulatory behavior. However, at some point it is necessary to return copulation to its social context.

An example of an effect of social facilitation on copulatory activity is provided in the observations of Bingham (Paper 12) on chimpanzees. Bingham noted, "It was in exhibitionist activities that some of the most pronounced explorations in sexual behavior appeared" (p. 136). Larsson (1956) compared the copulatory behavior of rats copulating with either three males and three females in one cage, one pair in each of three adjacent cages, or one pair copulating with no other copulating animals in the room. Animals copulating in groups attained more ejaculations per hour and ejaculated in less time than isolated animals. The presence of animals copulating in an adjacent cage produced a lesser effect. Permitting males to mate with one versus five females produced little effect on the parameters of copulatory behavior (Tiefer, 1969). Several species have been studied in the editor's laboratory under conditions in which two males have access to a single female. In deer mice, as in other species, there were but minor alterations in the parameters of copulatory behavior compared to those in the one-male–one-female situation. The socially dominant males achieved significantly more ejaculations than the subordinate males (Dewsbury, 1979b).

The results are relevant to field conditions. In wild Norway rats, it frequently has been observed that several males pursue a single female (for example, Robitaille and Bovet, 1976). Using electrophoretic techniques, it has been determined that litters of deer mice often are sired by more than one male (Birdsall and Nash, 1973). In one study of elephant seals, it was found that 4 percent of the males achieved 85 percent of the recorded ejaculations with females, indicating the importance of social dominance (LeBoeuf and Peterson, 1969).

An excellent example of the role of social factors in sexual behavior is provided by Beach and LeBoeuf (Paper 33). They found that in a series of mating tests, female dogs displayed consistent preferences for individual males. They emphasize that a state of estrus does not imply a state of automatic receptivity to all potential partners but may be greatly complicated by such individual preferences. Beach and LeBoeuf cite relevent literature in other species, especially nonhuman primates, wherein similar preferences have been manifest. The work of Herbert (1968) on rhesus monkeys provides an excellent example.

REFERENCES

Barfield, R. J., and B. D. Sachs, 1968, Sexual behavior: Stimulation by painful shock to skin in male rats, *Science* **161**:392–395.

Beach, F. A., 1942, Analysis of factors involved in the arousal, maintenance, and manifestation of sexual excitement in male animals, *Psychosomat. Med.* **4**:173–198.

Beach, F. A., 1956, Characteristics of masculine "sex drive," *Nebraska Symp. Motiv.* **4**:1–32.

Beach, F. A., and R. E. Whalen, 1959, Effects of ejaculation on sexual behavior in the male rat, *J. Comp. Physiol. Psychol.* **52**:249–254.

Beach, F. A., M. W. Conovitz, F. Steinberg, and A. C. Goldstein, 1956, Experimental inhibition and restoration of mating behavior in male rats, *J. Genet. Psychol.* **89**:165–181.

Bermant, G., 1961, Response latencies of female rats during sexual intercourse, *Science* **133**:1771–1773.

Bermant, G., 1964, Effects of single and multiple enforced intercopulatory intervals on the sexual behavior of male rats, *J. Comp. Physiol. Psychol.* **57**:398–403.

Birdsall, D. A., and D. Nash, 1973, Occurrence of successful multiple insemination of females in natural populations of deer mice (*Peromyscus maniculatus*), *Evolution* **27**:106–110.

Bolles, R. C., H. M. Rapp, and G. C. White, 1968, Failure of sexual activity to reinforce female rats, *J. Comp. Physiol. Psychol.* **65**:311–313.

Caggiula, A. R., and R. Eibergen, 1969, Copulation of virgin male rats evoked by painful and peripheral stimulation, *J. Comp. Physiol. Psychol.* **69**:414–419.

Carlsson, S. G., and K. Larsson, 1962, Intromission frequency and intromission duration in the male rat mating behavior, *Scand. J. Psychol.* **3**:189–191.

Dewsbury, D. A., 1971, Copulatory behavior of old-field mice (*Peromyscus polionotus subgriseus*), *Animal Behav.* **19**:192–204.

Dewsbury D. A., 1979a, Factor analysis of measures of copulatory behavior in three species of muroid rodents, *J. Comp. Physiol. Psychol.* **93**:868–878.

Dewsbury, D. A., 1979b, Copulatory behavior of deer mice (*Peromyscus maniculatus*). II. A study of some factors regulating the fine structure of behavior, *J. Comp. Physiol. Psychol.* **93**:161–177.

Dewsbury, D. A., and S. K. Bolce, 1970, Effects of prolonged postejaculatory intervals on copulatory behavior of rats, *J. Comp. Physiol. Psychol.* **72**:421–425.

Dewsbury, D. A., 1981, Effects of novelty on copulatory behavior: The Coolidge effect and related phenomena, *Psychol. Bull.*, **89**:464–482.

Drewett, R. F., 1973, Sexual behavior and sexual motivation in the female rat, *Nature* **242**:476–477.

Fisher, A. E., 1962, Effects of stimulus variation on sexual satiation in the male rat, *J. Comp. Physiol. Psychol.* **55**:614–620.

Fowler, H., and R. E. Whalen, 1961, Variation in incentive stimulus and sexual behavior in the male rat, *J. Comp. Physiol. Psychol.* **54**:68–71.

Gerall, A. A., 1958, Effect of interruption of copulation on male guinea pig sexual behavior, *Psychol. Rep.* **4**:215–221.

Hård, E., and K. Larsson, 1970, Effects of delaying intromissions on the male rat's mating behavior, *J. Comp. Physiol. Psychol.* **70**:413–416,

Herbert, J., 1968, Sexual preferences in the rhesus monkey *Macaca mulatta* in the laboratory, *Animal Behav.* **16**:120–128.

Hsiao, S., 1965, Effect of female variations on sexual satiation in the male rat, *J. Comp. Physiol. Psychol.* 60:467–469.

Jenkins, M., 1928, The effects of segregation on the sex behavior of the white rat as measured by the obstruction box method, *Genet. Psychol. Monog.* **3**:455–471.

Kagan, J., 1955, Differential reward value of incomplete and complete sexual behavior, *J. Comp. Physiol. Psychol.* **48**:59–64.

Larsson, K., 1956, Conditioning and sexual behavior in the male albino rat, *Acta Psychol. Gothoburg.* **1**:1–269.

Larsson, K., 1958, Aftereffects of copulatory activity of the male rat: I., *J. Comp. Physiol. Psychol.* **51**:325–327.

Larsson, K., 1959, Effects of prolonged postejaculatory intervals on the mating behavior of the male rat, *Z. Tierpsychol.* **16**:628–632.

LeBoeuf, B. J., and R. S. Peterson, 1969, Social status and mating activity in elephant seals, *Science* **163**:91–93.

Moss, F. A., 1924, A study of animal drives, *J. Exp. Psychol.* **7**:165–185.

Nissen, H. W., 1929, The effects of gonadectomy, vasectomy, and injections of placental and orchic extracts on the sex behavior of the white rat, *Genet. Psychol. Monog.* **5**:451–550.

Peirce, J. T., and R. L. Nuttall, 1961, Self-paced sexual behavior in the female rat, *J. Comp. Physiol. Psychol.* **54**:310–313.

Pepelko, W. E., and M. T. Clegg, 1965, Studies of mating behavior and some factors influencing the sexual response in the male sheep *Ovis aries*, *Animal Behav.* **13**:249–258.

Robitaille, J. A., and J. Bovet, 1976, Field observations on the social behavior of the Norway rat, *Rattus norvegicus* (Berkenhout), *Biol. Behav.* **1**:289–308.

Sachs, B. D., 1978, Conceptual and neural mechanisms of masculine copulatory behavior, in *Sex and Behavior: Status and Prospectus*, T. E. McGill, D. A. Dewsbury, and B. D. Sachs, eds., Plenum, New York, pp. 267–295.

Sachs, B. D., and R. J. Barfield, 1976, Functional analysis of masculine copulatory behavior in the rat, *Adv. Study Behav.* **7**:91–154.

Sachs, B. D., R. Macaione, and L. Fegy, 1974, Pacing of copulatory behavior in the male rat, *J. Comp. Physiol. Psychol.* **87**:326–331.

Schwartz, M., 1956, Instrumental and consummatory measures of sexual capacity in the male rat, *J. Comp. Physiol. Psychol.* **49**:328–333.

Seward, J. P., and G. H. Seward, 1940, Studies on reproductive activities of the guinea pig. IV. A comparison of sex drive in males and females, *J. Genet. Psychol.* **57**:429–440.

Sheffield, F. D., J. J. Wulff, and R. Backer, 1951, Reward value of copulation without sex drive reduction, *J. Comp. Physiol. Psychol.* **44**:3–8.

Stone, C. P., R. G. Barker, and M. I. Tomlin, 1935a, Sexual drive in potent and impotent males as measured by the Columbia obstruction apparatus, *J. Genet. Psychol.* **47**:33–48.

Stone, C. P., M. I. Tomlin, and R. G. Barker, 1935b, A comparative study of sexual drive in adult male rats as measured by direct copulatory tests and by the Columbia obstruction box. *J. Comp. Psychol.* **19**:215–241.

Tiefer, L., 1969, Copulatory behavior of male *Rattus norvegicus in a multiple-female exhaustion test*, *Animal Behav.* **17**:718–721.

Toates, F. M., and C. O'Rourke, Computer simulation of male rat sexual behavior, *Med. Biol. Eng. Comput.* **16**:98–104.

Whalen, R. E., 1961, Effects of mounting without intromission and intromission without ejaculation on sexual behavior and maze learning, *J. Comp. Physiol. Psychol.* **54**:409–415.

Wilson, J. R., R. E. Kuehn, and F. A. Beach, 1963, Modification in the sexual behavior of male rats produced by changing the stimulus female, *J. Comp. Physiol. Psychol.* **56**:636–644.

29

Reprinted from pp. 1–5, 63–68 of *Comp. Psychol. Monog.* **4**:1–68 (1927)

A STUDY OF SEX BEHAVIOR IN THE WHITE RAT BY MEANS OF THE OBSTRUCTION METHOD[1]

L. H. WARNER, Ph.D.

Psychological Laboratory, Columbia University

ACKNOWLEDGMENTS

This investigation was made in the Animal Laboratory of the Department of Psychology, Columbia University, under the direction of Professor C. J. Warden. The writer is glad to have this opportunity to express his warm appreciation to Professor Warden for his constant interest in the problem and for his encouragement and criticism throughout the progress of the experiment. Dr. G. N. Papanicolaou gave generously of his time in examining and classifying the microscopic records and in reading and criticizing the section devoted to the physiological conditions

---

[1] This report covers one major topic in a project on animal drives (under the general direction of Professor C. J. Warden), supported by the Council for Research in the Social Sciences of Columbia University.

in the female animal related to sex behavior. To him the writer expresses sincere gratitude. The writer is also indebted to Professors R. S. Woodworth and H. E. Garrett for their critical reading of the manuscript.

I. INTRODUCTION

The general purpose of this investigation was the study of the sex behavior of the normal male and female albino rat. More specifically, the attempt was made to determine the effect upon such behavior of varying the physiological conditions most intimately related to it.

The female mammal is subject to periodic changes in the reproductive tract which are, apparently, paralleled by definite changes in the behavior of the animal toward a sex object. In the female white rat ovulation is supposed to occur at rather regular intervals of from four to six days. Observation of behavior has shown that the animal "comes into heat," i.e., it will accept copulation, during only a few hours at a given time, and that these periods of heat, or oestrum, also recur rather regularly at intervals of from four to six days. At other times the female will avoid mating by fleeing or fighting.

Although the relationship between the changes in the reproductive tract and the animal's behavior has been noted it has not been worked out in great detail. It is not known, for example, whether the onset and termination of active sex behavior is abrupt or gradual. One reason for the lack of exact knowledge has been the lack of a method of observing any sex behavior other than that at the two extremes: Active mating and active avoidance of mating. A second reason has been the lack of an accurate method for the determination of the oestrous condition of the animal.

No rhythm analogous to the oestrous cycle (the periodic changes in the reproductive tract of the female) has been observed in the male rat. In the case of the male the effect upon behavior of variations in the physiological condition due to various periods of segregation involving sex deprivation has been studied. The data obtained show the speed of recovery of the copulative ability of the male after a period during which it has mated freely and may be said to be "satiated" sexually. So far as the writer knows no data bearing directly upon this aspect of male sex behavior have previously been reported.

The experimental study of the dynamic aspect of animal behavior has been neglected very largely in the past in favor of the study of learning and of sensory discrimination. "Motivation" has usually been studied merely as a phenomenon involved, more or less, in all learning problems. Likewise the theoretical consideration of the dynamic aspect of behavior has failed to interest psychologists as much as has the consideration of such problems as the elimination of random movements, the response to relationship between stimuli rather than to absolute, isolated stimuli, and other problems bearing upon learning and discrimination. Except for the fact that there is a tendency to discredit the use of the term "instinct" as descriptive of the source of energy in animal behavior, very little has been said to clarify matters. Recently a new word, "drive," has appeared in the literature, coined, no doubt, in the hope of avoiding the implications which cling to the terms previously used. It would seem that in human psychology the terms "motive," "interest" and the like may be considered satisfactory but that in the field of animal psychology the term drive is to be preferred. This term has one advantage, certainly. It has not yet been defined. Nor will the present writer attempt a final definition. The justification for the use of this term is based largely upon convenience, and the lack of a better one.

The term, drive, as it is used here, implies an analysis of the animal's behavior from two standpoints: first, with regard to the animal's physiological condition; second with regard to the type of external stimuli capable of dominating the behavior of the organism and of altering this physiological condition ("satisfying" the drive).

For example, we would say that the hunger drive is in operation in the case of an animal which is physiologically in that condition which follows deprivation of food for a given time and whose behavior is oriented toward that class of stimuli, olfactory, visual, etc., which have been in the past associated with the food consuming activity of the animal. Likewise we would say that the sex drive was operative in a female whose physiological condition is that of oestrum and which is responding positively to stimuli customarily associated with the male of the same species.

Of these two factors implied by the term, drive, the first is essential, the second non-essential. Drive, as we use the term, is related most closely to the animal's physiological condition, to its internal stimulation. Thus the type of behavior which we classify as that related to the hunger drive may be observed even in the absence of food or of external stimuli usually associated with food. In the course of the experiments here reported it was found that a female rat in oestrum shows more drive in the absence of the male than does a female in dioestrum (the period of sexual inactivity) even though the male be present. The origin of the drive is then largely internal, arousing activity in the animal, but also sensitizing the animal especially toward a certain type of external stimulation—the type in each case depending upon the nature of the internal condition. A female in oestrum, and recently fed, if offered the choice between food and a male will probably choose the latter. Thus in the absence of definite knowledge of the physiological conditions of the animal something regarding its nature may be deduced from observation of the type of stimuli toward which the animal is especially sensitive.

The purpose of this discussion is, mainly, to emphasize the fact that the term, drive, as we use it, is in no sense explanatory but is merely descriptive of the observed facts.

The scattered experiments bearing upon animal drives have made use of four methods. The first studies the effect of various incentives, positive and negative, upon the rate of learning. Simmons has reported such an experiment (26). The second correlates the physiological condition of the animal with the amount of "spontaneous" activity. Wang, for example, has found a definite relation between the oestrous cycle in the white rat and the amount of general activity by means of this method

(34). The third method has been called the choice method since it makes note of the behavior of the animal when faced with the stimuli related to two different drives. Tsai has compared the hunger and the sex drives by this method (33). The fourth method has been designated the obstruction method since it involves observation of the behavior of the animal when the attainment of the stimuli related to the drive is hindered by some obstruction or obstacle. The literature has been covered recently in connection with a detailed description of the apparatus used in the present experiment (11).

Moss is the first and only previous worker to have studied sex behavior by means of the obstruction method. His work (19) deserves the credit due all pioneer investigations but possesses, naturally, some of the defects of such work. Its chief value consists in the first application of the principle of measuring the strength of the tendency of an animal to approach a certain object by introducing an obstruction between the animal and the object. An obstruction might conceivably be any stimulation to which the animal normally reacts negatively.

The quantitative data Moss reports are of small value. His application of the method leaves a number of technical difficulties unsolved. So far as these relate to his apparatus they have been considered in the article cited above which includes also a description of our own efforts to overcome them. His procedure is more difficult to criticize since it is not described in great detail. Two important points, however, should be noted. We are given no further information regarding the condition of the females when tested (or when serving as stimulus animals during the testing of the males) than that they were in heat. On what basis this determination was made, we do not know. Whether judged from a physiological or from a behavior point of view the state known as "heat" is anything but a fixed and definite condition. It is better described as a continuous process presenting a different picture at different stages.

[*Editor's Note*: Material has been omitted at this point.]

V. SUMMARY OF CONCLUSIONS

A. Male

1. The tendency of a male rat to approach a female rat in oestrum may be measured in terms of the number of times it will cross an electrical obstruction to reach the female within a given period of time. That the behavior of such males was dominated by the presence of the female in the present experiment was indicated by the results of a control group tested under conditions exactly like those of one of the test groups except that there was no female present in the incentive compartment. The males in this control group crossed the obstruction to the incentive compartment decidedly and reliably less often than did the animals in the comparable test group.

2. The tendency of a male rat to cross an electrical obstruction to a female rat in oestrum is at its low point immediately after a period (two hours in this case) during which the male has had access to and mated frequently with a female in oestrum. Recovery of the tendency from this low point is rapid during the first six hours after such a period of mating, almost as rapid during the second six hours after which it has reached a point only slightly below the maximum manifestation of this tendency which is found in an animal twenty-four hours after a period of mating.

3. Intervals of sex deprivation longer than one day do not increase the tendency of the male to cross. The data at hand suggest that this tendency decreases slightly from this point on to a deprivation interval of twenty-eight days, the longest interval studied. This decrease is not statistically reliable, however, and the only conclusion upon this point is that there is no increase in the tendency after the first day.

B. Female

1. The tendency of a female rat to approach a male rat may be measured in terms of the number of times it will cross an electrical obstruction to reach the male within a given period of time. That the behavior of the females in oestrum was dominated by the presence of the male in the present experiment was indicated by the results of a control group run under conditions exactly like those of one of the test groups except that there was no male present in the incentive compartment. The females in this control group crossed the obstruction to the incentive compartment decidedly and reliably less often than did the animals in the comparable test group.

2. There is a very definite relationship between the histological character of the vaginal secretion of a female rat and its tendency, at the moment, to cross an electrical obstruction to reach a male, i.e., between the oestrous rhythm and behavior toward a sex object. The group of animals, which judging by the vaginal smear, were tested during the early part of the stage during which cornified cells only are found in the secretion displayed a decidedly and reliably greater amount of crossing and other activity oriented toward the male than did the group whose vaginal secretion was characterized by the presence of epithelial cells and leucocytes. In other words, activity directed toward a male is confined rather rigidly to a single period during the oestrous cycle, the period usually known as oestrum.

3. From the standpoint of behavior the onset of oestrum is sudden while its cessation is relatively gradual. The behavior of the group representing the transition into oestrum indicates that the group is not homogeneous but contains certain animals

whose behavior resembles that commonly seen in dioestrum, and others whose behavior resembles that commonly seen in oestrum. When oestrum sets in it is apparently with practically full force. The groups representing transition out of oestrum show, in general behavior midway between that seen in oestrum and that seen in dioestrum. The transition out of oestrum appears to be gradual.

C. Comparison of the male and the female

1. Since the strength of the tendency of the rat to cross an electric obstruction to reach a sex object is related to quite different physiological factors in the two sexes, the only fair comparison of the sexes from this standpoint is found in the comparison of those groups in which the physiological condition is such that the tendency was at its maximum for each sex. For the male this group is that tested after one day of sex deprivation. For the female it is the group of animals tested when in the early cornified stage. The behavior record shows that of these two groups the female group was the more active although the difference is not statistically reliable. When it is considered that the incentive stimulus in the case of the male (a female in oestrum) was in most cases extremely active, displaying that type of activity commonly preceding mating whereas the stimulus animal for the female (a male) was decidedly less active and less frequently made such advances, the assumption seems justified that were it possible to equate the activity of the incentive stimuli the female group would be found to show a reliably greater amount of activity directed toward a sex object.

2. The behavior data indicate that sex activity is initiated rather more by the external stimulus situation in the male than in the female and rather more by internal stimulation in the female than in the male. The female rat in oestrum displays more activity in the form of crossing the electrical obstruction to the incentive compartment *even though that compartment be empty* than does the female in dioestrum *even though the compartment contains a male*. A male rat which has not mated for twenty-four hours (the interval which, of those studied, found sex activity at its maximum) will cross electrical obstruction less often

to an empty incentive compartment than will a male of any of the other sex deprivation intervals studied (0, 6, 12 hours, 4, 7, 28 days) to an incentive compartment containing a female in oestrum.

REFERENCES

(1) ALLEN, E.: a. The oestrous cycle in the mouse. Amer. Jour. Anat., 1922, xxx, 297–348.

ALLEN, E.: b. Racial and familial cyclic inheritance and other evidence from the mouse concerning the cause of oestrous phenomena. Amer. Jour. Anat., 1923, xxxii, 293–304.

(2) BISCHOFF, TH. L. W.: Entwicklungsgeschichte des Meerschweinchens. Giessen, 1852.

(3) BLAIR, E. W.: Contraction rate of the uterine musculature of the rat with reference to the oestrous cycle. Proc. Amer. Assoc. Anat., Anat. Rec. 1922, xxiii, 9–10.

(4) CORNER, G. W.: a. A review of some recent work on the mammalian reproductive cycle. Jour. Mammal., 1921, ii, 227–231.

CORNER, G. W.: b. Cyclic variation in uterine and tubal contraction waves. Amer. Jour. Anat., 1921, xxxii, 345–351.

(5) GREENMAN, M. J., AND DÜHRING, F. L.: Breeding and care of the albino rat for research purposes. The Wistar Institute of Anatomy and Biology, Philadelphia: 1923.

(6) HARTMAN, C.: The oestrous cycle in the opossum. Amer. Jour. Anat., 1923, xxxii, 353–395.

(7) HEAPE, W.: The sexual season of mammals and the relation of the "Prooestrum" to menstruation. Quar. Jour. Micr. Sci., 1900, xliv, 1–70.

(8) HENSEN, V.: Beobachtungen uber die Befruchtung und Entwickelungen des Kaninchens und Meerscheinchens. Zeit. f. Anat. u. Entwick., 1876, i, 213–272; 353–423.

(9) HOLDEN, F.: A study of the effect of starvation upon behavior by means of the obstruction method. Comp. Psych. Mon., 1926, iii, no. 17, 1–45.

(10) ISHII, O.: a. Observations on the sexual cycle of the guinea-pig. Biol. Bull., 1920, xxxviii, 237.

ISHII, O.: b. Observations on the sexual cycle in the white rat. Anat. Rec., 1922, xxviii, 311.

(11) JENKINS, T. N., WARNER, L. H., AND WARDEN, C. J.: Standard apparatus for the study of animal motivation. Jour. Comp. Psych., 1926, vi, 361.

(12) KEYE, J. D.: Periodic variation in spontaneous contraction of uterine muscle, in relation to the oestrous cycle and early pregnancy. Bull. Johns Hopkins Hosp., 1923, xxxiv.

(13) KÖNIGSTEIN, H.: Die Veränderungen der Genital Schleimhaut während der Gravidität und Brunst bei einigen Nagern. Arch. f. Physiol., 1907, cxix, 553–570.

(14) LATASTE, F.: a. Transformation périodique de l'épithélium du vagin des rongeurs. Mem. de la Soc. de Biol., 1892, xliv, 765–769.

LATASTE, F.: b. Rhythme vaginal des Mammifères. Ibid., 1893, xlv, 135–146.

(15) Loeb, L.: a. The cyclic changes in the ovary of the guinea-pig. Jour. Morph., 1911, xxii, 37–70.

Loeb, L.: b. The correlation between the cyclic changes in the uterus and the ovaries of the guinea-pig. Biol. Bull., 1914, xxvii, 1–44.

Loeb, L.: c. The mechanism of the sexual cycle with special reference to the corpus luteum. Amer. Jour. Anat., 1923, xxxii, 305–343.

(16) Long, J. A., and Evans, H. M.: The oestrous cycle in the rat and its related phenomena. Memoirs of the University of California, 1922, vi, 1–148.

(17) Marshall, F. H. A.: Physiology of reproduction. London: Longmans, Green & Co., 1922.

(18) Morau, H.: Des transformations épithéliales de la muqueuse du vagin de quelques rongeurs. J. de l'Anat. et de la Physiol., 1889, xxv, 275–297.

(19) Moss, F. A.: A study of animal drives. Jour. Ex. Psych., 1924, vii, 165.

(20) Papanicolaou, G. N.: Oestrous in mammals from the comparative point of view. Amer. Jour. Anat., 1923, xxxii, 284–292.

(21) Rein, G.: Beiträge zur Kenntniss der Reifungserscheinungen und Befruchtungsvorgänge am Säugethierei. Arch. f. Mikr. Anat., 1883, xxii, 233–270.

(22) Retterer, E.: a. Sur la morphologie et l'évolution de épithélium au vagin mammifères. Mém. Compt. rend. Soc. de biol., 1892, xliv, 101–107.

Retterer, E.: b. Évolution de l'épithélium du vagin. Ibid., 1892, 566–568.

Retterer, E.: c. Sur les modifications de la muqueuse utérine l'époque du rut. Ibid., 637–642.

(23) Rubaschkin, W.: Über die Reigungs- und Befruchtungs-prozesse des Meerschweincheneies. Anat. Hefte, Wiesbaden, 1905, xxix, 507–553.

(24) Seckinger, D. D.: Spontaneous contractions of the fallopian tube of the domestic pig with reference to the oestrous cycle. Bull. Johns Hopkins Hosp., 1923, xxxiv.

(25) Selle, P. M.: Changes in the vaginal epithelium of the guinea-pig during the oestrous cycle. Amer. Jour. Anat., 1922, xxx, 429–449.

(26) Simmons, R.: The relative effectiveness of certain incentives in animal learning. Comp. Psych. Mon., 1924, ii, no. 7.

(27) Slonaker, J. R.: a. The effect of pubescence, oestruation, and menopause on the voluntary activity in the albino rat. Amer. Jour. Physiol., 1924, lxviii.

Slonaker, J. R.: b. Analysis of the daily activity of the albino rat. Ibid., 1925, lxxiii.

Slonaker, J. R.: c. Long fluctuations in voluntary activity in the albino rat. Ibid., 1926, lxxvii.

(28) Smith, H. P.: The ovarian cycle in mice. Jour. Roy. Mic. Soc., 1917, p. 252.

(29) Sobotta, J.: Die Befruchtung und Furchung des Eies der Maus. Arch. f. Mikr. Anat., 1895, xlv, 15–93.

(30) Stockard, C. R.: The general morphological and physiological importance of the oestrous problem. Amer. Jour. Anat., 1923, xxxii, 227–283.

(31) Stockard, C. R., and Papanicolaou, G. N.: a. The existence of a typical oestrous cycle in the guinea-pig, with a study of its histological and physiological changes. Amer. Jour. Anat., 1917, xxii, 225–283.

Stockard, C. R., and Papanicolaou, G. N.: b. The vaginal closure membrane, copulation, and the vaginal plug in the guinea-pig, with further consideration of the oestrous rhythm. Biol. Bull., 1919, xxxvii, 222–243.

(32) Stone, C. P.: a. The congenital sexual behavior of the young male albino rat. Jour. Comp. Psych., 1922, ii.

Stone, C. P.: b. Further study of sensory function in activation of sexual behavior in the male rat. Jour. Comp. Psych., 1923, iii.

Stone, C. P.: c. The awakening of copulation ability in the male albino rat. Amer. Jour. Physiol., 1924, lxviii.

Stone, C. P.: d. Delay in the awakening of copulatory ability in the male albino rat incurred by defective diets. Jour. Comp. Psych., 1924, iv and v.

Stone, C. P.: e. The initial copulatory response of female rats reared in isolation from the age of twenty days to the age of puberty. Jour. Comp. Psych., 1926, vi, 73–83.

(33) Tsai, C.: The relative strength of sex and hunger motives in the albino rat. Jour. Comp. Psych., 1925, v, 407.

(34) Wang, G. H.: Spontaneous activity and the oestrous cycle. Comp. Psych. Monog., 1923, ii, no. 6.

30

Reprinted by permission from *J. Comp. Physiol. Psychol.* **52**:245–248 (1959)

EFFECTS OF "SITUATIONAL ANXIETY" ON SEXUAL BEHAVIOR IN MALE RATS[1]

FRANK A. BEACH[2] AND HARRY FOWLER

Yale University

It has been reported that the sexual responsiveness of male mammals may be depressed in an environmental setting previously associated with conflict or pain. "Neurotic" male dogs are slow to mate with estrous females presented in the experimental room where the "neurosis" was induced, although in the paddock copulation may occur promptly (Grantt, 1944). Male rats subject to electric shock each time they attempt to mate eventually cease to display copulatory reactions in the experimental cage; and in some individuals the loss or inhibition of sexual responsiveness persists for many weeks without any additional punishment (Beach, Conovitz, Steinberg, & Goldstein, 1956).

The present experiment was conducted to study the sexual behavior of male rats tested in an environment in which they had previously been subjected to a series of painful electric shocks. It was assumed that males which had recently been punished in a particular environment would experience fear or "situational anxiety" when they were returned to that same setting.

METHOD

Subjects

Thirty male rats from a mixed, hooded and albino strain were chosen as Ss on the basis of their mating performance in preliminary tests. Each animal was tested with receptive females from one to three times. As soon as a male copulated and reached one ejaculation, he was put aside and held for the experiment. This was the full extent of the Ss' pre-experimental, heterosexual experience.

Males were approximately 120 days old at the beginning of experimentation. Throughout the investigation they lived in individual cages with drinking water and Purina chow constantly available. The living cages were in the experimental room where the day-night light cycle was artificially reversed.

Apparatus

The apparatus consisted of two identical Conditioning Cages and several identical Neutral Cages. The Conditioning Cages were black boxes 15 in. long, 12 in. wide, and 12 in. high. The tops were of glass, and the floors were metal grids. The floors of the Conditioning Cages were wired in parallel with a matched impedance a.c. shock source of 400 v. in series with a $\frac{1}{2}$-meg. resistance.

The Neutral Cages were semicircular with white, metal sides, a glass front, and a solid floor covered with sawdust. They were 20 in. wide at the front, 12 in. deep, and 10 in. high.

Experimental Treatment

The 30 Ss were randomly distributed into three groups of 10 each. Phase I of the experiment lasted for ten days. On Days 1 to 5 members of Groups A and B were placed twice a day in the Conditioning Cage and given electric shocks. Every individual in these groups spent two 5-min. periods in the Conditioning Cage during each of which he received 100 randomly dispersed shocks of $\frac{1}{2}$-sec. duration. The daily periods were separated by at least 6 hr. On the first five days of Phase I the Ss in Group C were given the same exposure to the Conditioning Cage, but no electric shock was administered.

Sex tests were conducted on Days 6 to 10 of Phase I. During this time no shock was employed. Members of Groups A and C were tested in the Conditioning Cage. Subjects in Group B were tested in the Neutral Cage.

One day after the fifth sex test, Phase II began. On Days 11 to 15 of the experiment every S spent two 5-min. periods in the Conditioning Cage. The procedure was identical to that followed in Phase I with one exception. During Phase II the members of Group C, as well as those of Groups A and B, received electric shocks. On Days 16 to 20 sex tests were administered, Groups A and C being tested in the Conditioning Cage, and Group B in the Neutral Cage.

Administration of Sex Tests

All sex tests were conducted during the dark phase of the light-dark cycle. Stimulus animals were female rats which had been brought into heat by hormone treatment. Two days before testing 0.21 mg. of estradiol benzoate was injected subcutaneously. Six hours before a test 1.0 mg. of progesterone was given. Only fully receptive females were used in the tests.

Subjects were placed in the Conditioning Cage or the Neutral Cage 1 min. before the introduction of the female. If a male did not achieve intromission within 15 min. after presentation of the female, he was removed. If intromission did occur within this time limit,

[1] This research was supported in part by Research Grant M-943 from the USPHS.

The hormone preparations were generously supplied by Edward J. Henderson of Schering Corporation, Bloomfield, N. J.

[2] Now at the University of California, Berkeley.

TABLE 1

Comparative Performance of the Experimental Groups

Phase of the Experiment	Tests	Groups	N	M Mount Latency	M Intro-mission Latency	M Intro-mission Frequency	M Mount Frequency	Mdn Ejaculation Latency	Mdn Intercop-ulatory Interval
I	1 & 2	A	7	164.28	197.68	8.43	7.0	380.0	45.5
		B	8	215.93	228.12	11.87	9.2	801.2	65.8
		C	7	256.42	273.92	13.44	9.0	695.0	43.4
	4 & 5	A	7	78.57	109.64	11.67	15.5	480.0	36.8
		B	9	46.67	79.72	13.00	6.0	387.5	33.5
		C	7	86.42	113.57	11.83	6.5	410.0	37.3
II	1 & 2	A	8	162.50	194.68	9.56	7.2	426.2	43.0
		B	9	55.00	89.44	11.89	6.5	405.0	48.0
		C	9	286.11	343.61	10.17	8.5	535.0	48.6
	1 to 5	A	9	111.03	147.38	10.38	9.0	463.3	49.4
		B	10	114.84	170.82	12.20	9.7	500.0	43.1
		C	9	216.44	261.25	11.41	8.4	585.0	55.4

the male was left with the female until ejaculation occurred or until 20 min. had elapsed from the time of the first intromission. Tests in which males failed to achieve intromission, or achieved intromission but did not ejaculate within 20 min., were not included in the analysis of data.

During the test the observer recorded several measures of sexual behavior on a constant-speed kymograph. At the end of the test the kymographic record was analyzed to determine the frequency or duration of the various measures. The following measures were employed. *Intromission frequency:* number of mounts with intromission. *Mount frequency:* number of mounts without intromission. *Mount latency:* time from the introduction of the female until the occurrence of the first mount. *Intromission latency:* time from the introduction of the female until the male first achieved intromission. *Ejaculation latency:* time from the first intromission until the occurrence of ejaculation. *Intercopulatory interval:* time from the first intromission to ejaculation divided by the number of intromissions preceding ejaculation.

RESULTS

Behavior in Phase I

The principal results are summarized in Table 1. Groups A and C were both tested in the Conditioning Cage, but males in Group A had been shocked in that environment, whereas Group C males had not. There were no statistically significant differences between the two groups with respect to the length of delay before mating was initiated (mount and intromission latencies). Once mating had begun,

the average length of intervals between intromissions was not different (intercopulatory interval).

Males in Group A ejaculated after fewer intromissions than did members of Group C. This difference was pronounced in the first two tests ($p < .001$),[3] but disappeared in Tests 4 and 5. However, for all five tests taken together the difference remained significant ($p < .01$). Mount frequency was also lower for Group A during Tests 1 and 2 ($p = .01$), but no significant difference was found in Tests 4 and 5. Ejaculation latency was significantly shorter for males in Group A, but this merely reflected the fact that fewer intromissions were required to reach ejaculation.

Members of Group B were shocked in the Conditioning Cage and tested in the Neutral Cage. Their scores did not differ significantly from those of Group C on any of the measures. The differences between Groups B and A males in intromissions, mounts, and ejaculation latencies were not significant ($.05 < p < .10$ in all three instances). Differences between Groups A and B for the remaining measures were also not statistically significant ($p < .10$).

[3] The Mann-Whitney U Test and the Wilcoxon matched-pairs signed-rank tests were employed to ascertain the significance of a difference between independent and related samples, respectively.

Behavior in Phase II

The mating tests in Phase II were preceded by five days of shock for all animals. This might be thought of as "reconditioning" for Groups A and B, but Group C had not been shocked previously. To evaluate the effects of five days of shock in the Conditioning Cage upon the behavior of Group C, it is appropriate to compare average scores of this group for Tests 4 and 5 of Phase I with those of Tests 1 and 2 of Phase II. Such a comparison reveals that the frequency of intromissions preceding ejaculation was reduced after the shock ($p = .025$). If all five tests in Phase II are used in such a comparison, the difference remains ($p = .05$). Both the intromission and mount latencies were increased in the first two tests of Phase II ($p = .025$ for both comparisons).

Although the preceding shock reduced the number of intromissions to produce ejaculation in Group C males, the effect was less pronounced than it had been for males in Group A during Phase I. Group A males had had very limited sexual experience before being exposed to shock in the Conditioning Cage. In contrast, Ss in Group C had had many opportunities to copulate in the Conditioning Cage before being shocked there. This may explain the fact that Group C males showed more intromissions in Phase II than did Group A males in Phase I. The difference over the first four tests (Phase I for Group A and Phase II for Group C) was significant at the .05 level of confidence.

To determine whether the effects of punishment in the Conditioning Cage were re-established in Group A, it is necessary only to compare the scores of this group in the last two tests of Series I (when the effects of the first conditioning had largely worn off) with those in the first two tests of Series II. These comparisons revealed a decrease in the number of intromissions ($p = .025$) and mounts ($p = .05$), and an increase in intromission latency ($p = .05$). The reduction in intromissions was not as pronounced as it had been in the earlier mating tests. The frequency of intromissions preceding ejaculation in the first two tests of Phase I was significantly lower than the comparable score in Tests 1 and 2 of Phase II ($p = .05$). This may be related to the fact that conditioning in Phase II was preceded by

five mating tests in which the Ss gained sexual experience in the Conditioning Cage.

Comparisons between the sexual behavior of Group B (shocked in the Conditioning Cage and tested in the Neutral Cage) in Phase I and Phase II revealed no statistically significant differences.

DISCUSSION

The results of this experiment suggest that "situational anxiety" tends to reduce the amount or duration of sexual stimulation necessary to produce ejaculation in male rats. In this connection it is of interest to note that under some circumstances, anxiety is associated with premature ejaculation in human males. *Ejaculatio praecox* is one form of impotence associated with certain types of neurosis (Abraham, 1927).

In an earlier study from our laboratory (Beach et al., 1956), male rats were given an electric shock each time they mounted the receptive female. When the shock was very intense males eventually ceased to make any copulatory attempts. With lower levels of shock, mating was not inhibited, but the number of intromissions preceding ejaculation was reduced. It seems probable that this effect was the same as that operating in the present study.

Why should fear or anxiety produce premature ejaculation? The answer is not at all clear, but certain observations on normal male rats suggest one possible interpretation. When a fully rested male copulates with a receptive female, the first ejaculation occurs after an average of 10 to 12 intromissions. After an interval of approximately 5 min. mating is resumed, and the second ejaculation is usually preceded by only six or seven intromissions (Beach & Jordan, 1956; Larsson, 1956).

This "facilitative" effect of the initial orgasm is maximal for about 15 min., and wears off gradually over a period of approximately 1 hr. (Beach & Whalen, 1959). It has been suggested (Beach, 1956) that the event of ejaculation is triggered by neural discharge from a hypothetical Ejaculatory Mechanism, and that the occurrence of the first orgasm sensitizes or excites this mechanism, lowering its threshold for a limited period of time. If copulation is resumed while the state of sensitization persists, a second ejaculation will necessitate less genital stimulation than did the first.

Since ejaculation involves the autonomic system, and since parts of this system are believed to be hyperactive in conditions of anxiety, it may be that anxiety potentiates or renders hyperexcitable the mechanisms involved in sexual orgasm.

SUMMARY

Three groups of male rats were placed in a Conditioning Cage twice a day for five days. Members of Groups A and B were given electric shock in the Conditioning Cage, whereas Group C was not punished. Daily sex tests without shock were given for the next five days. Males in Groups A and C were tested with receptive females in the Conditioning Cage, and Group B was tested in a dissimilar Neutral Cage.

The mating performance of Groups B and C was essentially similar. That of Group A was unique in that these males tended to ejaculate with less than the normal amount of genital stimulation. This effect was pronounced in the first two tests, which occurred within two days after the last exposure to shock. It was not evident in the fourth and fifth sex tests, after more time had elapsed and after the males had mated several times in the experimental environment.

Following the fifth sex test all Ss were placed in the Conditioning Cage twice daily for five days. Upon these occasions all males including the previously unpunished Group C received electric shock. The final day of conditioning was followed by five daily mating tests, Groups A and C being tested in the Conditioning Cage, and Group B in the Neutral Cage.

The mating behavior of Group B remained unchanged. Group A and C males showed a significant reduction in the number of intromissions preceding ejaculation during the first two sex tests after punishment. The data suggested that sexual experience in the Conditioning Cage tended to reduce the effects of punishment upon subsequent sexual performance in that environment.

REFERENCES

ABRAHAM, K. *Ejaculatio praecox. Selected papers.* London: Institute Psycho-analysis and Hogarth Press, 1927.

BEACH, F. A. Characteristics of masculine "sex drive." In M. R. Jones (Ed.), *Nebraska symposium on motivation, 1956.* Lincoln, Neb.: Univer. Nebraska Press, 1956. Pp. 1–31.

BEACH, F. A., CONOVITZ, M. W., STEINBERG, F., & GOLDSTEIN, A. C. Experimental inhibition and restoration of mating behavior in male rats. *J. genet. Psychol.,* 1956, **89,** 165–181.

BEACH, F. A., & JORDAN, L. Sexual exhaustion and recovery in the male rat. *Quart. J. exp. Psychol.,* 1956, **8,** 121–133.

BEACH, F. A., & WHALEN, R. E. Effects of ejaculation on sexual behavior in the male rat. *J. comp. physiol. Psychol.,* 1959, **52,** 249–254.

GANTT, W. H. *Experimental basis for neurotic behavior.* New York: Hoeber, 1944.

LARSSON, K. Conditioning and sexual behavior in the male albino rat. Stockholm: Almqvist and Wiksell, 1956.

Received February 3, 1958.

31

Reprinted from *Animal Behav.* 7:23–25 (1959)

THE EFFECT OF RESTRAINT UPON COPULATORY BEHAVIOUR IN THE RAT

By KNUT LARSSON

Department of Psychology, University of Göteborg

Ejaculation in the rat is preceded by a series of intromissions each separated from its predecessor by a short intercopulatory interval (ICI). When these intervals are artificially prolonged the number of intromissions required before ejaculation is effectively decreased (Gerall, 1958; Larsson, 1956; Rasmussen—unpublished), an extension to 1 minute duration reducing the frequency of intromission by half its normal value. Enforced intervals of up to 5 minutes give similar results, but short restraints lasting 15 seconds are ineffective. These findings show that the time factor plays an important part in regulating the processes of excitation which underly this aspect of mating behaviour, but so far the precise relationship between the duration of the ICI and number of intromissions necessary to achieve ejaculation has not been determined. The present experiment describes the effect of varying this interval from 0·1 minute to 2 minutes.

Materials and Methods

Animals

Observations were made on the mating behaviour of eleven male rats approximately 12 months old whose mean weight was 372g. (range 346-404g.). All had previously taken part in similar experiments. The females used for mating were treated with 10μg. oestradiol benzoate 36 hours before being presented to the male.

Conduct of Tests

The animals were observed in open cages placed on the laboratory floor. After the male had been given time to adapt to the mating cage a female was introduced and was gently withdrawn as soon as intromission had occurred. The female was then reintroduced after a specified interval of time and again withdrawn immediately after the second intromission. This procedure was repeated until ejaculation had taken place, after which the male was rested for 4-5 days before further testing.

Measurement of Intercopulatory Intervals

A stopwatch graduated in 1/100 sec. was used to time (1) the interval between the presentation of the female to the male and the entry of the penis into the vagina (copulatory latency), and (2) the enforced intercopulatory interval, measured from the moment of intromission to the next presentation of the female. Intercopulatory intervals of 0·0 (*ad lib.*), 0·1, 0·2, 0·4, 0·6, 0·8, 1·0 and 2·0 minutes were enforced in this way, the performance of each rat being observed in relation to each ICI in accordance with a prearranged schedule.

Results

Normal Duration of the Intercopulatory Interval

The duration of the ICI in the unrestrained rat is illustrated in Fig. 1, which shows that 19 per cent. of all intervals were shorter than 0·1 minutes, 61 per cent. shorter than 0·2 minutes, and 85 per cent. shorter than 0·4 minutes. Only one interval was longer than 0·6 minutes.

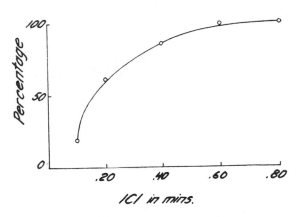

Fig. 1. The cumulative distribution curve showing length of the ICI in animals copulating *ad lib*. The abscissa gives the length of the ICI in hundredths of a minute, the ordinate the percentage of intervals not exceeding each duration.

Effect of Enforced ICI on Frequency of Intromission

The relationship between the number of intromissions preceding ejaculation and the duration of an enforced ICI is given in Fig. 2. When the animals were allowed to copulate *ad lib.*, the mean number of intromissions was 7·8. Prolongation of the ICI from 0·2 to 0·4 minutes resulted in a decrease in the number of intromissions from 8·9 to 5·2, a still greater reduction being recorded when the intervals were extended

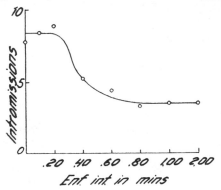

Fig. 2. The effect of restraint on the number of intromissions preceding ejaculation. The abscissa gives the length of the enforced ICI, and the ordinate the mean number of intromissions preceding ejaculation.

to 0·6 and 0·8 minutes. Thereafter, no further decrease took place even when the interval was prolonged to 2 minutes. Analysis of variance showed the overall decrease in the number of intromissions to be statistically significant ($P > 0.001$) while the individual decrease resulting from an enforced ICI of 0·4 minutes was significant at the 5 per cent. level (single-tail t-test).

Effect of Copulatory Latency

The length of the interval between intromissions is slightly longer than that recorded as ICI owing to the latency between the presentation of the female to the male and the occurrence of intromission. The mean copulatory latency for the first intromission of animals allowed to copulate *ad lib.* was 0·1 minute, and for rats in which an ICI of 0·1 and 0·2 minutes was enforced the latencies were 0·07 and 0·06 minutes respectively. When the ICI exceeded 0·2 minutes the male became highly excited, hurrying to the female and copulating with the shortest possible latency. Premature ejaculation was sometimes seen. On two occasions when the

female was dropped into the mating cage the male took a few steps forward, stopped suddenly and ejaculated in the air 20-30 cm. away from the female.

Discussion

These results emphasise the importance of the variable of time for the elicitation of mating behaviour. Successive intromissions may be considered as building up an excitatory state which culminates in the ejaculatory reflex. Each intromission makes its individual contribution to this rising excitation, the extent of which, however, depends on the time relationships of the sequential sensory inputs. When a very short interval separates two copulations the excitatory effect of the first intromission has not reached its maximal height before that of the second supervenes. The full effect of each intromission does not appear to be realised for about 0·5 minutes and since, in the unrestrained rat, the intromissions succeed each other at very short intervals, the maximal effects of each are never attained. When, however, intercopulatory intervals are artificially prolonged and the full effect of each intromission can be utilised and the number of copulations necessary to cause ejaculation radically lowered, even to the extent that the reflex can appear in the absence of sensory stimulation from the glans penis. The optimum duration of the enforced ICI would seem to lie between 0·6 and 0·8 minutes, and the failure of longer intervals to cause any further decrease in the number of intromissions suggests that after 1·0 second the excitatory effect of each intromission begins to wane.

Summary

Intervals, ranging from 0·1 to 2·0 minutes, have been enforced between the several intromissions of penis into vagina which comprise the copulatory behaviour of the rat.

The number of intromissions required for ejaculation is reduced as a result of increasing this intercopulatory interval. This reduction becomes significant when an interval of 0·4 minutes is enforced and is even more marked after an interval of 0·6 minutes.

In such circumstances the animals become excited, show a reduced copulatory latency, and may exhibit premature ejaculation.

Acknowledgments

This study has been supported by the Magnus Bergwall Foundation. The hormone products used in this experiment were generously supplied by harmacia Corporation, Uppsala, Sweden.

REFERENCES

Gerall, A. A. (1958). Effect of interruption of copulation on male guinea pig sexual behaviour. *Psychol. Rep.*, **4**, 215-221.

Larsson, K. (1956). *Conditioning and sexual behaviour in the male albino rat.* Stockholm: Almqvist & Wiksell. p. 269.

Rasmussen, E. W. The effect of an enforced pause between each coitus on the number of copulatory necessary to achieve ejaculation in the albino rat. *Unpublished manuscript.*

Accepted for publication 23rd *September,* 1958.

32

Reprinted by permission from J. Comp. Physiol. Psychol. 45:508–510 (1952)

PSYCHOLOGICAL MODIFICATION OF FATIGUE FOLLOWING ORGASM (EJACULATION) IN THE MALE GUINEA PIG[1]

JEROME A. GRUNT AND WILLIAM C. YOUNG

Department of Anatomy, University of Kansas School of Medicine

The behavior of the male guinea pig at the time of copulation differs from that of many mammals in that when a male is placed with a receptive female, usually only a single ejaculation occurs (2). During the hour following copulation with ejaculation, and probably for longer, little interest is displayed in the female. The male also tends to be quiet in the cage. We became interested in ascertaining if there are circumstances under which there can be a restimulation of sexual activity. The most likely stimulus, it seemed, would be access to a second female in heat. Observations were planned that would test this possibility.

METHOD

Ten males that could generally be depended on to achieve copulation with ejaculation in 10 min. were selected for observation in three types of 60-min. tests. Five of the ten were also observed in a fourth type of 60-min. test.

In every case ejaculation was achieved within the first 14 min. and, as is the case with the guinea pig, usually within 10 min. Test 1: A female in heat (1) was placed in the male's cage, and the pair was observed for an hour. Test 2: An estrous female was replaced after 30 min. by a second estrous female that was observed with the male for the remaining 30 min. of the test. Test 3: An estrous female was replaced after 30 min. by a female that was not in heat. Test 4: Five males observed in Tests 1, 2, and 3 were used in a fourth variation of the test. An estrous female was removed from the cage after 30 min., but returned in the same time that was required for the replacement of the first by the second female in Tests 2 and 3.

The behavior of the male was recorded and scored by a standardized procedure described elsewhere (2). It is sufficient to note here that the score is determined from the amount of each measure of behavior that is obviously directed toward the female (sniffling and nibbling, nuzzling, mounting, intromission, ejaculation), and from the length of the interval between the beginning of the test and ejaculation. The possible score varies from 0 for tests during which the male displays no behavior that would be regarded as sexual to 20 for tests in which ejaculation occurs within the first 15-sec. period.

For purposes of scoring, the hour observation period was divided into six time intervals. The first of the six time intervals was the period from the 0-minute through the time of ejaculation. The second was the period from the time of ejaculation through the twentieth minute. The third, fourth, fifth, and sixth time intervals each consisted of 10 min.

RESULTS

The behavior of the males during the tests is revealed by the trend of the average scores as the tests progressed (Fig. 1).[2] After the sharp drop in activity

[1] This investigation was supported in part by a research grant from the Division of Research Grants and Fellowships of the National Institutes of Health, United States Public Health Service, in part by the Committee for Research in Problems of Sex, National Research Council, and in part by the University of Kansas Research Fund.

[2] The mean and range of scores for each of the six time intervals during the 60-min. tests are available by ordering Document 3675 from the American Documentation Institute, 1719 N St., N. W., Washington 6, D. C., remitting $1.00 for microfilm (images 1 in. high on standard 35 mm. motion picture film) or $1.00 for photocopies (6 by 8 in.) readable without optical aid.

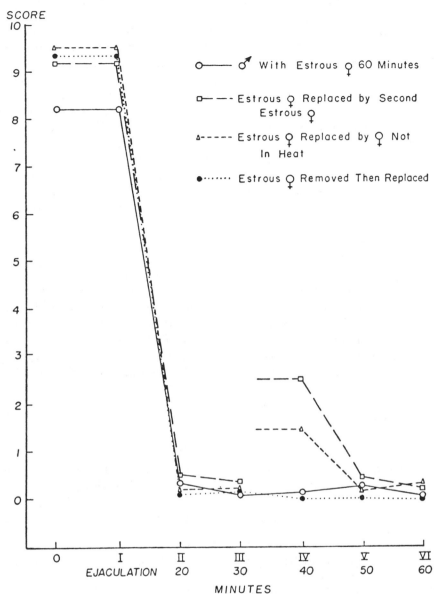

FIG. 1. SEXUAL BEHAVIOR OF MALE GUINEA PIGS IN VARIOUS TEST CONDITIONS

TABLE 1

MEAN FREQUENCIES OF TYPES OF BEHAVIOR DIRECTED TOWARD THE FEMALE DURING THE LAST THREE TIME INTERVALS

TEST CONDITION	SNIFFING & NIBBLING	NUZZLING	MOUNTING	INTROMIS-SION	EJACULA-TION
Estrous female for 60 min..............	3.5	0.5	0.1	0.0	0.0
Estrous female replaced by second estrous female after 30 min..........	4.0	4.0	2.0	0.6	0.1
Estrous female replaced after 30 min. by a female not in heat.................	3.5	4.4	0.9	0.0	0.0
Estrous female removed after 30 min., then replaced........................	1.2	0.0	0.0	0.0	0.0

which followed ejaculation, there was no significant change during the remainder of the tests in which the same female was left with the male or in which the same female was removed and replaced immediately after the first 30 min. Only when the first female was replaced by a second was there increased activity. The average score for the tests in which an estrous female was replaced by a second estrous female after 30 min. was 2.636 for the fourth time interval. The average score for the tests in which an estrous female was replaced after 30 min. by a female not in heat was 1.500 for the fourth time interval. As tested by the method of mean gains, both scores are significantly higher than those for the third time intervals. The critical ratios of both scores are likewise significant when compared to the fourth time interval scores of 0.150, for tests in which an estrous female was placed with the male for the entire 60 min., or of 0.060, for tests in which an estrous female was removed and immediately replaced. All scores tested for significance at the 5 per cent level or better.

Not only was more sexual behavior displayed following the introduction of a second female, but it was of a higher degree (Table 1). A second ejaculation occurred in one test and intromission was seen in several tests when the second female was in heat. The frequency of mounting was increased when the first female was replaced by the second, even though the second was not in heat. Superficially, the introduction of an estrous female for the second 30 min. of the test would seem to have been more stimulating than the introduction of a female not in heat, but the comparison is not reliable. A male confined with a female not in heat would not achieve as high a degree of sexual behavior as he would with an estrous female because in the former instance intromission and ejaculation would be precluded.

SUMMARY

The sexual activity of the male guinea pig can be stimulated during the first hour after ejaculation by the replacement of the first female by a second female. The stimulation, although definite and statistically significant, generally falls far short of that shown in response to the introduction of the first female.

REFERENCES

1. YOUNG, W. C., DEMPSEY, E. W., & MYERS, H. I. Cyclic reproductive behavior in the female guinea pig. *J. comp. Psychol.*, 1935, 19, 313–335.
2. YOUNG, W. C., & GRUNT, J. A. The pattern and measurement of sexual behavior in the male guinea pig. *J. comp. physiol. Psychol.*, 1951, 44, 492–500.

Received October 13, 1951.

33

Reprinted from *Animal Behav.* **15**:546–558 (1967)

COITAL BEHAVIOUR IN DOGS. I. PREFERENTIAL MATING IN THE BITCH*

By FRANK A. BEACH & BURNEY J. LeBOEUF†

Department of Psychology, University of California, Berkeley

Investigators of copulatory behaviour in lower mammals usually describe the female's sexual condition as 'receptive' or 'nonreceptive'. Differing degrees of receptivity have of course been recognized and studied, but they have been treated as variations in the female's overall level of sexual responsiveness which affect her reactions to all males. Consistent differences in a particular female's readiness to mate with different masculine partners have not been reported for any infraprimate species. The purpose of the present report is to describe such differences in one group of dogs.

Method

Subjects

These were five male and five female purebred beagles obtained from a large colony maintained at the University of California at Davis when they were 40 to 50 days of age. The same animals were used throughout the experiment, at the end of which time they were approximately 2 years old.

Conditions of Maintenance

From the beginning of the investigation all ten animals lived together in a ¾-acre field which was enclosed by an 8-ft sheet-metal fence. The dogs had constant access to one of ten indoor living-units where they could take shelter and where drinking water was available. The subjects were fed communally and were in daily contact with one or more of several caretakers who were also members of the team of experimenters. In view of the results to be reported it is worth emphasizing that the subjects grew up together, and that the only additions to the pack consisted of off-spring produced as a result of the mating tests.

Beginning when the animals were approximately 5 months old all females were examined

*This investigation was supported in part by Public Health Service Research Grant No. MH 04000 from the National Institute of Mental Health. The experiment was carried out at the University of California Field Station for Behavior Research.

†Present address of B. LeBoeuf: Dept. of Psychology, University of California, Santa Cruz.

periodically for signs of oestrus. At the first indication of vulvar swelling and/or vaginal bleeding a bitch was confined to one of the living units. Each of these consisted of an outdoor and an indoor section connected by a guillotine door which could be opened and closed from the service corridor that ran behind all ten units and gave access to the inner section of each. The outer section was enclosed on both sides by metal walls, but the front end, which gave upon the field, was constructed of heavy wire mesh. Animals in the outer section of a living unit had visual, auditory and olfactory access to the field and any other dogs in it. A guillotine door in the outer end of this section could be raised from the inside corridor allowing dogs to pass into the field or to enter the unit from the field.

In addition to serving for the confinement of females in heat, the living units were used to accommodate pregnant and lactating bitches and puppies too young to be released in the field. They were also employed to confine all animals not in use while a mating test was in progress. At such times the eight animals not being tested in the field were shut in the inner sections of living units to minimize the distraction they might introduce into the testing situation.

Conditions of Testing

Mating tests were carried out in one of two settings. In the early phases of the experiment the male and female to be observed were set free in the ¾-acre field and the experimenter was stationed in an elevated observation tower situated at one end of the row of living units. From this position he had visual access to nearly all parts of the field. This situation offered the advantage of extreme freedom of movement for the dogs, but it had the disadvantage of occasional periods when the animals were not visible because they moved behind a tree or into a small hollow at one end of the field. Furthermore, during the winter months fog sometimes limited the observer's visual contact with the animals.

Experience gained during the initial tests

using the entire field was exceedingly valuable, but after a number of such tests had been conducted it was decided to reduce the area so that dogs could be kept under closer observation every minute of the test period. Accordingly one portion of the field, approximately 1700 sq. ft in extent, was fenced off and a small observation shed was constructed in the middle of one of the sides of this enclosure. In this situation the animals had ample room to move about and the observer could keep them under close scrutiny at all times.

When tests were conducted in the large field the standard procedure was to release the male while the female was confined in the outer section of one of the living units. As mentioned earlier, all other dogs were shut in the inside sections of the units. Under these conditions the male almost invariably detected the presence of the female at once and responded by barking and pawing at the wire separating him from the bitch. When the test was to begin one experimenter raised the outer door which would allow the bitch to leave the living unit and enter the field.

If a female was ready to copulate with the waiting male she promptly joined him and he usually began mounting within less than 1 min. If the bitch was not yet in full oestrus, or if the male was one with which she did not want to mate, she characteristically refused to leave the unit. When this occurred some males immediately entered the outer section of the living unit and attempted to initiate copulatory contact. To prevent this reaction it was necessary for one of the experimenters to station himself in the living unit with the bitch and to force her out into the field at the beginning of the test.

This procedure involved a number of undesirable elements. For example, some males mounted the bitch within 1 or 2 sec after she had literally been tossed through the doorway and, as explained later in this report, once a female has been mounted her tendency to reject the male is markedly reduced. For this and other reasons the treatment of the two dogs was reversed. The bitch was released first, and after she had been in the field for 5 to 10 min the male was allowed to join her. Under these conditions the female was free to accept or reject the male and evidence for the existence of individual preferences became much more clear.

Scheduling of Tests

As indicated earlier, females were withdrawn from the pack and housed in individual units as soon as the advent of oestrus was detected. This usually antedated the onset of sexual receptivity by several days and an attempt was made to test each bitch for 3 or more days before and after the period during which she would allow males to copulate.

Except for a few cases of excessive bleeding produced by coition on the preceding day, every bitch was tested at least once each day throughout her period of oestrus. The schedule called for obtaining one completed mating every day as long as the female would permit copulation. If the first mating of a given day was successful a bitch was not tested again until 24 hr later. If the first male failed to carry out a complete mating he was replaced by a second male. This procedure was repeated until mating took place or until all five males had been tested at least once.

Daily tests were conducted until a female had failed to mate with any male for 3 successive days, at which time she was judged to have passed completely out of oestrus.

This treatment invariably resulted in pregnancy and females were confined a few days before parturition and kept in one of the living units until the puppies were weaned. The bitch was then returned to live in the field until her next oestrous period, at which time the testing for copulatory behaviour was begun again. The results to be reported are based upon tests conducted in the course of two natural oestrous periods for each female.

It will be evident that under this plan the timing of mating tests was determined by the natural reproductive rhythm of the females. As a result there were extended periods when no tests were conducted, and there were other times when as many as three bitches were in heat simultaneously. Accordingly it was not feasible to follow any predetermined schedule for testing each male. Nevertheless attempts were made to test every male with every female as frequently as possible; and over a period of nearly 2 years all possible pairings were achieved often enough to yield a reasonably consistent picture of the mating preferences of each bitch, and to indicate the general sexual responsiveness and effectiveness of each male.

Testing Procedure

The first response of male dogs given access to an oestrous bitch is to sniff and lick her vulvar area (Plate XI, Fig. A; Table 1, code 3)

If the female is disposed to accept the male she permits the investigation and he mounts *a posteriori* with his forelegs clasping her sides just anterior to the hipjoint (Plate XI, Fig. B; Table I, code 5). As soon as he is mounted the male begins to execute pelvic thrusts which direct the penis toward the general vicinity of the vaginal aperture (Table I, code 6). When the glans penis enters between the labia majora this partial intromission evokes the deep thrust of insertion as the male simultaneously performs three responses. He forces his ventral pelvic region strongly against the female, pulls his forelegs backward into the junction between the female's rear legs and body, and depresses his tail.

The insertion thrust is followed immediately by a period of very rapid pelvic thrusting during which intromission is maintained and the male's rear legs move spasmodically in what appears sometimes to be poorly coordinated stepping movements while at other times the behaviour looks more like kicking and both hindfeet may leave the ground simultaneously. High speed photography reveals that in some cases the rapid thrusting is accompanied by oscillation or swivelling movements of the male's pelvis or entire hind quarters. The entire pattern is here referred to as 'behaviour of intromission' (Table I, code 7) and it usually lasts for 15 to 20 sec.

Insertion precedes the development of full genital tumescence, and erection is completed during the rapid thrusting phase. This involves swelling of the bulbus glandis which is a segment of the penile shaft located nearest the base. The fully distended bulb fits so tightly inside the vulva that its circumference exceeds that of the vaginal opening and as a result withdrawal of the penis normally is prevented until detumescence begins. The relation of the pair while intromission is thus maintained is referred to as a 'lock' or 'tie' (Table I, code 8).

After the period of rapid thrusting is completed and a lock has been established the male dismounts or he may be dislodged by the bitch. In either event he subsequently lifts one rear leg over the female's back and assumes a quadrupedal posture facing away from the bitch while the penis, reflected backward between his legs, remains in the vagina (Plate XI, Figs C & D). This position or some variant of it is maintained until the male's erection begins to subside which usually occurs in 10 to 15 min, although some pairs may remain tied for an hour or longer.

In the present experiment females were permitted only one lock a day, and if this occurred on the first test the result was a single test for that day. If the first male failed to lock, the test usually lasted for 10 to 15 min depending upon the observer's estimate as to the probability of a successful copulation. Failure to lock sometimes occurred because the male displayed little interest in the bitch and exhibited few or no mounting reactions. In other tests the same result occurred because the male was unable to achieve intromission even though the female permitted repeated mountings. A third type of negative test occurred when the bitch refused to accept certain males. Under any of these circumstances tests usually were terminated within 10 to 15 min or as soon as the observer judged that mating was not going to occur.

The various criteria for continuing or discontinuing testing were based upon experience gained during preliminary tests and upon a knowledge of the individual characteristics of each of the ten dogs. In the course of initial tests it became evident that if a female rejects a male during the first 5 min of a test, she will continue to reject him for a much longer period of time; and a male that fails to mount the female within the first 5 min is very unlikely to do so if the pair is left together for 30 min or even longer.

Throughout each test all behavioural observations were dictated into a recorder. The first records consisted of continuous narrative accounts, but gradually there emerged a recognition of the repertoire of reasonably discrete acts by the male and female which combine to constitute the patterns of successful or unsuccessful mating, or of rejection. These acts were categorized and the dictated record became one of a sequence of responses by each animal with the time of occurrence and duration for each. After these records were transcribed they became the raw material for analysis of the behaviour under examination.

Analysis of Behavioural Records

The first step in analysing the records was to inspect the transcript for each daily test and to code each act by the male and each response shown by the bitch. The coded behavioural units are shown in Table I.

The second step in the analysis consisted of assembling all of the daily tests for any one pair

PLATE XI

Copulatory Behaviour in the Dog.

A. Pre-mounting investigation of the female by the male.
B. Mounted male just prior to insertion.
C & D. Positions commonly assumed during the lock.

Table I. Criteria for Scoring Mating Test

Male		Female	
behaviour	code	behaviour	code
Approach: moves to within 2 ft of female	1	*Attack:* Pursue, chase, drive, bite or any combination	−3
Attempted Investigation: Brings nose within 6 in. of tail region, or makes momentary contact with that area	2	*Threat:* Snarl, growl, bark, snap, lunge or any combination	−2
Investigation: Sniffs and/or licks anogenital area for at least 5 sec	3	*Avoid:* Withdraw, whirl around, throw off, fall or sit down	−1
		Ignore: No recorded response, pays no attention	0
Pre-mount Contact: Attempted mount, paws on back, chin over back, etc.	4	*Stand/Investigation:* Maintain location and orientation (tail usually deviated) for at least 5 sec or until male desists	+1
Mount: Rear position, clasp, no thrusting	5	*Stand/Mount:* Maintain location and orientation (tail usually deviated) for 5 to 10 sec or until male dismounts if mount less than 5 sec	+2
Clasp and Thrust: Rear position only	6	*Stand Long/Mount:* Like stand/mount but for more than 10 sec	+3
Behaviour of Intromission: Insertion, rapid thrusting, pelvic oscillation, stepping and treading	7	*Stand/Insertion:* Maintain location and orientation for verified insertion of more than 10 sec if no lock. Any length O.K. with lock	+4
Lock: Scored whenever pair remains tied after mount is discontinued	8		

throughout an entire oestrous period and transferring the coded behavioural units to an Interaction Matrix like the ones shown in Fig. 1. A tally in the appropriate position indicates the nature of each male-female interaction. From a completed matrix many calculations are made, and those which are pertinent to the present study will be explained in connection with the presentation of results.

Results

As noted earlier, mating tests were begun before the onset of sexual receptivity, whenever possible, and continued after the bitch had ceased to receive any male. The following analysis is based upon tests conducted during the period when the female was able and willing to copulate. The beginning of this period was defined as the first day on which the female locked, and its termination was defined as the last day on which a lock occurred.

Spacing of Oestrous Periods and Frequency of Tests

Table II shows the dates of the first and last lock for each female in the two oestrous periods on which this report is based. The interval

between periods ranged from 8 to 13 months, and for three of the five bitches 12 months or more elapsed. This observation is relevant to questions concerning the permanence of mating preferences and consistency of sexual performance.

The number of times each female was tested with each male is shown in the table, and considerable variation is evident. This was because some males quickly locked with a given female whereas others frequently failed to do so. In the latter instance the same pair was likely to be tested more often in an attempt to allow for a successful mating. For example, Peggy was consistently resistant to Ken and these two animals were tested together fairly frequently to discover whether the female's rejection would endure. This same bitch was tested next most often with John whom she rarely rejected but who had great difficulty in achieving intromission and therefore proved an unsuccessful copulator. During her second oestrous period Dewey developed an ataxia of the hindlimbs after having locked several times. For this reason she could not be tested with each male as frequently as desired,

Male _Broadus_   Dates _Jan. 5-16, 1966_   Locks _4_
Female _Peggy_   Tests _4_   Intros. _9_

♀

♂	-3	-2	-1	0	1	2	3	4	Σ	
1			/						1	Per Cent Lock _100_
2				/					1	Rejection Coefficient _0_
3			//		⦀⦀⦀ (THL THL II THL)				19	
4			////	THL	////	THL II			20	
5			/	//	//	///	THL III		16	
6			/			///	THL THL THL		19	
7							THL ////		9	
Σ			9	7	24	13	23	9	85	

Male _Ken_   Dates _Jan. 6-16, 1966_   Locks _2_
Female _Peggy_   Tests _9_   Intros. _2_

♀

♂	-3	-2	-1	0	1	2	3	4	Σ	
1	///	////							7	Per Cent Lock _22_
2	/	THL ///							9	Rejection Coefficient _82_
3	THL THL	THL //	/	//	//				22	
4	///	THL ////	//	/		/	/		17	
5		//	/			/	/		5	
6					THL	THL THL			15	
7							//		2	
Σ	17	30	4	3	2	7	12	2	77	

Fig. 1. Interaction matrices for one bitch with two males during one entire heat period.

Table II. Dates of Periods of Receptivity and Number of Times Each Pair Was Tested

Female	Oestrus	Date of first and last locks	Number of tests with each male				
			Broadus	Eddie	John	Clark	Ken
Peggy	I	18 to 27 Jan. 1964	3	3	6	3	4
	II	7 to 18 Mar. 1965	6	5	4	5	8
Spot	I	16 to 22 May 1964	3	3	3	5	4
	II	5 to 12 June 1965	3	2	4	4	2
Blanche	I	4 to 13 May 1964	7	6	5	6	7
	II	11 to 18 Nov. 1964	2	5	5	6	5
Kate	I	16 to 24 May 1964	3	3	3	3	3
	II	17 to 27 Mar. 1965	4	5	4	4	5
Dewey	I	12 to 21 May 1964	4	4	3	5	3
	II	19 to 26 June 1965	4	4	1	5	0

Differences in Female Responsiveness to Different Males

As indicated, the results of early tests during the females' first oestrous period made it quite plain that some bitches would readily mate with some males and vigorously reject others. The kinds of reactions involved are indicated in the rating scale shown in Table I. In an attempt to quantify the negative responses of each female to each male an index was calculated using information contained in the Interaction Matrix for that pair. The number of male acts coded 1 to 4 was totalled and the percentage of these acts to which the female responded with 'attack' (code −3) or 'threat' (code −2) was defined as the *rejection coefficient* or RC.

The results of such an analysis of all tests conducted during the first and second oestrous periods for each female are shown in Table III.

Values presented in this table reveal clear-cut differences in the acceptability of individual males. Broadus was never rejected by three females and very rarely rejected by the remaining two. In contrast, Ken was fully accepted only by Dewey and met with fairly strong rejection on the part of all others. Clark was consistently unpopular with all females except Kate who rejected him infrequently. It is apparent that the same male might be quite acceptable to one female but unacceptable to another. For example, Eddie was never rejected by Peggy; but when tested with Blanche his rejection coefficient was 58.

Examination of mean scores for the five females shows that Kate was the least rejecting and Blanche the most. The most marked discrimination was shown by Peggy who never rejected Broadus or Eddie, but showed an RC of 81 when tested with Ken.

Table III. Rejection Coefficients for Each Female with Each Male (Two Oestrus Periods)

Male	Female					Mean
	Peggy	Spot	Blanche	Kate	Dewey	
Broadus	0	4	6	0	0	2
Eddie	0	12	58	12	35	23
John	10	10	15	2	34	14
Clark	33	61	55	5	46	40
Ken	81	33	59	30	0	41
Mean	25	24	39	10	23	

Table IV. Rejection Coefficients in First and Second Oestrus Periods

| Male | Peggy | | Spot | | Blanche | | Kate | | Dewey | | Mean | |
	I	II	I	II	I	II	I	II	I	II	I	II
Broadus	0	0	5	0	0	20	0	0	0	0	1	4
Eddie	0	0	10	14	62	43	0	13	18	84	17	31
John	23	2	0	15	19	7	0	2	24	100	13	25
Clark	29	33	50	85	60	51	5	17	67	91	42	55
Ken	62	89	30	67	51	72	42	29	0	—	37	64
Mean	23	25	19	36	38	38	9	12	22	69		
Rho	+1·00		+0·70		+0·50		+0·90		+0·80*		+0·90	

*N = 4 (excludes Ken)

In order to determine whether the preferences revealed by these rejection scores were consistent and long lasting we compared the rejection coefficients based on tests in the first oestrous period with those for the same pairs tested during the female's second oestrous. The results appear in Table IV. It can be seen that the magnitudes of the coefficients vary considerably. In fact only for Kate and Peggy are the correlations statistically significant at the 5 per cent level of confidence. Nevertheless scores shown in the table lead us to believe that strong preferences shown in the first oestrous period tended to reappear in the second. For example, in both periods Spot was much more negative toward Clark and Ken than to the other three males, and Blanche was consistent in rejecting Eddie, Clark and Ken to a more marked degree than Broadus or John.

In calculating the rejection coefficient we used only those responses by the female which had been coded −2 ('Threat') or −3 ('Attack'). Possibly negative reactions coded −1 ('Avoid') were not included because of our inability to interpret them. Females are often mounted by males which they force to dismount and then allow to remount immediately with no other evidence of rejection. In some instances this occurs when the male is not appropriately oriented to the female and is thrusting from one side. In other cases it seems to occur because the male has been mounted and thrusting for a long time but is not close enough to the female to effect insertion. Upon still other occasions a female may be hyperactive, and although she vigorously solicits and permits mounting she may whirl around and dislodge the male before he has time to clasp her firmly or to execute pelvic thrusts. There is no question but that some responses scored −1 constitute a form of rejection; but it is equally certain that many of them do not reflect unwillingness to copulate, as evidenced by the fact mentioned earlier that the female at once initiates a renewal of sexual contact. For all of these reasons we consider our rating of −1 for the female to be an ambiguous category.

Inasmuch as we have arbitrarily defined rejection in terms of responses shown by the female to premounting activity on the part of the male, it is reasonable to enquire into the behaviour shown by females after the males have achieved a mount. To investigate this question we combined all of the male's acts coded 5, 6 or 7 and calculated the percentages of these totals to which the female's responses were rated +1 or higher. The results of these calculations are summarized in Table V.

It is evident that individual differences in this measure are much less marked than those involved in rejection of premounting acts. The general tendency for all five females was to respond permissively or cooperatively after the male had achieved a mount. Differences in the mean scores for the individual females are small, ranging from 85 per cent in the cases of Peggy and Dewey to 90 per cent in the case of Spot. This difference of 5 per cent between individuals contrasts sharply with a range of 29 per cent in the case of rejection coefficients for the same five females.

Table V. Percentage of Male Acts Involving Mounting which Were Responded to Positively by the Female

Male	Female					Mean
	Peggy	Spot	Blanche	Kate	Dewey	
Broadus	93	93	95	95	100	95
Eddie	87	93	74	87	91	86
John	88	91	87	91	63	84
Clark	88	83	93	87	68	84
Ken	71	86	88	78	100	85
Mean	85	90	87	87	85	

As far as the males are concerned there is again clear evidence of an attenuation of individual differences. Broadus is still the most generally accepted dog, but mean scores for the remaining four males differ from one another by no more than 2 per cent. We feel that this indicates the appropriateness of restricting the concept of rejection to reactions shown by the bitch before the male has mounted. This restriction applies, of course, only to the period during which the male remains mounted. Once he dismounts, the bitch may show rejection if he attempts to mount again.

One very interesting exception to the last statement occurs if a male achieves insertion but withdraws without locking. When this takes place the female frequently ceases to show any rejection and instead reacts very positively to the male, persistently investigating his penis, prancing about in front of him with forelegs flexed, and often mounting him repeatedly with a strong rear clasp and pelvic thrusting movements. In our experience it appears that no matter how violently a female has rejected a male, if insertion occurs and is then terminated without locking she usually becomes extremely receptive and cooperative for the remainder of that particular test. However, the facilitating effects of temporary vaginal stimulation do not carry over from one test to another, and the next day the male's acceptability as a sexual partner returns to its original level.

Effects of the Female's Rejection Behaviour

In considering the effects of rejection by the female upon the behaviour of the male, one obvious possibility is that repeated rejections might deter the male from showing any further responses to the female. He might, so to speak, 'lose interest', and cease altogether to interact with a bitch that repeatedly threatened or attacked him. To investigate this possibility we

Table VI. Mean Interactions per Test*

Male	Female					Mean
	Peggy	Spot	Blanche	Kate	Dewey	
Broadus	12·5 (112)	16·7 (100)	10·2 (92)	17·7 (124)	8·4 (67)	13·5
Eddie	11·0 (88)	20·5 (101)	17·5 (194)	39·4 (315)	7·2 (58)	19·0
John	24·4 (244)	27·5 (191)	16·8 (168)	44·6 (312)	15·5 (62)	26·8
Clark	6·9 (55)	13·0 (117)	17·8 (214)	31·4 (220)	18·3 (183)	17·5
Ken	16·0 (192)	19·6 (98)	13·2 (159)	32·0 (255)	8·0 (24)	17·8
Mean	14·2	19·5	15·1	33·0	11·5	

*Values in parentheses represent the sum of interactions in all tests.

calculated the average frequency of interactions per test for each pair and compared the resulting scores with the rejection coefficients for the same pairs. The results are summarized in Table VI.

In interpreting the values shown in this table it should be recalled that a low interaction frequency could be caused by several different factors. It could result from avoidance of the bitch by the male; but it could also result from the prompt achievement of a lock in a high porportion of the mating tests. If a male mounted a female once, achieved intromission and locked, the interaction score on that test would be 1, since no interactions were scored after the lock had been established. A high frequency of interaction could reflect acceptance by the female and inefficiency of the male, or rejection by the bitch and persistence by the male.

With these possible sources of ambiguity in mind it can be noted that mean values shown in Table VI reveal fairly large individual differences between males as well as between females. Broadus engaged in an average of 13·5 interactions per test, whereas the comparable score for John was 26·8. Among the females, Dewey's average interaction frequency was 11·5, while Kate's was 33·0.

Inspection of the individual records suggests that frequency of interaction was not controlled by the tendency of the female to reject the male. Interaction frequencies were not closely related to rejection coefficients. This becomes clear if one compares values shown in Tables III and VI. For example Broadus and Peggy had an RC of 0 and an average interaction frequency of 12·5. In comparison Ken, whose RC with Peggy was 81, engaged in an average of 16·0 interactions per test with this same female. Ken was never

rejected by Dewey, and the mean interaction frequency for this pair was 8·0. If we compare Eddie's interaction scores and rejection coefficients with each of the five females there is no obvious relationship between the two measures. Finally, a comparison of the relative rankings of the five males in terms of mean frequency of interactions with all five females and in terms of mean rejection coefficients shows that only one male retains the same rank on both scales. Broadus was the least rejected male and the male with the lowest average frequency of interactions. In contrast, John was the next-to-least rejected but had the highest mean interaction frequency.

Evidence summarized in Table V indicates that once a male has succeeded in mounting a female she tends to respond permissively as long as that mount is maintained. This led us to suspect that rejection by the bitch is revealed in her reactions to the male's premounting activities. To check this impression we noted the total number of interactions occuring in all tests for each pair (values shown parenthetically in Table VI) and determined the percentage of these interactions in which the male's behaviour was coded 5 or higher, i.e. the percentage in which a mount was achieved. The results are shown in Table VII.

The table reveals that males differed with respect to the proportion of all interactions that progressed at least as far as mounting the female. Comparison of these data with those shown in Table III makes it evident that there was a relationship between the female's rejection behaviour and the male's mounting. The rank orders for the five males on these two measures correlate highly (Rho = + 0·90). Nevertheless

Table VII. Percentage of Total Interactions in which Male's Behaviour was Coded 5 or Higher

Male	Female					Mean	Rank Order	
	Peggy	Spot	Blanche	Kate	Dewey		Least rejection	Highest % >4
Broadus	49	73	63	66	76	65	1	1
Eddie	59	75	47	62	60	61	3	3
John	71	67	59	63	48	62	2	2
Clark	44	30	27	75	34	42	4	5
Ken	18	66	60	57	87	58	5	4
Mean	49	62	51	65	61			

a low percentage of mounting reactions does not always reflect rejection by the female. Broadus was never rejected by Peggy, but less than half of his responses to her involved mounting. Blanche showed an RC of 59 when tested with Ken, but he achieved a mount in 60 per cent of his interactions with her. It would seem that a male may mount a female relatively infrequently even though she does not reject him. Whether or not rejection reduces mounting frequency depends on the female involved. Males succeeded in mounting some females quite frequently despite being rejected. For example Blanche's record shows that rejection coefficients and successful mounts for the five males were not highly correlated (Rho = +0·30). In contrast, the same correlation for Dewey was + 0·88.

So far it has been shown that rejection does not profoundly affect frequency of interaction but does decrease the frequency of mounting. The next obvious question is whether it influences the probability that a lock will occur. It should be noted that when a female successfully prevents a high percentage of a male's attempts to mount her, this can reduce the male's copulatory success, but does not automatically do so. To achieve a lock only one mount is necessary provided this mount results in insertion, and it has already been shown that females do not often react negatively to a male while he is mounted. To examine possible relationships between the female's rejection behaviour and the male's copulatory success a different comparison is needed and this is embodied in Table VIII.

Comparisons between Tables III and VIII fail to reveal any simple and clear-cut relationship between rejection by the female and lock success by the male. When males are compared in terms of rank orders on the two measures it is seen that Broadus ranked first as least rejected and first as achieving the highest percentage of locks. Eddie ranked third on both measures and Clark ranked fourth on both. However, John, the second most popular male, was the least efficient copulator, whereas Ken, the most frequently rejected individual, was scored second only to Broadus in copulatory efficiency.

Part of the explanation for this discrepancy lies in the fact that though he was not often rejected, John rarely achieved insertion and therefore could not lock. Most females permitted him to mount frequently, as is indicated in his score in Table VII. Ken, in contrast, was allowed to mount somewhat less often, but when he succeeded in doing so he frequently was able to achieve insertion and to establish a lock. A third male, Clark, was often rejected and was next to last in lock frequency because, as shown by his score in Tables VI and VII, when he was rejected he soon ceased interacting and therefore achieved relatively few mounts.

Behaviour in Tests Before and After the Locking Period

As stated under *Methods*, for the purposes of this investigation 'behavioural oestrus' or the period of 'receptivity' was defined as beginning on the day a female first locked with any male and as ending on the day of her last lock. In one sense this is a rather stringent criterion and the limits involved are probably narrower than those which would have resulted from use of vaginal smears as an indicator of oestrus. The

Table VIII. Percentage of Tests with Lock

Male	Female					Mean	Rank order	
	Peggy	Spot	Blanche	Kate	Dewey		Least rejection	Most lock
Broadus	89	67	88	85	63	78	1	1
Eddie	62	60	18	37	75	50	3	3
John	0	0	30	14	75	14	2	5
Clark	75	11	8	85	20	40	4	4
Ken	33	60	42	62	100	59	5	2
Mean	52	40	37	57	57			

implication of this qualification is that females were not in complete anoestrus 1 or 2 days before or 1 or 2 days after we obtained their first and last locks. For this reason behaviour shown during tests conducted during what we have termed the 'pre-lock' and 'post-lock' periods cannot be taken as representative of responses which would occur at times more distantly removed from the period of 'heat'. Nevertheless there were clear-cut behavioural differences in addition to the absence of locking.

Although we attempted to test every female for several days before the first lock and several days after the last, we were unsuccessful in the case of Dewey, who locked during the first test of her first oestrous period. The other four females received from 1 to 5 tests before locking and 1 to 3 tests after locking during both oestrus periods. Omitting the record of Dewey, and combining the pre- and post-lock scores for the remaining four females we have calculated the rejection coefficients for all pairs.

For all males and all females the number of interactions per test was lower before and after than during the lock period. The over-all mean was 20·5 interactions per test during the lock period and 12·2 during the pre- and post-lock combined. This decrease characterized the behaviour of sixteen of the twenty pairs, the four exceptions being Peggy and Broadus, Kate and Eddie, Blanche and Broadus, and Blanche and Ken. Mean frequency of interactions per test for each of the five males during the pre- and post-lock as compared with the lock period showed no relationship as far as rank order was concerned. No individual maintained the same rank under the two conditions. The same lack of correlation of ranks was seen in the case of the four females.

Rejection coefficients for all pairs are presented in Table IX which shows an increase in average RC for three bitches and no change for Kate who was the least rejecting female during the lock phase. During pre- and post-lock tests Peggy was more rejecting towards four males and less so toward Ken, but as in the case during the lock tests she was least rejecting toward Broadus and Eddie and most negative towards Clark and Ken. Spot showed increased rejection towards three males, decreased negativity toward one and no change in the case of the fifth partner. This bitch was, however, still least rejecting to Broadus and most rejecting to Clark and Ken.

In the case of Blanche rejection coefficients were higher for three males, lower in the case of one animal and unchanged in one. In contrast to her behaviour during the lock phase, Blanche strongly rejected Broadus before and after the lock period. Nevertheless her marked aversion for Clark and Ken was evident under both conditions. Before and after the period of locking Kate showed less rejection towards two males, more towards two and no change towards one. For this female the order of preference was changed, but her overall level of acceptance was so high that the significance of the shift is difficult to interpret. Under both conditions Broadus was immune from rejection, but Ken, the most frequently rejected male during lock tests was never rejected before or after lock tests.

Table IX. Rejection Coefficients During 'Lock Period' Compared with Pre- and Post-lock Periods

Male	Female									
	Peggy		Spot		Blanche		Kate		Mean	
	Lock	P&P	Lock	P&P	Lock	P&P	Lock	P&P	Lock	P&P
Broadus	0	13	4	15	6	66	0	0	2·5	23·5
Eddie	0	3	12	12	58	23	12	31	20	17
John	10	21	10	6	15	41	2	19	9	22
Clark	33	89	61	70	55	55	5	0	50	38
Ken	81	65	33	50	59	71	30	0	51	46
Mean	25	37	24	31	31	52	10	10		

Table IX shows that, although some females, e.g. Peggy and Spot, maintained more or less the same order of male preference in pre- and post-lock tests as they had displayed during the lock phase, the other two bitches were less consistent. Accordingly it cannot be concluded that the preferences existing in behavioural oestrus were also present before or after this period.

The data are sparse and the sources of variation numerous, but it seems safe to conclude that when females are not in a condition to lock the frequency of interactions with males tends to decrease, although as suggested earlier, this frequency is appreciably higher than would be expected in tests conducted at times more distant from the period of locking. Such decrease as does occur cannot be accounted for purely in terms of an increased tendency of the bitch to reject the male. There is undoubtedly an increase in this function, but there is at the same time a decrease in the male's tendency to interact with the female regardless of her attitude toward him.

Discussion

The findings here reported lead to two important conclusions. (1) Female dogs in physiological oestrus may copulate readily with certain masculine partners and actively resist the mating attempts of other males. (2) It is inadvisable to assume for any species that females must be either 'receptive' or 'nonreceptive' if receptivity is taken to signify indiscriminate and equal acceptance of every conspecific male that attempts to copulate.

It has long been apparent that feminine selectivity characterizes the mating patterns of several species of nonhuman primates including rhesus monkeys (Kempf, 1917; Carpenter, 1942), baboons (De Vore, 1965) and chimpanzees (Yerkes & Elder, 1936; Yerkes, 1939; Young & Orbison, 1944). In contrast, published descriptions of copulatory behaviour in laboratory rodents such as the rat (Stone, 1932; Beach, 1943), guinea pig (Young et al., 1939), hamster (Beach & Rabedeau, 1959) and inbred mouse (McGill, 1962) characteristically refer to a female's conditions as either 'receptive' or 'nonreceptive'. Variations in female receptivity have of course been recognized, and there are reports of consistent differences between strains in guinea pigs (Goy & Young, 1957), between individuals in rats (Ball, 1937) and even in the same female rat at different stages of the oestrous cycle (Beach & Kuehn, 1963). However, in all such studies the 'receptivity' of a given female at a given time has, by implication, been treated as a constant. The possibility of differences in responsiveness to different males has not been explicitly considered. Certainly it has not been systematically investigated and until this has been done the possibility of selective receptivity cannot be ignored.

On the basis of the published evidence one is tempted to conclude that dogs resemble female rats, mice, hamsters and guinea pigs to the extent that females of all these species will copulate only under the influence of ovarian hormones, but that the dog is different from rodents in that some oestrous females respond selectively toward different males. Such a difference, in kind or in degree may well exist, but its existence remains to be established experimentally.

Summary

Five male and five female beagles were raised together from puppyhood in a large field and were tested for copulatory behaviour when the females came into oestrus. Mating tests were repeated 8 to 13 months later during a second oestrous period. Females exhibited clear-cut preferences for particular males as sexual partners. Feminine rejection behaviour ranged from simple avoidance to active attack. Some females were more selective than others, but all showed discriminatory responses. Some males were rarely rejected by any bitch, whereas others were generally unpopular.

It is suggested that any concept of sexual receptivity as an endogenously controlled condition leading to indiscriminated acceptance of all conspecific masculine partners must be evaluated separately for each species. The absence of preferential responsiveness on the part of the females should not be assumed a priori. Its existence or nonexistence can be established only by direct investigation.

REFERENCES

Ball, J. (1937). A test for measuring sexual excitability in the female rat. Comp. psychol. Monogr., **14,** 1–37.

Beach, F. A. (1943). Effects of injury to the cerebral cortex upon the display of masculine and feminine mating behavior by female rats. J. comp. Psychol., **36,** 169–199.

Beach, F. A. (1947). A review of physiological and psychological studies of sexual behavior in mammals. Physiol. Rev., **27,** 240–307.

Beach, F. A. & Kuehn, R. (1963). Quantitative measurements of sexual receptivity in female rats. Behaviour, **21,** 282–299.

Beach, F. A. & Rabedeau, R. (1959). Sexual exhaustion and recovery in the male hamster. *J. comp. Psychol.*, **52**, 56–61.

Carpenter, C. R. (1942). Sexual behavior of free ranging rhesus monkeys (*Macaca mulatta*). II. Periodicity of estrous, homosexual, autoerotic and nonconformist behavior. *J. comp. Psychol.*, **33**, 143–162.

De Vore, I. (1965). Male dominance and mating behavior in baboons. In *Sex and Behavior* (ed. by F. A. Beach), pp. 266–289. New York: John Wiley & Sons.

Fuller, J. L. & Du Buis, E. M. (1962). The behavior of dogs. In *The Behaviour of Domestic Animals* (ed. by E. S. E. Hafez). pp. 415–452. Baltimore: Williams and Wilkins.

Goy, R. W. & Young, W. C. (1952). Strain differences in the behavioral responses of female guinea pigs to alpha-estradiol benzoate and progesterone. *Behaviour*, **10**, 340–354.

Harrop, A. E. (1960). *Reproduction in the Dog*. London: Baillière, Tindall & Cox.

Kempf, E. J. (1917). The social and sexual behavior of infra-human primates with some comparable effects in human behavior. *Psychoanalyt. Rev.*, **4**, 127–153.

McGill, T. E. (1962). Sexual behavior in three inbred strains of mice. *Behaviour*, **19**, 341–350.

Stone, C. P. (1932). Sex drive. In *Sex and Internal Secretions* (ed. by E. Allen). Baltimore: Williams and Wilkins.

Yerkes, R. M. (1939). Social dominance and sexual status in the chimpanzee. *Quart. Rev. Biol.*, **14**, 115–137.

Yerkes, R. M. & Elder, J. H. (1936). Oestrus, receptivity and mating in the chimpanzee. *Comp. psychol. Monogr.*, **13**, 1–176.

Young, W. C., Dempsey, E. W., Hagquist, C. W. & Boling, J. L. (1939). Sexual behavior and sexual receptivity in the female guinea pig. *J. comp. Psychol.*, **27**, 49–68.

Young, W. C. & Orbison, W. D. (1944). Changes in selected features of behavior in pairs of oppositely sexed chimpanzees during the sexual cycle and after ovariectomy. *J. comp. Psychol.*, **37**, 107–143.

(*Received* 20 *December* 1966; *revised* 14 *February* 1967; *Ms. number:* A536)

Part IV

EVOLUTION

Editor's Comments
on Papers 34 and 35

34   ARONSON
Hormones and Reproductive Behavior: Some Phylogenetic Considerations

35   NEVO
Mole Rat Spalax ehrenbergi: *Mating Behavior and Its Evolutionary Significance*

In studying the evolution of behavior, we attempt to use available information to reconstruct the probable evolutionary history of the behavioral pattern under study. It is necessary to study a group of species and then to discern some orderly trend that would suggest probable evolutionary history. There appear to be few viable generalizations one can make regarding the evolutionary history of patterns of sexual behavior, at least among vertebrates.

Cross-species comparisons can be made at either the species level or the phyletic level (King, 1963). When making species-level comparisons, one considers a group of closely related species— those from a single genus, family, or superfamily. At the phyletic level one attempts to understand the broader sweep of evolutionary history by comparing a wider range of diverse animal species. In general, psychologists have tended to work at the phyletic level whereas some biologists have worked at the species level.

Dewsbury (1972) wrote a phyletic-level review of the available literature on the patterns of copulatory behavior in male mammals with respect to the presence of locking, intravaginal thrusting, multiple intromissions, and multiple ejaculations. He was able to detect no continuous trend. Rather, similar patterns appear to have evolved in many different taxa, probably in response to similar selective pressures.

Beach (1942, 1947a, 1947b, 1948) proposed some preliminary, phyletic-level generalizations concerning the evolution of sexual behavior and its physiological control. In essence, Beach noted a difference in the degree of development of the cerebral cortex among vertebrates in general and in the "lower" mammals (rodents and rabbits) and "higher" mammals (carnivores and primates) in particular. As one moves from these lower to higher forms, Beach proposed that there is (1) greater variability of stimuli effective in initiating sexual activity, (2) greater variability in motor patterns, and (3) less direct control of behavior by gonadal hormones. In essence, in the higher mammals sexual behavior was viewed as more under the control of the cerebral cortex and less under the control of hormones than in lower forms. In addition, Beach proposed that in males there is greater dependence on the cerebral cortex, less complete dependence on hormones, and greater risk of distraction than in females.

Aronson (Paper 34) reviews the control of reproductive behavior by hormones among the vertebrates and evaluates Beach's proposals critically. Aronson concludes that there is no systematic relationship, either within mammals or within vertebrates. Although there are many differences within and among species, there is no progressive order.

According to King, comparisons at the species level are more likely to reveal important biological principles than are those at the phyletic level. An example of this approach is provided in Paper 15 by Nevo. Nevo compared four different forms of mole rats—forms that are probably different sibling species. He found that mating behavior appears to be one factor in ensuring reprodutive isolation between forms with different chromosome numbers. Aggression was more frequent and copulations less frequent in heterogametic than in homogametic matings.

Dewsbury (1975) reviewed data from thirty-one species of muroid rodents studied in his laboratory. He concluded, "There appears to be no simple progressive pattern of evolution with respect to copulatory behavior. In broad perspective, rodent copulatory patterns have some commonalities with those of primates, carnivores, insectivores, and others. Within the muroids studied there is appreciable variation within genus" (p. 752). He found no support for the proposal of Bignami and Beach (1968) that there has been a differential evolution of male copulatory behavior among myomorph versus muroid rodents in association with differences in the female estrous cycles in the two suborders.

What are the implications for the study of sexual behavior of the existence of appreciable species differences in behavior together with a lack of orderly evolutionary progression? We cannot exclude the possiblity that the variation may be either random or an artifact of some as yet undetermined factor. More likely,

293

copulatory patterns are evolutionarily, though not ontogenetically, labile, changing readily as species evolve. It appears that analogous patterns have appeared again and again, probably in response to the operation of similar selective pressures. When one finds this kind of pattern, the study of adaptive significance and of ecological factors in adaptation becomes especially important.

REFERENCES

Beach, F. A., 1942, Central nervous mechanisms involved in the reproductive behavior of vertebrates, *Psychol. Bull.* **39**:200–226.

Beach, F. A., 1947a, Evolutionary changes in the physiological control of mating behavior in mammals, *Psychol. Rev.* **54**:297–315.

Beach, F. A., 1947b, A review of physiological and psychological studies of sexual behavior in mammals, *Physiol. Revs.* **27**:240–307.

Beach, F. A., 1948, *Hormones and Behavior*, Harper, New York.

Bignami, G., and F. A. Beach, 1968, Mating behavior in the chinchilla, *Animal Behav.* **16**:45–53.

Dewsbury, D. A., 1972, Patterns of copulatory behavior in male mammals, *Quart. Rev. Biol.* **47**:1–33.

Dewsbury, D. A., 1975, Diversity and adaptation in rodent copulatory behavior, *Science* **190**:947–954.

King, J. A., 1963, Maternal behavior in *Peromyscus*, in *Behavior in Mammals*, H. L. Rheingold, ed., Wiley, New York, pp. 58–93.

34

Reprinted from *Comparative Endocrinology*, A. Gorbman, ed., John Wiley, New York, 1959, pp. 98–120.

Hormones and Reproductive Behavior: Some Phylogenetic Considerations

LESTER R. ARONSON

Department of Animal Behavior

American Museum of Natural History

New York, New York

Problems in evolution have constituted a major line of research and advanced thought at the American Museum of Natural History since its inception. It is not surprising, therefore, that a Museum scientist and student of behavior, F. A. Beach (1942), was first to devote serious attention to the phylogenetic relationships of reproductive behavior. By 1947 Beach had formulated his evolutionary hypotheses concerning the relationships between the central nervous system, gonadal hormones and sexual behavior. These hypotheses will be reviewed in part, and discussed in the light of recent advancements in this field.

THE THEORETICAL PICTURE

Central to Beach's (1947, 1948) theory are the concepts of sexual arousal and the sexual arousal mechanism which he later (Beach, 1956) labeled the S.A.M. In male mammals the S.A.M. functions to increase sexual excitement to the point where the copulatory threshold is·attained. Approach to the receptive female, mounts, intromissions, and ejaculations follow more or less in this order, though specific copulatory patterns may differ. The stimuli which activate the arousal mechanism are multisensory. They may include olfactory, auditory, visual, chemical, tactile, pressure, and genital sensations

The researches of the author and collaborators cited in this article were supported in part by grants from the Committee for Research in Problems of Sex, National Academy of Sciences, National Research Council.

(Beach, 1942*a*, 1951). These are integrated in the central nervous system to the extent that a deficit in one stimulus modality can be overcome by a higher level in others.

The motor functions of sexual behavior are mediated by central neural structures below the forebrain, while the cerebral neocortex serves to integrate the sexual stimuli of the S.A.M. The cortex, in turn, facilitates or activates the motor mechanisms.

Gonadal hormones are believed to sensitize the S.A.M., thus lowering the copulatory threshold. Moreover, endocrine deficits can be compensated for, up to a certain point, by increased sensory stimulation (Beach, 1942*b*). Likewise, cortical deficits can be compensated for, in part, by high levels of androgen (Beach, 1942*b*). The regular order in which the major elements of the mating pattern in male rodents disappear following castration (first ejaculation, then intromission, and finally mounting) reflects differences in the arousal thresholds of the neural mechanisms mediating these patterns.

In the evolution of vertebrates, the forebrain attains an increasingly dominant position. This change is commonly referred to as encephalization. The term refers especially to the development of the cerebral cortex in mammals, and to the progressively greater role of the neocortex in the sensory and motor functions of higher mammals. According to Beach (1947) this increase in the functional position of the forebrain includes greater control over the sexual responses. In mammals, lower neural mechanisms mediating the various elements of the sexual pattern become more and more dependent upon activation by the neocortex. In considering the phylogenetic progression (Beach, 1948) from the lower mammals (rodents and lagomorphs), through an intermediate group (especially carnivores and ungulates) to primates and man, (1) the variety of external stimuli which evoke the sexual responses increases, (2) the sexual responses become more variable, and (3) the direct importance of gonadal hormones to sexual behavior tends to decrease. It is this last generalization that particularly concerns us here.

The major lines of evidence which Beach used to develop this hypothesis are as follows: (1) In adult male rodents and lagomorphs, sexual behavior drops off rapidly after castration, whereas in primates and man the complete sexual pattern may last for months or even years. (2) In female mammals other than primates, estrous behavior is closely assocated with the cyclical activity of the ovaries, and can no longer be elicited when the ovaries are removed. In monkeys, apes, and man, on the other hand, receptivity of the female is largely independent of the condition of the ovaries and may continue after ovariotomy. (3) Extra-sexual stimuli, previous sexual experiences, and social situations are much more important factors in initiating sexual activity in primates than in other mammals. (4) Infantile sex play is much more frequent and complete in primates than in nonprimates. (5) Sexual behavior in both male and female prepuberally castrated primates is much more pronounced than in lower mammals.

If, with encephalization of the forebrain, and with elaboration of the neocortex, direct participation of the gonadal hormones in the elicitation of sexual behavior decreases, it follows that in the lower vertebrates, with little or no cortical representation, we should expect a very direct action of gonadal hormones in mediating sexual behavior. Beach has not offered an answer to this question, presumably because of the limited available evidence. However, since this problem is basic to our understanding of hormone-behavior relationships, I will attempt an evaluation with this point in mind.

FISHES

Among the Elasmobranchs, Cyclostomes and other primitive vertebrates I have found no information bearing upon this question. In teleosts, much of the evidence has been derived from the castration of adult individuals. A few investigators have claimed a rapid decline in sexual behavior when the gonads are removed. Thus Bock (1928), Ikeda (1933), and Baggerman (1957) report nearly complete cessation of nest-building behavior following castration in the three-spined stickleback, Gasterosteus aculeatus. Likewise, a castrated male Salmo salar "showed no interest in females," but surgically sterilized control males exhibited normal spawning activity (Jones and King, 1952). Among females, spayed jewel fish, Hemichromis bimaculatus, and Siamese fighting fish, Betta splendens exhibited no sexual activities when placed with ripe males (Noble and Kumpf, 1936), and I have obtained similar negative results with spayed females of the West African mouthbreeding cichlid, Tilapia macrocephala.

On the other hand, considerable sexual activity has been observed after gonadectomy in other species or under other conditions. Thus Noble and Kumpf (1936) reported typical courtship, fertilization movements, and brooding behavior in male jewel fish for as long as 202 days after castration. Likewise, spayed swordtails, Xiphophorus helleri, remained "sexually attractive to males." Unfortunately, this study was published only in abstract, and there is no indication in this brief report whether the gonadectomized individuals were ever examined for completeness of the operation. This is an item of paramount importance in studies of gonadal relationships in fishes, since testicular and ovarian remnants can be overlooked easily during the operation, and these remnants may induce a variable amount of gonadal regeneration. Moreover, small gonadal fragments, in fact as little as two per cent of the whole gonad, are sufficient to maintain normal secondary sex characters (reviewed by Forselius, 1957: 523).

A male or female *Tilapia macrocephala,* separated by a glass partition from a spawning female, may build his or her own nest. In a group of 15 males so kept opposite females, nests were found on the male side 17 per cent of the times that the female spawned. When castrated males or intact females were placed opposite spawning females, nests were found in an equivalent percentage of spawnings, but spayed females placed opposite intact females only occasionally built nests (Aronson, 1951). In this experiment nests were found built up to 14 months after castration.

In *Tilapia,* courtship, spawning and parental patterns of males and females are qualitatively alike, but there are decided differences in the frequency with which various elements of the pattern are displayed (Aronson, 1949). When two females are paired, they frequently spawn. Under such circumstances the female that does not deposit eggs will perform all of the elements of the mating pattern at a rate close to that of males. When a spayed female was paired with a normal female, the operated fish exhibited a level of nest-building and nest-cleaning activity within the range of variability of intact males. However, several other elements of courtship appeared infrequently or not at all.

When males of the gobiid fish *Bathygobius soporator* were isolated in small aquaria they promptly attacked any other introduced male of the species. On the other hand, the resident male courted gravid and nongravid females that were introduced, but responses to the latter were shorter in duration. Following castration, the resident males became nondiscriminatory and courted introduced males, gravid and nongravid females in like manner. The vigorous fanning movements of courtship became spasmodic in the castrates, but characteristic gasping and snapping movements were more frequent than in the intact controls. In castrated males, behavior patterns associated with spawning (e.g., darkening, nest-rubbing, erection of genital papilla, and fertilization movements) were similar to those of intact fish. In this experiment Tavolga (1955) interpreted the failure of the castrated males to make the usual discriminations as due to changes in perceptual capacities.

Most recently we have studied the behavior of another cichlid fish, the blue acara, *Aquidens latifrons*. Observations have been made of eight spawnings of three males, and these records (table 1) have been compared with observations of four postcastrational spawnings by the same three males. Up to 1½ months after operation all elements in the mating pattern were still present, most of these showing little change in frequency of occurrence or duration from their own preoperative levels. There was, however, a decided increase among the castrates in frequency of nest-passing behavior (rubbing genital papilla over nest), as well as increases in several other items of behavior, particularly during the observation periods one day after spawning. On the other hand, there was a noticeable decline after operation in nest-building behavior (scooping up gravel with mouth).

Thus far we have been considering oviparous species in which the sexual acts are intimately associated with deposition of eggs by the female. In viviparous forms such as the platyfish, *Xiphophorous maculatus*, the sequence of reproductive events is much more akin to that of mammals. Here during copulation, the transfer of sperm is accomplished by contact of the gonopodium (modified anal fin) with the genital aperture of the female (Clark, Aronson and Gordon, 1954). The frequency of occurrence of the major items in the mating pattern was studied before and after castration in 24 males (table 2). Copulations (i.e., prolonged gonopodial contact wtih the female usually resulting in insemination), thrusts (i.e., momentary gonopodial contact with the female), swings (i.e., undirected lateral and forward movement of gonopodium), and sidling (i.e., movement of male alongside of female) dropped in frequency or duration in tests after operation. In the remaining items of behavior no changes could be ascertained. A further analysis of the data is presented in table 3 for three of the above patterns, namely, thrusting, swinging, and sidling. These acts persisted in some individuals up to nine months after castration, a time which represents the main part of the test period. The data in table 3 also show that the duration of persistence after castration is highly variable among individuals, and that within any group the patterns drop out in different sequences (Chizinsky and Aronson, unpublished).

The variable effects of castration on sexual behavior as outlined in this section may, perhaps, be due in part to species differences. In addition, variable technical procedures used, for example in looking for gonadal remnants, and differences in testing arrangement may also be responsible. Despite these uncertainties, the evidence as it

299

TABLE 1

AVERAGE BEHAVIOR OF THREE MALE BLUE ACARAS BEFORE AND AFTER CASTRATION

		Total No. of Spawnings	Nest Passing		Nest Building		Body Quivering		Mouthing		Slate Cleaning		Nest Cleaning		Fanning			Guarding		
			♂/♀*	No./Min.	♂/♀	No./Min.	♂/♀	No./Min.	♂/♀	No./Min.	♂/♀	No./Min.	♂/♀	No./Min.	♂/♀	No./Min.	Dur./Min.	♂/♀	No./Min.	Dur./Min.
½ hr. Pre.†	Intact	3	0.2	0.2	2.0	0.2	0.3	1.0	0.2	0.3	0.5	2.0	0.7	0.4						
	Castrated	7	1.0	0.5	0.6	0.2	0.3	1.0	0.2	0.3	0.5	1.0	0.2	0.2						
Spawning	Intact	4	0.4	0.9	6.0	0.1	1.0	0.1	4.0	0.3	0.3	0.04	1.0	0.1						
	Castrated	8	0.7	2.0	1.0	0.03	3.0	0.3	5.0	0.6	0.4	0.1	5.0	0.4						
½ hr. Post.‡	Intact	4	0.3	0.2	1.0	0.3	2.0	0.2	0.2	0.1	0.4	0.04	2.0	0.6	0	0	0	0.5	2.0	6.0
	Castrated	8	0.3	0.7	0.5	0.1	0.6	0.05	2.0	4.0	0	0	2.0	0.3	0.2	0.02	0.08	1.0	0.3	5.0
1 day Post.§	Intact	3			3.0	0.6	0.3	0.04	1.0	0.3	0.1	1.0	3.0	0.3	0.4	0.2	0.3	3.0	0.9	12.0
	Castrated	6	0.3	0.7	2.0	0.5	2.0	0.2	13.0	0.7	1.0	0.2	2.0	2.0	0.2	0.1	2.0	2.0	0.5	10.0

* Frequency.
† ½ hour record before first egg appeared.
‡ ½ hour record after first fanning.
§ A 20-minute record 1 day later.

accumulates seems to support increasingly the following conclusions: (1) Removal of testes or ovaries causes an eventual decline in some or all elements of sexual behavior. (2) In all species studied, the decline is more pronounced in the female than in the male. (3) Among various species certain elements of the pattern are affected more

TABLE 2

SEXUAL BEHAVIOR OF MALE PLATYFISH BEFORE AND AFTER CASTRATION*

Behavior†	No. of Fish Showing Pattern		McNemar Test for Signif.	Average No. or Duration of Activity per Test		Wilcoxon Test for Signif.	
	Preop.	Postop.		Preop.	Postop.		
Copulations (No.)	6	2	N.S.‡	0.11	0.004	T < 0.025 > 0.01	S.
Thrusts (No.)	23	14	S.§	4.5	1.3	T < 0.01	S.
Swings (No.)	24	12	S.	8.1	2.2	T < 0.01	S.
Swings (duration)				1.24 sec.	1.1 sec.	T > 0.025	N.S.
Sidles (duration)	24	20	N.S.	66.9 sec.	27.6 sec.	T < 0.01	S.
Pecks (No.)	24	23	N.S.	22.9	14.2	T > 0.025	N.S.
Backs (No.)	24	22	N.S.	2.8	3.9	T > 0.025	N.S.
Approaches‖				89.5 sec.	118.6 sec.	T > 0.025	N.S.

* Unpublished data of Chizinsky and Aronson.

† Based on 24 fish, with 10 preoperative tests for each one, and with an average of 10 postoperative tests for each one.

‡ Not significant
§ Significant } at .05 level.

‖ Time from beginning of test to first approach.

rapidly and more drastically than others. (4) In males of some species, parts of the sexual pattern may persist for long periods after the gonads have been removed and after the androgen level has presumably dropped. Thus, it seems likely that in fishes, sexual behavior of experienced male adults is relatively independent of the direct action of testicular hormones.

In many species of fish it is easy to observe that sexual behavior makes its first appearance when the gonads are maturing and the secondary sex characteristics are developing. Also, male behavior has been induced in immature male and female platyfish and guppies with androgenic compounds (Eversole, 1941; Cohen, 1946; Tavolga, 1949). From such observations and experiments one may conclude that the development or organization of the mating pattern is directly contingent upon the presence of increasing levels of gonadal hormone. Until more adequate tests are performed, however, as for example

TABLE 3

PERSISTENCE OF THRUSTING, SWINGING AND SIDLING AFTER
CASTRATION OF ADULT MALE PLATYFISH*

Male No.	Weeks† Tested	Thrusting (weeks)	Swinging (weeks)	Sidling (weeks)
2	22	10	19	22
5	28	28	28	28
17	40	36	40	40‡
30	22	22	0	22
32	15	5	2	5
35	8	7	7	7‡
38	41	35	19	33
40	41	7	2	22‡
52	8	4	0	7
56	10	0	10	1
61	10	10	10	10
78	8	0	0	0
		Sham Operates§		
47	10	9	10	10
48	10	10	10	10
71	8	8	8	8
95	9	8	9	9
36‖	28	19	19	28‡

* Unpublished data of Chizinsky and Aronson.

† No. of weeks after castration that males were tested.   Approximately one test every other week.

‡ Fish checked histologically for gonadal remnants.

§ Similar abdominal incision.

‖ Partial castrate—gonadal remnants found.

by studying prepuberal castrates, this must remain an open question.

Pickford (1952, 1954) reported that sexual responses in male *Fundulus* may be controlled directly by pituitary action, and Wilhelmi, Pickford and Sawyer (1955) have identified the responsible agents as oxytocin and vasopressin from the neurohypophysis.   Additional evidence suggesting a direct action of pituitary hormones on sexual behavior has been reviewed by Aronson (1957) and by Pickford and Atz (1957).   One should not overlook the possibility that direct actions such as these may represent a primitive condition in vertebrates.   On the other hand, the same concept was also suggested a number of years ago for mammals (Smith, 1930), but it has never received additional support.

The effects of castration as reviewed above might be taken as evidence opposing the evolutionary theory of sexual behavior stated previously. Yet, it is important to consider that the forebrain of teleosts has evolved in a unique fashion, as a specialized offshoot from the main line of vertebrate evolution. In some teleostean families the forebrain is a small relatively simple structure. At the opposite end of this evolutionary progression we find families in which the forebrain is a complex and highly differentiated body which in some species includes cortex-like arrangements (Papez, 1929; Meader, 1939; Aronson, 1957). Those species cited above in which sexual behavior seems relatively independent of gonadal secretion are at least intermediate in this evolutionary progression of the teleostean forebrain. In other words, we must also recognize the possibility that, incorporated within the complex teleostean forebrain, there may be a mechanism equivalent to that indicated in higher mammals, whereby some elements of the reproductive pattern have been released from the functional control of the gonadal hormones.

AMPHIBIANS AND REPTILES

The amphibians and reptiles will be treated more briefly, since rather little has been added to this topic since the reviews by Beach (1948, 1951). Several investigators (Nussbaum, 1905; Noble and Greenberg, 1941; Reynolds, 1943) have observed that following gonadectomy, mating behavior is eliminated in male frogs and in lizards of both sexes. Noble and Aronson (1942) tied recently castrated male *Rana pipiens* to the backs of ovulating females. The latter oviposited in normal fashion but the action of the female did not elicit ejaculatory movements by the operated males. The clasp reflex will not develop in frogs if castration is performed well before the breeding season (Schrader, 1887; Steinach, 1910; Baglioni, 1911). When castration is performed during the breeding season the clasp reflex persists for a considerable time (Golz, 1869; Tarcharnoff, 1887; Busquet, 1910).

Castration of the male toad *Bufo arenarum* caused a gradual regression of the clasp reflex, which was nearly complete by 60 days. Croaking behavior, however, was unaffected by the operation (Burgos, 1950; Houssay, 1954).

As in fishes, the development of the sexual pattern may be directly dependent on gonadal hormones, since treatment of immature male frogs and lizards with androgens will induce mating responses (Noble and Greenberg, 1940, 1941; Blair, 1946). Similarly, estrous behavior can be induced in female *Anolis* with estrogen treatment (Greenberg and Noble, 1944).

Concerning the clasp reflex, it may be said that once the pattern becomes organized, it is less dependent upon gonadal hormones for its continued performance, at least for the remainder of the spawning season. Also, as in fishes, the possibility of a direct pituitary action is suggested by the experiments of Rey (1948) and Burgos (1950), who found that the simultaneous action of testosterone and hypophyseal hormones is necessary to restore the clasp reflex in intact and castrated *Bufo vulgaris* and *Bufo arenarum* males.

BIRDS

It is evident from the work of many investigators that castration in birds reduces the frequency of crowing, courtship, and copulatory behavior (Lipschütz and Withelm, 1929; Scott and Payne, 1934; Van Oordt and Jung, 1936). The effects are particularly pronounced when castrations are performed at an early age (Goodale, 1913; Guhl, 1949). However, as Collias (1950) notes, it is not always easy to evaluate studies on castration in birds because of difficulties in achieving complete removal of the gonads and in detecting small gonadal remnants at autopsy.

Although sexual activity is reduced by removal of the testes, it is not eliminated in all individuals, for some males continued mating (Goodale, 1918) even when complete gonadectomy seemed certain (Benoit, 1929). In this respect, Carpenter's work (1933, 1933a) is most important, since he used large numbers of experimental animals, quantitative techniques for recording behavior, and regular monthly observations up to nine months in one group. When the experiments were terminated he examined histologically all tissues suspected of containing testicular nodules. Seven out of fourteen completely castrated birds copulated after operation, but a gradual reduction in sexual activity took place, so that only four were still copulating when the last observations were made at eight months after the testes were removed. Eleven of these castrates continued billing after the operation and in some cases even a pronounced increase in billing was observed in later tests.

In contrast with the events in males, ovariotomy apparently eliminates all sexual activity shortly after operation (Goodale, 1913; Noble and Worm, 1940; Allee and Collias, 1940; Davis and Domm, 1943).

As in lower vertebrates, sexual precocity can be induced in young males by treatment with testosterone propionate (Noble and Zitrin, 1942; Hamilton and Golden, 1939) and in young females by treatment with alpha-estradiol benzoate (Noble and Zitrin, 1942).

Davis (1957) reported that male starlings maintained "fighting and song" for a month after castration and had a clear social rank which was not affected by adequate doses of testosterone. He suggests as a possible explanation a direct action of gonadotrophic hormones on behavior.

These data, although limited in scope, make it evident that birds also conform to a general vertebrate pattern which is becoming better defined. Thus, we note the following: (1) Removal of the gonads depresses sexual behavior in females more rapidly and effectively than in males. (2) The gonadal hormones seem to be more important in males for the organization of the sexual pattern than for its continued performance. (3) Some elements of the pattern are affected by removal of the gonads, to a greater extent than others, but in this there is considerable individual variation. (4) Some elements of the mating pattern increase in frequency of performance after castration. (5) Some elements may be directly influenced by hormones of hypophyseal origin.

MAMMALS

In reference to mammals, I shall confine my remarks mainly to recent experiments in our laboratory, particularly to some aspects of the studies on male cats by Rosenblatt and Aronson (1958, 1958a). We discovered early in our experiments that castration of sexually experienced male cats produces highly variable behavioral effects. On the one hand, an experimental male, Spike, continued to copulate in tests over 3½ years, and to mount the female in tests over 4½ years (fig. 1). White, at the other extreme, never achieved intromission after castration, while his mounting was sporadic and soon ceased altogether. Further study of many other animals showed that we could divide them into three groups with respect to the retention and eventual decline in sexual responses. In type A animals (fig. 2), intromissions lasted for many months or even years, whereas the capacity for mounting continued almost indefinitely. In type B males, capacity for intromissions lasted for about two months, but mounts without intromission increased in frequency and continued for a year or more. Many of the tests were also characterized by a large number of very brief mounts. In type C animals, intromissions dropped out rapidly, whereas mounting was weak and sporadic and rapidly disappeared. It is apparent that the extreme effects of castration described recently by Green, Clemente, and de Groot (1957) refer to type C animals.

305

The long persisting type A animals are particularly interesting in that the changes occurring after castration can be traced through and analyzed. Thus we recognized four postoperative periods with specific behavioral characteristics. In Spike (fig. 3) the first period of 5 weeks was characterized by no obvious changes from his preoperative per-

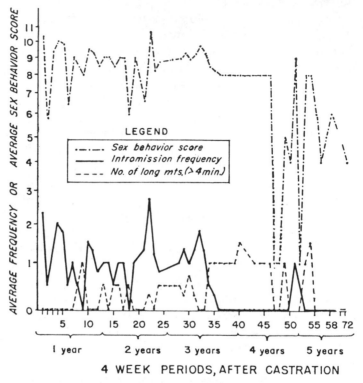

Figure I. Condensed sex behavior record of castrated male cat Spike. Ordinate graphed on logarithmic scale. [From Rosenblatt and Aronson (1958).] The sex score represents a comprehensive index of sexual behavior. For details of this scoring method, see p. 302 of the above reference.

formance. Period 2 extended from the 5th to the 110th week, during which time there occurred a gradual increase in mounting time, a marked decrease in duration of intromission, and an increase in the number of long mounts not terminating in intromission. In the third period, from the 110th to the 138th week, prolonged mounts not terminating in intromissions were frequent, and in some tests intromission did not take place. In the final period intromissions dropped out, but prolonged mounts were common, and often lasted the full 20 minutes of the testing period.

In prepuberally castrated males brought into sexual behavior by treatment with testosterone propionate, these same three types of decline were observed (fig. 4) when the hormone treatment was withdrawn (Rosenblatt and Aronson, 1958a).

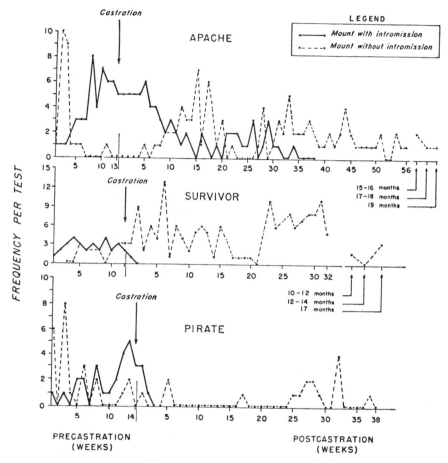

Figure 2. Comparisons of intromissions and mounts of three male cats, Apache, Survivor, and Pirate, typifying the three modes of decline of sexual behavior after castration. [From Rosenblatt and Aronson (1958).]

Several questions and ideas are suggested by the above observations. In mammals there need be no concern about the completeness of castrations. However, the explanation has often been advanced that in cases like our long persisting type A animals, androgens of sufficiently high level to facilitate the mating responses are being secreted by the adrenal cortex. Recent studies in hamsters (Warren and Aronson,

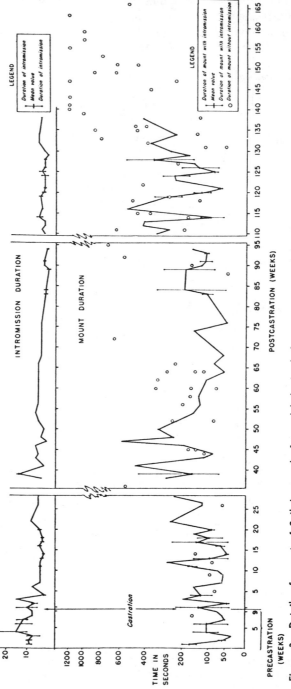

Figure 3. Details of a part of Spike's record of sexual behavior before castration and for three years after castration. Ordinate graphed on logarithmic scale. [From Rosenblatt and Aronson (1958).]

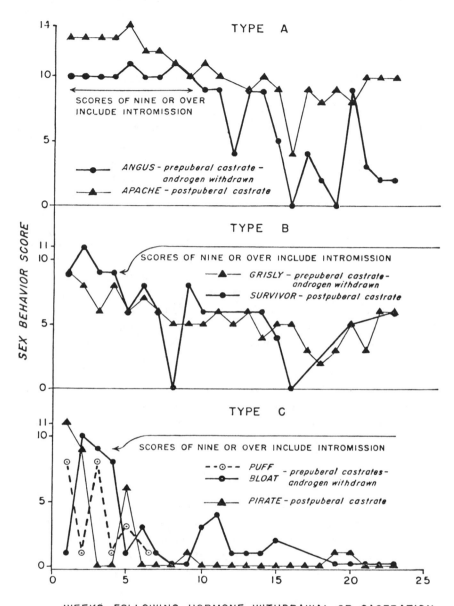

Figure 4. Decline in sexual behavior of androgen-treated prepuberally castrated male cats, compared with the decline of males after postpuberal castration. These graphs illustrate the three types of decline. [From Rosenblatt and Aronson (1959).]

1952, 1956, 1957), in cats (Cooper and Aronson, 1958), and in dogs (Schwartz and Beach, 1954) militate strongly against such an explanation.

Are the three types of decline unique for cats? We think not. We have examined comparable data for rats (Stone, 1927; Beach and Holz, 1946) and hamsters (Beach and Pauker, 1949, with additional data furnished by Pauker) and find striking evidence of several types of reproductive decline (Rosenblatt and Aronson, 1958). The clinical literature on castrations as reviewed by Beach (1947, 1947a, 1948) suggests a similar situation in man. The data for the lower vertebrates, although admittedly limited, also point to a comparable situation. We must recognize, of course, that among species with very varied reproductive structures and patterns, the number and detailed characteristics of such types are likely to be very different from corresponding types in cats.

The analysis of behavior by types underlines some important concepts which can easily be overlooked when only group trends are considered. Thus, in our cats the increase in frequency and duration of mounts following the cessation of intromission in type A and B males, and the characteristic behavioral changes in period 2 of type A males, are attributable to special physiological conditions stemming from the depletion of androgen and resulting in weakness or loss of erection. However, other parts of the total pattern are mostly unaffected. As another example, the repetitive short mounts which occur frequently in the later stages of type B males indicate on the one hand that the sensory aspects of the female perceived at a distance remain unchanged, and on the other hand that stimulation resulting from contact with the female is insufficient to sustain the mount. Here it would seem that the sensory receptive mechanism, particularly on the ventral surfaces of the male's body may be altered in these males by lack of testicular hormone. In type C animals, the rapid decline of interest in the female suggests changes in the distance sensory receptors and in the perceptive mechanism, to the extent that the female can no longer elicit positive responses from the male.

The conclusions of Soulairac (1952, 1952a) are in accord with this concept of the relative independence of the different components of sexual behavior. From this research, as summarized by Larsson (1956), Soulairac attempts to show that sexual behavior in the rat is controlled by nervous and endocrine processes. The behavior is described in terms of distinct reflexes and temporal relationships, which are regulated by three specific physiological mechanisms: "(1) the ejaculatory reflex by the endocrine balance; (2) the copulatory reflex by the cerebral cortex; and (3) the refractory period by enzymatic processes in the nervous tissue."

Although alterations in the level of gonadal hormones undoubtedly cause changes in central nervous processes, and may account for decreases or increases in the copulatory thresholds, peripheral changes and changes in systems other than the central nervous system must not be overlooked. Castration not only causes marked effects on genital structures, as for example changes on the surface and structure of the penis (Retterer and Lelievre, 1912, 1913; Pezard, 1949; Beach and Levinson, 1950), but many other morphological characteristics and physiological processes are also affected which may be directly or only indirectly related to the reproductive processes. These include cutaneous changes (Hartley, Grad, and Leblond, 1951), changes in olfactory thresholds (Le Magnen, 1952) and in the composition and strength of certain muscles (Wainman and Shipounoff, 1941). These widespread influences of the gonadal hormones seem to be characteristic of all vertebrates and their relevance to the functioning of the sexual pattern has been recognized by Beach (1951), Grunt and Young (1953) and Schneirla (1956). In addition, as shown by Valenstein and Young (1955) and Rosenblatt and Aronson (1958, 1958a), sexual experiences gained prior to castration will markedly affect the course of events after operation, and represent important factors both in the organization and retention of sexual behavior. If the retention or decline in sexual behavior after castration is the resultant of changes in a great many structures and processes, it is not surprising that the timing and sequence of events is as variable as it appears to be in many vertebrates.

We can define puberty in the male cat as starting at 3 to 4 months of age, for it is then that the hormone sensitive spines on the glans penis begin to develop (Aronson and Cooper, unpublished). Animals castrated at this age and tested for sex behavior in adulthood were surprisingly unresponsive (table 4). There were no intromissions; only 1 out of 13 males mounted, and the animal that mounted did not show typical stepping or thrusting movements. One additional male obtained occasional neck grips (Levy, Aronson and Rosenblatt, 1956; Rosenblatt and Aronson, 1958a). Although it is difficult to compare levels of sexual behavior in different species, we feel that this record for prepuberally castrated cats is distinctly inferior to results obtained for comparable groups of rats (Beach, 1942; Beach and Holz, 1946), guinea pigs (Sollenger and Hamilton, 1939), and hamsters (Warren and Aronson, 1957). The experiments of Rosenblatt and Aronson (1958a) also show that androgens and sexual experience are indis-

TABLE 4

SEXUAL PERFORMANCE OF PREPUBERALLY CASTRATED MALE CATS
DURING TEN TESTS IN ADULTHOOD (332–503 DAYS)*

Animal	No. of Mounts	No. of Dorsal Neckgrips	No. of Ventral or Lateral Neckgrips	No. of "Approaches"
Angus	14	9	7	17
Grisly	0	4	15	9
Charlie III	0	0	1	0
Charlie II	0	0	2	1
Bloat	0	0	1	3
Sleepy	0	0	6	10
Spotty	0	0	0	2
Fighter	0	0	4	1
Snowy	0	0	4	2
Puff	0	0	1	2
Charlie IV	0	0	7	3
Rusty	0	0	0	0
Charlie I	0	0	0	0

* Data from Rosenblatt and Aronson (1958a).

pensable for the organization of the mating pattern at puberty. But this behavior, once organized, becomes partially independent of the hormone.

In considering the phylogenetic relationship of testicular hormone to mating behavior in primates, considerable weight has been given to the performance of one prepuberally castrated chimpanzee, Don, who showed a high level of sexual behavior (Clark, 1945). However, Rosenblatt and Aronson (1958a) recently reviewed the published protocols and find no entry indicating that Don ever copulated *before* testosterone therapy began. Thus, in regard to prepuberal castrates, we feel that much more evidence is needed before a phylogenetic trend can be recognized.

CONCLUDING REMARKS

There is no question that the relations of the gonadal hormones to sexual behavior vary as we compare one species with another. These differences are particularly pronounced if we compare postpuberally castrated or spayed rodents with primates. However, the evidence does not reveal any clear-cut steps between these two mammalian orders, and below mammals an evolutionary progression is even less obvious.

In the past we depended extensively on the use of a single criterion, such as "sex drive," "hyper- or hyposexual behavior," "sex score" as used by Young and his collaborators in sex studies in guinea pigs (Young and Grunt, 1951), and "sexual arousal" as used by Beach. I fully concur in the concept that the forebrain contains an excitatory apparatus, even in lower vertebrates (Aronson, 1945, 1948), and that the gonadal hormones are likely to affect this mechanism, but evidence is accumulating that this is only one part, and possibly not even the most significant part, of the total organization of sexual behavior.

If we are going to discover phylogenetic trends in the relations of hormones to behavior, which evidently exist and certainly should be sought, it will be necessary to pay more attention to processes and structures controlling individual elements of the sexual acts. Investigations stemming from hypotheses searching out the direct and indirect action of the hormones on elements of the pattern seem more likely to reveal evolutionary trends than does research based on the conceptions of sexual behavior affected only as a unitary whole.

SUMMARY

In teleosts, amphibians, reptiles, birds, and mammals, experiments involving prepuberal and postpuberal castrations and the treatment of immature individuals with gonadal hormones reveal the following characteristics which seem to be common to all classes of vertebrates: (1) Removal of the gonads tends to depress sexual behavior but more so in females than in the males. (2) The gonadal hormones are more important for the organization of sexual behavior than for its continued performance. (3) Some elements of the sexual pattern are affected to a greater extent than others by the removal of hormonal stimulation, but considerable individual variability is recognized. (4) Some elements of the mating pattern may increase in frequency of performance at least for a time after castration. (5) Some elements of the mating pattern may be directly influenced by hormones of hypophyseal origin. This last matter needs considerably more investigation.

Experiments on mammals, and particularly on male cats, show that sexual behavior comprises several relatively independent processes which respond differentially to hormonal action or depletion. This accounts for the pronounced individual variation in response to changes in hormonal level. For this reason it seems more appropriate to analyze sexual behavior in terms of the several mechanisms comprising the total pattern rather than in terms of a single criterion, such as "sex drive," or "level of sexual arousal."

Although there are differences among species in the relations of gonadal hormone to sexual behavior, particularly noticeable when rodents and primates are compared, distinct phylogenetic trends are not discernible. It is suggested that the evolution of the endocrine mechanisms controlling reproductive behavior can be approached more effectively by the analysis of individual components of the pattern than by the consideration of sexual behavior as a single, well-circumscribed entity.

REFERENCES

Allee, W. C., and N. Collias, 1940, The influence of estradiol on the social organization of flocks of hens, *Endocrinology,* 27: 87–94.

Aronson, L. R., 1948, Problems in the behavior and physiology of a species of African mouthbreeding fish, *Trans. New York Acad. Sci.,* Ser. II, 2: 33–42.

———, 1949, An analysis of reproductive behavior in the mouthbreeding cichlid fish, *Tilapia macrocephala* (Bleeker), *Zoologica,* 34: Pt. 3, 133–158.

———, 1951, Factors influencing the spawning frequency in the female cichlid fish, *Tilapia macrocephala, Amer. Mus. Novit.* (1484), 1–26.

———, 1957, Reproductive and parental behavior, in *The Physiology of Fishes,* edited by M. E. Brown, Chap. III, Part 3, 271–304.

———, and G. K. Noble, 1945, The sexual behavior of Anura. 2. Neural mechanisms controlling mating in the male leopard frog, *Rana pipiens, Bull. Amer. Mus. Nat. Hist.,* 86: 87–139.

Baggerman, B., 1957, An experimental study of the timing of breeding and migration in the three-spined stickleback, *Arch. Néerl. Zool.,* 12: 1–213.

Baglioni, S., 1911, Zur Kenntnis der Zentrentätigkeit bei der sexuellen unklammerung der Amphibien, *Zentralbl. f. Physiol.,* 25: 233–238.

Beach, F. A., 1942, Central nervous mechanisms involved in the reproductive behavior of vertebrates, *Psychol. Bull.,* 39: 4:200–226.

———, 1942(a), Analysis of the stimuli adequate to elicit mating behavior in the sexually-inexperienced male rat, *Jour. Comp. Psychol.,* 33: 163–207.

———, 1942(b), Analysis of factors involved in the arousal, maintenance and manifestation of sexual excitement in male animals, *Psychosom. Med.,* 4: 173–198.

———, 1947, A review of physiological and psychological studies of sexual behavior in mammals, *Physiol. Rev.,* 27: 2:240–307.

———, 1947(a), Hormones and mating behavior in vertebrates, in *Recent Progress in Hormone Research,* edited by G. Pincus, pp. 27–63.

———, 1948, *Hormones and Behavior,* New York, Harper & Brothers, 368 pp.

———, 1951, Instinctive behavior: reproductive activities, in *Handbook of Experimental Psychology,* edited by S. S. Stevens, pp. 387–434.

———, 1956, Characteristics of masculine "sex drive," in *Nebraska Symposium on Motivation,* edited by M. R. Jones, pp. 1–32.

———, and M. Holz, 1946, Mating behavior in male rats castrated at various ages and injected with androgen, *Jour. Exp. Zool.,* 101: 91–142.

———, and G. Levinson, 1950, Effects of androgen on the glans penis and mating behavior of castrated male rats, *Jour. Exper. Zool.,* 114: 159–171.

———, and R. S. Pauker, 1949, Effects of castration and subsequent androgen administration upon mating behavior in the male hamster (*Cricetus auratus*), *Endocrinology,* 45: 211–221.

Benoit, J., 1929, Le déterminisme des caractères sexuels secondaires du coq domestique, *Arch. Zool. Expér. Gen.,* 49: 217–499.

Blair, A. P., 1946, The effects of various hormones on primary and secondary sex characters of juvenile *Bufo fowleri, Jour. Exper. Zool.,* 103: 365–400.

Bock, F., 1928, Kastration und sekundäre Geschlechtsmerkmale bei Teleostiern, *Z. wiss. Zool.,* 130: 455.

Burgos, M. H., 1950, Regulación hormonal de los caracteres sexuales secundarios en el sapo macho, *Rev. Soc. Argent. Biol.,* 26: 359.

Busquet, H., 1910, Existence chez la grenouille male d'un centre médullaire permanent présidant à la copulation, *Compt. rend. soc. de biol.*, 62: 880–881.

Carpenter, C. R., 1933, Psychobiological studies of social behavior in Aves. I. The effect of complete and incomplete gonadectomy on the primary sexual activity of the male pigeon, *Jour. Comp. Psychol.*, 16: 25–57.

———, 1933(a), Psychobiological studies of social behavior in Aves. II. The effect of complete and incomplete gonadectomy on secondary sexual activity, with histological studies, *Jour. Comp. Psychol.*, 16: 59–96.

Clark, G., 1945, Prepuberal castration in the male chimpanzee, with some effects of replacement therapy, *Growth*, 9: 327–339.

Clark, E., L. R. Aronson, and M. Gordon, 1954, Mating behavior patterns in two sympatric species of xiphophorin fishes: their inheritance and significance in sexual isolation, *Bull. Amer. Mus. Nat. Hist.*, 103: art. 2: 135–226.

Cohen, H., 1946, Effects of sex hormones on the development of the platyfish *Platypoecilus maculatus*, *Zoologica*, 31: 121.

Collias, N. E., 1950, Hormones and behavior with special reference to birds and the mechanisms of hormone action, in *Steroid Hormones*, edited by E. S. Gordon, 277–329.

Cooper, M., and L. R. Aronson, 1958, The effect of adrenalectomy on the sexual behavior of castrated male cats, *Anat. Rec.*, 131: 3: 544.

Davis, D. E., 1957, Aggressive behavior in castrated starlings, *Anat. Rec.*, 128: 537.

———, and L. V. Domm, 1943, The influence of hormones on the sexual behavior of domestic fowl, in *Essays in Biology*, California, U. Calif. Press, 171–181.

Eversole, W. J., 1941, The effects of pregneninolone and related steroids on sexual development of the fish (*Libestes reticulatus*), *Endocrinology*, 28: 603.

Forselius, S., 1957, Studies of anabantid fishes. III. Zool. Bidrag Fran Uppsala, 32: 379–597.

Goltz, F., 1869, *Beitrage zur Lehre von den Functionen der Nervencentren des Frosches*, Berlin, Hirschwald.

Goodale, H. D., 1913, Castration in relation to the secondary sexual characters in brown leghorns, *Am. Nat.*, 47: 159–169.

———, 1918, Feminized male birds, *Genetics*, 3: 276–299.

Green, J. D., C. D. Clemente, and J. De Groot, 1957, Rhinencephalic lesions and behavior in cats, *Jour. Comp. Neurol.*, 108: 505–536.

Greenberg, B., and G. K. Noble, 1944, Social behavior in the American chameleon (Anolis carolinensis Voigt), *Physiol. Zoöl.*, 17: 392–439.

Grunt, J. A., and W. C. Young, 1953, Consistency of sexual behavior patterns in individual male guinea pigs following castration and androgen therapy, *Jour. Comp. Physiol. Psychol.*, 46: 138–144.

Guhl, A. M., 1949, Heterosexual dominance and mating behavior in chickens. *Behaviour*, 2: 106–120.

Hamilton, J. P., and W. R. C. Golden, 1939, Responses of the female to male hormone substances with notes on the behavior of hens and newly-hatched chicks, *Endocrinology*, 25: 737–748.

Houssay, B., 1954, Hormonal regulation of the sexual function of the male toad. *Acta Physiol. Latino-Amer.*, 4: 2–41.

Ikeda, K., 1933, Effect of castration on the secondary sexual characters of anadromous three-spined stickleback, *Gasterosteus aculeatus*. *Japan. Jour. Zool.*, 5: 135.

Jones, J. W., and G. M. King, 1952, The spawning of the male salmon parr (*Salmo salar* Linn. juv.), *Proc. Zool. Soc. London*, 122: 615.

Larsson, K., 1956, Conditioning and sexual behavior in the male albino rat, *Acta Psych. Gotheburgensia*, I: 1–269.

Levy, M., L. R. Aronson, and J. Rosenblatt, 1956, Effects of prepuberal adrenalectomy on the development of sexual behavior in male cats maintained on DCA —a comparison with intact and prepuberally castrated males, *Anat. Rec.* 125: 587.

Lipschütz, A., and O. Withelm, 1929, Castration chez pigeon, *Jour. Physiol. et Path. Gen.*, 27: 46–54.

Meader, R. G., 1939, The forebrain of bony fishes, Proc. Koninkl. Ned. Akad. Wetenschap., 42: 3.

Noble, G. K., and L. R. Aronson, 1942, The sexual behavior of Anura. I. The normal mating pattern of *Rana pipiens*, *Bull. Am. Mus. Nat. Hist.*, 80: 127–142.

——, and B. Greenberg, 1940, Testosterone propionate, a bisexual hormone in the American chameleon, *Proc. Soc. Exper. Biol. & Med.*, 44: 460–462.

————, 1941, Induction of female behavior in male *Anolis carolinensis* with testosterone propionate, *ibid.*, 47: 32–37.

Noble, G. K., and K. F. Kumpf, 1936, The sexual behavior and secondary sexual characters of gonadectomized fish. *Anat. Record*, 67(suppl.): 113.

——, and M. Wurm, 1940, The effect of testosterone propionate on the black-crowned night heron, *Endocrinology*, 26: 837–850.

——, and A. Zitrin, 1942, Induction of mating behavior in male and female chicks following injections of sex hormones, *Endocrinology*, 30: 327–334.

Nussbaum, N., 1905, Innere Sekretion und Nervenienfluss, *Ergeb. Anat. u. Entwicklungsgeschichte*, 15: 39–89.

Oordt van, G. J., and G. C. A. Jung, 1936, Die hormonal Wirkung der Gonaden auf Sommer—und Prachtkleid. III. Der Einfluss der Kastration auf männliche Kampfläufes (*Philomachus pugnax*), *Arch. Entwicklungsmech. Organ.*, 134: 112–121.

Papez, J. W., 1929, *Comparative Neurology*, New York, Crowell, 378 pp.

Pauker-Warren, R., and L. R. Aronson, 1952, The relation of the adrenal cortex to mating behavior in the golden hamster, *Anat. Rec.*, 113: 4: 546–547.

Pickford, G. E., 1952, Induction of a spawning reflex in hypophysectomized killifish, *Nature*, 170: 807.

——, 1954, The response of hypophysectomized male killifish to purified fish growth hormone, as compared with the response to purified beef growth hormone, *Endocrinology*, 55: 274.

——, and J. W. Atz, 1957, *The Physiology of the Pituitary Gland of Fishes*, New York Zoological Society, New York.

Rey, P., 1948, Sur le déterminisme hormonal du réflexe d'embrassement chez les males de Batraciens anoures, *Jour. Physiol.* (Paris), 40: I, 292A–293A.

Reynolds, A. E., 1943, The normal seasonal reproductive cycle in the male Eumeces fasciatus together with some observations on the effects of castration and hormone administration, *Jour. Morphol.*, 32: 331–371.

Rosenblatt, J., and L. R. Aronson, 1958, The decline of sexual behavior in male cats after castration with special reference to the role of prior sexual experience, *Behaviour*, 12: 4: 285–338.

————, 1958a, The influence of experience on the behavioral effects of androgen in prepuberally castrated male cats, *Jour. Animal Behaviour*, 6: 3, 4: 171–182.

Schrader, M. E. G., 1887, Zur Physiologie des Froschgehirns, *Pflüger's Arch.*, 41: 75–90.

Schwartz, M., and F. A. Beach, 1954, Effects of adrenalectomy upon mating behavior in castrated male dogs, *Am. Psychol.*, 9: 467–468.

Scott, H. M., and L. F. Payne, 1934, The effect of gonadectomy on the secondary sexual characters of the bronze turkey (*M. gallaparo*), *Jour. Exp. Zoöl.*, 69: 123–136.

Smith, P. E., 1930, Hypophysectomy and a replacement therapy in the rat, *Am. Jour. Anat.*, 45: 205.

Sollenberger, R. T., and J. M. Hamilton, 1939, The effect of testosterone propionate upon the sexual behavior of castrated male guinea pigs, *Jour. Comp. Psychol.*, 28: 81–92.

Soulairac, A., 1952, La signification physiologique de la période réfractaire dans le comportement sexuel du rat mâle, *Jour. Physiol.*, 44: 99-113.

———, 1952(a), Analyse expérimentale des actions hormonales sur le comportement sexuel du rat mâle normal, *Jour. Physiol.*, 44: 327–330.

Steinach, E., 1910, Geschlechtstrieb und echt sekundare Geschlechtsmerkmale als Folge der innersekretorischen Funktion der Keimdrusen. II. Über die Entstehung des Umklammerungsreflexes bei Froschen, *Zentralbl. Physiol.*, 24: 551.

Stone, C. P., 1927, The retention of copulatory ability in male rats following castration, *Jour. Comp. Psychol.*, 7: 369–387.

Tarchanoff, J. R., 1887, Zur Physiologie des Geschlechtsapparates des Frosches. *Arch. ges. Physiol. Pflüger's*, 40: 330–351.

Tavolga, M. C., 1949, Differential effects of estradiol, estradiol benzoate and pregneninolone on *Platypoecilus maculatus, Zoologica*, 34: 215.

Tavolga, W. N., 1955, The effects of gonadectomy and hypophysectomy on the prespawning behavior in males of the gobiid fish *Bathygobius soporator, Zoologica*, 28: 218.

Warren, R. P., and L. R. Aronson, 1956, Sexual behavior in castrated-adrenalectomized hamsters maintained on DCA, *Endocrinology*, 58: 3: 293–304.

———, 1957, Sexual behavior in adult male hamsters castrated-adrenalectomized prior to puberty, *Jour. Comp. and Physiol. Psychol.*, 50: 5: 475–480.

Wilhelmi, A., G. E. Pickford, and W. Sawyer, 1955, Initiation of the spawning reflex response in *Fundulus* by the administration of fish and mammalian neurohypophyseal preparations and synthetic oxytocin, *Endocrinology*, 57: 243.

Young, W. C., and J. A. Grunt, 1951, The pattern and measurement of sexual behavior in the male guinea pig, *Jour. Comp. Physiol. Psychol.*, 44: 492–500.

317

35

Reprinted from *Science* **163**:484–486 (1969)

Mole Rat Spalax ehrenbergi: Mating
Behavior and Its Evolutionary Significance

Eviatar Nevo
Laboratory of Genetics,
Hebrew University, Jerusalem, Israel

Abstract. *Mating behavior of the subterranean mole rat,* Spalax ehrenbergi, *consists of three distinct stages—agonistic, courtship, and copulation.* Spalax *sexual behavior reflects certain cricetid affinities, some features general in rodents, and others presumably related to its subterranean, territorial life. Within four groups of* Spalax ehrenbergi, *each with different numbers of chromosomes, recently found in Israel, mating behavior seems to provide partial reproductive barriers. Selective matings between chromosome forms may complement a cytologic isolating mechanism to prevent widespread natural hybridization.*

The mole rats, genus *Spalax*, are subterranean rodents comprising the monotypic family Spalacidae. Little has been reported on the biology of natural populations of *Spalax* (*1*), nor is its classification satisfactory (*2*). The currently accepted taxonomy (*3*) recognizes three species, *Spalax microphthalmus, S. leucodon,* and *S. ehrenbergi,* distributed from southeast Europe through the Middle East to North Africa.

Phenotypic variation in *Spalax* is relatively small, in contrast to its considerable genetic variation. Chromosome variation in *S. leucodon* (*4*) and *S. ehrenbergi* (*5*) is remarkable. Four forms of *S. ehrenbergi* occur in Israel and neighboring areas having diploid numbers ($2n$) 52, 54, 58, and 60, respectively (*5*). These mole rats are distributed clinally and parapatrically from north to south, along biogeographic regions of increasing aridity, probably reflecting adaptive systems. The $2n = 52$ form ranges in the Upper Galilee Mountains; $2n = 54$ form is found in the Golan Heights and Mount Hermon; $2n = 58$ form is found in central Israel; and $2n = 60$ form is found in Samaria, Judea, and northern Negev.

The relative rarity of natural hybrids, karyotypic homozygosity, and the mating trials reported here suggest that the four forms are probably sibling species. If this hypothesis is substantiated, Pleistocene speciation of *Spalax*, masked by convergent morphological adaptations to subterranean life, might have been operating on a larger scale than hitherto assumed. We now report mating behavior of *Spalax* which suggests taxonomic affinities and the operation of ethologic barriers to reproduction between chromosome forms.

Mating experiments were conducted with adult animals of all chromosome forms, collected across Israel from August through November 1967. Each mole rat was kept in a separate cage to prevent casualties. Experiments were conducted in aquaria (70 by 35 cm) with a 5-cm layer of sawdust. For tests, animals were transferred to each other's cage, male to female and vice versa. Observations were recorded on movies, color slides, and tape recorder. When copulation occurred it was usually interrupted before ejaculation to save females for further tests. Description of copulation is based on three matings which were allowed to proceed to termination. Of the three females, one became pregnant but died several days later; the other two were not pregnant.

Seventy-seven mating trials were conducted on 17 females and 13 males, including animals of different chromosome forms. Only nine females were receptive. Receptivity was determined on the basis of behavioral, not vaginal, estrus. Copulation was achieved in 15 trials, involving 8 of 20 homogametic and 7 of 45 heterogametic tests. Homogametic tests involved partners of the same chromosome forms, whereas

318

heterogametic tests involved different chromosome forms.

Three distinct stages characterize the mating of *Spalax*: (i) agonistic (hostile), involving both attacks and fleeings; (ii) courtship; and (iii) copulation.

When a male and a female met, an initial characteristic defensive posture was assumed by each (Fig. 1A). This was followed by a series of attacks and fleeings. Confrontation involved exposure of incisors ready to bite. The male usually leaped and bit first, frequently in his partner's nasal region. Sometimes the male growled in a low-pitched voice, while the female cried in a high-pitched voice similar to that of the young. Receptive females did not attack but posed in a defensive posture. Nonreceptive females sometimes kept attacking and fleeing, calling vigorously and biting the male severely if he kept chasing them. Pregnant females uttered a low-pitched, trembling call, like ringing bells, which may be an avoidance signal. Between defensive displays either sex, but particularly males, roamed around, digging, eating, scratching, teeth-chattering, and climbing the cage of the cage.

The agonistic stage was usually prominent, yet it varied widely in pattern, intensity, and duration; rarely was it omitted. Bitings were significantly more frequent in heterogametic than in homogametic matings ($\chi^2_{(1)} = 8.85$; $.01 > P > .001$) (Table 1). The agonistic stage lasted from 0 to 15 minutes, averaging 8.3 minutes in 12 trials that achieved copulation. When either sex was not ready to mate, it lasted for hours, sometimes ending in casualties. When both were ready to mate, a dramatic switchover took place and courtship ensued.

Courtship was usually initiated by the male, though sometimes a female in full estrus was the initiator. Courtship started with partners facing each other with withdrawn incisors. They nuzzled each other gently, uttering a series of barely audible, trembling calls; vociferous cries were uttered only by females. The male caressed the female's head, sides, and entire body. The female remained sometimes passive while the male pulled her skin or pushed her body. Sometimes, the female kept crying, licked the male, and turned sidewise to him; she then presented her back so he could sniff and lick her genitalia (Fig. 1B). Sexually vigorous males actively licked, caressed, pulled, pushed.

Table 1. Mating behavior of *Spalax ehrenbergi* in homogametic and heterogametic matings.

| Type of mating | Tests (No.) | Matings with agonistic behavior | | Copulations (No.) |
		Bitings	No bitings	
Homogametic	20	4	15	8
Heterogametic	45	25	15	7

bit gently, and stepped over the female's head. Frequently the male shivered and called tremblingly in front of the female. Copulation did not ensue unless both partners were ready to mate.

The courtship stage was not continuous; partners parted and met time and again. During pauses, both sexes sometimes licked their own genitalia. Upon reencounter, feeble fights sometimes occurred, but courting soon resumed with increased efforts. Receptive females reacted promptly by licking the male and crying in a high-pitched voice.

Fig. 1. Mating behavior of *Spalax ehrenbergi*. (A) Agonistic display; (B) courtship; and (C) copulation.

In seven heterogametic matings, receptive females solicited courtship, but males did not respond.

The courtship stage varied in pattern, intensity, and duration both within and between chromosome forms. Within the latter, individual variation was remarkable. Some males tried to mate hastily before the female was fully receptive. Others displayed remarkable courting abilities and patience. Most males fell in the middle range, performing the male routine adequately to induce females to copulate. Courtship lasted from 8 to 55 minutes, averaging 21.5 minutes in 12 trials that achieved copulation.

Mounting, which starts the copulation stage, is attempted infrequently at first to test the female's receptivity. Some sexually excited males tried to copulate with the female's head. Normally, however, mounting started once the female presented her back to the male. She may have sniffed his genitalia before accepting him. Sometimes she moved away soliciting more courtship. Active courtship was resumed between the first mounting attempts. A high-pitched cry emitted by the female, as in courtship, continued frequently throughout the copulation stage.

First mounts were usually unsuccessful with no attempted intromission. The male mounted the female from behind, clasping her lumbar region with his front limbs (Fig 1C). He then exercised a series of rapid probing pelvic thrusts mostly without intromission. A fully receptive female raised her hindquarters and maintained this posture until the male dismounted, thus permitting intromission to take place. Frequently both sexes licked their genitalia thoroughly after the male dismounted, particularly after intromission. Between mountings a female often lay on her back while the male palpated her. Frequency of mountings increased with time. Rate of thrusting was reduced as the male achieved intromission, but each thrust was deeper and more energetic. Intromissions were brief but repeated. Copulation lasted 20, 45, and 90 minutes in three tests, the last one involving more than 60 mounts. This male kept copulating with a thoroughly exhausted female, and when she lowered her pelvis to prevent further intromission he "copulated" with her head and sides. An average mount lasted 10 seconds (range, 3 seconds to several minutes). Sometimes the female crawled during copulation carrying the male on her back. After a varied number of intromissions the male gripped the female strongly, inserted his penis deeply, ejaculated, and then left her. He paid no attention to her after ejaculation. Usually he started eating eagerly, being indifferent to her even if she continued to solicit courtship. Yet no further agonistic behavior ensued. Additional copulations were observed sometimes within several hours.

Copulation was highly stereotyped in each sex and was the least variable stage. Nevertheless, individual variation was apparent in the intensity and subtle copulatory activities. The percentage of trials terminating in copulation was significantly higher in homogametic than in heterogametic matings ($\chi^2_{(1)} = 4.66$, $.05 > P > .025$) (Table 1).

Certain aspects of the mating behavior of *Spalax* reflect cricetid affinities, whereas others are common to several unrelated rodents (6). The copulatory behavior of *Spalax* recalls that of *Mesocricetus auratus* in the rear mounting, long duration, repeated mounts, and preejaculatory intromissions. Similar agonistic precopulatory behavior has been reported in *Microtus californicus* and other microtines. A prolonged courtship also occurs in *Microtus californicus* and *Cricetus cricetus*. Finally, remarkable individual variation characterizes also the mating behavior of *Microtus californicus*. Varied vocalizations function in sexual behavior of many unrelated species. Yet, despite cricetid affinities the complex of activities comprising the mating ritual of *Spalax* exhibits specializations presumably linked to its subterranean, highly territorial habits.

Elaborate breeding mounds are built by females in nature during the breeding season of *Spalax* from December to April (7). Encounters between the territorial males and females of *Spalax* elicit aggressive behavior in the laboratory, and presumably also in nature, regardless of sex or season. The sexes are found together in breeding mounds only during the winter reproductive season. Otherwise, males have separate, much simpler mounds than the breeding mounds of females. Copulation in nature is presumably effected within the breeding mounds. It is suggested that the elaborate courtship of *Spalax* may have evolved in order to overcome hostility which characterizes the encounter of the sexes and to induce females to copulate.

Mating behavior is of cardinal importance as an ethologic isolating mechanism (8), and may have played an important role in speciation of *Spalax*, concomitant with cytologic mechanisms (5). First, aggression is much more pronounced in the agonistic stage of heterogametic than homogametic matings, when frequency of bitings is used as an index of aggression. Second, copulations proved significantly more frequent in homogametic than in heterogametic matings (Table 1). If these trends are substantiated on a larger scale by assaying all possible combinations, they may indicate selective matings between chromosome forms. The ethologic barriers to reproduction may prove greater the larger the difference in chromosome numbers. They may complement a cytologic isolating mechanism operating to prevent widespread natural hybridization.

References and Notes

1. B. Bodnár, *Terre Vie* **4**, 323 (1934); E. Nevo, *Mammalia* **25**, 127 (1961); I. S. Ognev, *Mammals of the U.S.S.R. and Adjacent Countries*, vol. 5, *Rodents* (Moskva-Lenningrad, 1947) [Engl. Transl. Israel Program Sci. Transl., Jerusalem (1963), pp. 487–556]; I. R. Savić, thesis, University Belgrade (1965).
2. S. Schaub, in *Traité de Paléontologie*, P. P. Grassé, Ed. (Masson, Paris, 1958), p. 659; F. Petter, *Mammalia* **25**, 485 (1961); C. A. Reed, *J. Mammalogy* **39**, 386 (1958); G. B. Corbert, *The Terrestrial Mammals of Western Europe* (Foulis, London, 1966), pp. 210–212.
3. J. R. Ellerman and T. C. S. Morrison-Scott, *Checklist of Palaearctic and Indian Mammals 1758–1946* (British Museum, London, 1951), pp. 553–556.
4. J. Walknowska, *Folia Biol. (Krakow)* **11**, 293 (1963); B. Soldatović, S. Živković, I. Savić, M. Milošević, *Z. Säugetierk.* **32**, 238 (1967); P. Raicu, S. Bratosin, M. Hamar, *Caryologia* **21**, 127 (1968).
5. J. Wahrman, R. Goitein, E. Nevo, in *Comparative Mammalian Cytogenetics*, K. Benirschke, Ed. (Spring-Verlag, New York, 1969); ———, *Science*, in press.
6. C. A. Reed, *J. Comp. Physiol. Psychol.* **39**, 185 (1946).
7. E. Nevo, *Mammalia* **25**, 127 (1961).
8. N. Tinbergen, in *Sex and Behavior*, F. A. Beach, Ed. (Wiley, New York, 1965), pp. 1–33.
9. I thank Professors S. A. Barnett, F. A. Beach, E. Mayr, C. A. Reed, N. Tinbergen, and D. Zohary, Mrs. S. Blondheim, and Dr. Sarah Nevo for critical reading of the paper.

10 September 1968; revised 3 December 1968

Part V

FUNCTIONS

Editor's Comments
on Papers 36 Through 41

In investigating the functions of sexual behavior, we inquire as to the role behavior plays in the survival and reproduction of the organisms displaying it. Emphasis shifts from the study of the determinants of copulatory behavior to its consequences. We shall consider a number of studies that bear on the consequences of sexual activity, the ways in which these appear to function in affecting reproductive success, and, it is hoped, the selection pressures that led to the evolution of these behavioral patterns.

Dewsbury (1978a) proposed that there are essentially four methods with which animal behaviorists study adaptive significance: the behavior-genetic method; the method of adaptive correlation; the experimental method; and the method of within-species correlation. These methods can be used at the phyletic, species, or genetic (that is, within species) level (see Part IV).

A considerable body of genetic theory suggests that inferences about adaptive significance can be made on the basis of data generated from a diallel cross design. In this design, the F_1 crosses of three or more inbred strains are bred, and their scores are compared with those of the parental genotypes. Inferences about adaptive significance are based on the direction in which the F_1 scores fall relative to parental scores. Dewsbury (1975a) conducted a diallel cross study using four inbred strains of laboratory rats and their twelve F_1 crosses. The data indicated directional dominance and possible adaptive significance especially for ejaculation after relatively few mounts and intromissions and also for relatively fast mating.

Using the second method, the method of adaptive correlation, one measures the behavior and various other traits in a variety of species and works to generate meaningful correlations among traits. These correlations often suggest adaptive significance and are testable as additional species are studied. For example, Dewsbury (1975b), working with muroid rodents at the species level, found that species with locking patterns had relatively protected nest sites and suggested that a protected nest site may be necessary for the evolution of a locking copulatory pattern—a pattern that might increase susceptibility to predation during mating. Among simple-baculum muroid rodents, species that lock appear to have thicker glans penes with more prominent penile spines, a reduced complement of reproductive accessory glands, and no copulatory plug as compared to species that do not lock (Dewsbury, 1975b; Hartung and Dewsbury, 1978). Various writers have proposed a relationship between penile morphology and copulatory behavior at the phyletic level (Asdell, 1946; Walton, 1960).

The most powerful conclusions about adaptive significance have generally come from the use of experimental methods, often in conjunction with within-species correlation. This entails a shift of frame of reference from the research discussed earlier in the book in that behavior now becomes the independent variable, and its consequences the dependent variables.

We shall first consider a number of lines of research on the role of behavior in pregnancy initiation. One obvious function of sexual behavior is the transfer of sperm from male to female. In many species, however, additional processes must be stimulated if reproduction is to be successful. The most obvious case is that of "induced" or "reflex ovulation." In a variety of mammalian species from at least nine mammalian orders, females do not ovulate during each estrous cycle but rather ovulate only following copulation or similar vaginal stimulation (Jöchle, 1973). It is thus stimulation from sexual behavior that provides a critical link in reproduction. Inferences about adaptive significance might be based on the study of this important link (Conaway, 1971).

Greulich (Paper 36) conducted a classical study of the role of stimulation in triggering ovulation in cats. Greulich found that ovulation could be induced via stimulation of the genital tract with a glass rod. He thus demonstrated conclusively that ovulation is indeed copulation induced; seminal fluid is not essential. As ovulation is not inevitable following copulation, the possibility for study of the behavior-ovulation relationship exists. Vaginal stimulation appears essential for triggering ovulation in rabbits; mounting without intromission has little effect (Staples, 1967; see also Paper 22). Similarly, a single service triggers ovulation in alpacas, although mounting without insertion has little effect (Fernandez-Baca et al., 1970). The distinction between induced and spontaneous ovulators is not absolute; under some conditions species that are normally spontaneous ovulators can become induced ovulators (Zarrow and Clark, 1968).

In many species of muroid rodents ovulation is spontaneous but pseudopregnancy, including the secretion of functional amounts of progesterone from the corpora lutea and the preparation of the uterus for implantation, is contingent upon mating almost exactly as is ovulation in species with induced ovulation. A pioneering study of the role of behavior in pregnancy initiation in laboratory rats was conducted by Ball (Paper 37). She demonstrated a quantitative relationship between the amount of copulation and the probability of pseudopregnancy. Under her testing conditions, female rats required several ejaculations to reach maximal probabilities of pseudopregnancy.

Wilson et al., (Paper 38) studied the role of pre-ejaculatory intromissions (see Paper 4) in pregnancy initiation. The young females in this study showed maximal levels of pregnancy after a single complete ejaculatory series. However, if fewer than the normal number of pre-ejaculatory intromissions were received, the probability of pregnancy was greatly reduced. Thus pre-ejaculatory intromissions function in pregnancy initiation in rats. They do this by helping to trigger critical neuroendocrine reflexes and by facilitating sperm transport (Adler, 1969). The stimulus requirements of the female vary with different conditions. For example, older, multiparous females require multiple ejaculations to reach maximal levels of pregnancy (Davis et al., 1977), just as did Ball's females.

In Paper 39, Land and McGill report that in mice pre-ejaculatory intromissions, and the thrusts that accompany them, are without any apparent function in pregnancy initiation in house mice. The ejaculatory reflex alone is sufficient for pregnancy initiation (McGill, 1970). These data contrast with those from rats and provide an interesting example of species differences among closely related species. A major paradox in this area is the problem of why male house mice display such a complex pattern delivering much stimulation to the female when such stimulation appears to play no demonstrable role in pregnancy initiation. Similar studies have been completed with a variety of other species of rodents (Dewsbury, 1978b). It appears that in general the stimulus requirements of the females of different species are correlated with the persistence of male copulatory behavior in those species.

Other recent research has revealed that for pregnancy to be maximized, the male rat must remain immobile at the time of ejaculation so that the vaginal plug is lodged tightly in the female cervix (Matthews and Adler, 1977). Further, the female needs a period of several minutes without copulation following an ejaculation lest sperm transport be disrupted (Adler and Zoloth 1970).

The research discussed thus far has been focused on adaptations that are related to successful reproduction by all members of a species. Indeed, they are sometimes discussed incorrectly as adaptations "for the good of the species." More recently studies in evolutionary biology have moved to analysis at the level of the individual and to the study of differential reproduction by individuals. Natural selection works through the differential representation of some genes in future gene pools relative to that of others. Thus there is competition within species acting to maximize the contributions to the gene pool of particular individuals. Obviously, this is not a conscious competition on the part of individuals but rather behavior occurring as the result of the actions of natural selection. The study of sexual selection (for example, male-male competition and female choice) has become popular under the rubric of "sociobiology" (for example, Wilson, 1975). However, it has a long historical tradition (for example, Darwin, 1871; Noble, 1938).

In Paper 40, Sumption reports a study of multiple-sire mating in swine, finding that in a group-mating situation, there is considerable variability with regard to the mating frequencies of different individuals. Such variability is essential if natural selection is to work through differential reproduction and is central to Darwinian fitness. This paper also permits acknowledgment of the contributions to the study of sexual behavior of fields underrepresented in this book—those of the animal and agricultural sciences.

Although controversial in some species, a number of authors have sought to relate social dominance to differential mating frequencies and reproduction (for example, DeFries and McClearn, 1970; Dewsbury, 1979; Duvall et al., 1976; LeBoeuf and Peterson, 1969). Contrary to the views sometimes implicit in early work, the female is not a passive spectator in male-male competition. Cox and LeBoeuf (1977) suggested that female elephant seals incite male-male competition through loud vocalizations emitted when they are mounted. Parker and Pearson (1976) suggested that the mounting of males by females (see Paper 8) attracts dominant males for copulation.

The final paper in the volume, that of Parker (Paper 41), is both the sole contribution from the 1970s and the sole paper on invertebrates. However, the concepts discussed were judged sufficiently important to warrant the violation of two of the guidelines for inclusion in this book. From the viewpoint of a male, it is not important merely that his partner become pregnant but that she bear *his* offspring. Parker points out that under conditions in which several males may mate with a single female, selection will act to favor males that can somehow neutralize or displace the sperm deposited by males in prior matings, and also to favor males that act to prevent such displacement by males mating after them. Adaptations that function in the context of sperm competition appear widespread among insects.

Sperm competition appears to be important in mammals as well. The multiple ejaculatory pattern of male laboratory rats functions, at least in part, in preventing displacement by subsequent males (Lanier, et al., 1979). When two males have simultaneous access to a female, the male that ejaculates last and/or more frequently sires a disproportionate percentage of the offspring (Dewsbury and Hartung, 1980). The pattern of copulation-induced decrements of female receptivity that has been demonstrated in numerous species (for example, Carter and Schein, 1971; Crews, 1973; Goldfoot and Goy, 1970; Hardy and Debold, 1972) may function in this way. If a male can stimulate a decreased receptivity in a female by prolonged mating, the probability of another male mating subsequently is reduced. In guinea pigs, although not in rats, the copulatory plug functions to exclude other males (Martan and Shepherd, 1976). In all these cases strong selective pressures would be expected to act to favor those genes producing individuals able to gain differential reproduction in subsequent generations.

REFERENCES

Adler, N. T., 1969, Effects of the male's copulatory behavior on successful pregnancy of the female rat, *J. Comp. Physiol. Psychol.* **69**:613–622.

Adler, N. T., and S. R. Zoloth, 1970, Copulatory behavior can inhibit pregnancy in female rats, *Science* **168**:1480–1482.

Asdell, S. A., 1946, *Patterns of Mammalian Reproduction*, Comstock, Ithaca, N.Y.

Carter, C. S., and M. W. Schein, 1971, Sexual receptivity and exhaustion in the female golden hamster, *Horm. Behav.* **2**:191–200.

Conaway, C. H., 1971, Ecological adaptation and mammalian reproduction, *Biol. Reprod.* **4**:239–247.

Cox, C. R., and B. J. LeBoeuf, 1977, Female incitation of male competition: A mechanism in sexual selection, *Am. Naturalist* **111**:317–335.

Crews, D., 1973, Coition-induced inhibition of sexual receptivity in female lizards (*Anolis carolinensis*), *Physiol. Behav.* **11**:463–468.

Darwin, C., 1871, *The Descent of Man, and Selection in Relation to Sex*, John Murray, London.

Davis, H. N., G. D. Gray, and D. A. Dewsbury, 1977, Maternal age and male behavior in relation to successful reproduction by female rats (*Rattus norvegicus*), *J. Comp. Physiol. Psychol.* **91**:281–289.

DeFries, J. C., and G. E. McClearn, 1970, Social dominance and Darwinian fitness in the laboratory mouse, *Am. Naturalist* **104**:408–411.

Dewsbury, D. A., 1975a, A diallel cross analysis of genetic determinants of copulatory behavior in rats, *J. Comp. Physiol. Psychol.* **88**:713–722.

Dewsbury, D. A., 1975b, Diversity and adaptation in rodent copulatory behavior, *Science* **109**:947–954.

Dewsbury, D. A., 1978a, *Comparative Animal Behavior*, McGraw-Hill, New York.

Dewsbury, D. A., 1978b, The comparative method in studies of reproductive behavior, in *Sex and Behavior: Status and Prospectus*, T. E. McGill, D. A. Dewsbury, and B. D. Sachs, eds., Plenum, New York, pp. 83–112.

Dewsbury, D. A., 1979. Copulatory behavior of deer mice (*Peromyscus maniculatus*). II. A study of some factors regulating the fine structure of behavior, *J. Comp. Physiol. Psychol.* **93**:161–177.

Dewsbury, D. A., and T. G. Hartung, 1980, Copulatory behavior and differential reproduction of laboratory rats in a two-male, one-female competitive situation, *Animal Behav.* **28**:95–102.

Duvall, S. W., I. S. Bernstein, and T. P. Gordon, Paternity and status in a rhesus monkey group, *J. Reprod. Fert.* **47**:25–31.

Fernandez-Baca, S., D. H. L. Madden, and C. Novoa, 1970, Effect of different mating stimuli on induction of ovulation in the alpaca, *J. Reprod. Fert.* **22**:261–267.

Goldfoot, D. A., and R. W. Goy, 1970. Abbreviation of behavioral estrus in guinea pigs by coital and vagino-cervical stimulation, *J. Comp. Physiol. Psychol.* **72**:426–434.

Hardy, D. F., and J. F. DeBold, 1972, Effects of coital stimulation upon behavior of the female rat, *J. Comp. Physiol. Psychol.* **78**:400–408.

Hartung, T. G., and D. A. Dewsbury, 1978, A comparative analysis of copulatory plugs in muroid rodents and their relationship to copulatory behavior, *J. Mammalogy* **59**:717–723.

Jöchle, W., 1973, Coitus-induced ovulation, *Contraception* **7**:523–564.

Lanier, D. L., D. Q. Estep, and D. A. Dewsbury, 1979, The role of prolonged copulatory behavior in facilitating reproductive success in a competitive mating situation in laboratory rats, *J. Comp. Physiol. Psychol.* **93**:781–792.

LeBoeuf, B. J., and R. S. Peterson, 1969, Social status and mating activity in elephant seals, *Science* **163**:91–93.

McGill, T. E., 1970, Induction of luteal activity in female house mice, *Horm. Behav.* **1**:211–222.

Martan, J., and B. A. Shepherd, 1976, The role of the copulatory plug in reproduction of the guinea pig, *J. Exp. Zoology* **196**:79–84.

Matthews, M., and N. T. Adler, 1977, Facilitative and inhibitory influences of reproductive behavior on sperm transport in rats, *J. Comp. Physiol Psychol.* **91**:727–741.

Noble, G. K., 1938, Sexual selection among fishes, *Biol. Rev.* **13**:133–158.

Parker, G. A., and R. G. Pearson, 1976, A possible origin and adaptive significance of the mounting behavior shown by some female mammals in estrus, *J. Nat. Hist.* **10**:241–245.

Staples, R. E., 1967, Behavioral induction of ovulation in the estrous rabbit, *J. Reprod. Fert.* **13**:429–435.

Walton, A., 1960, Copulation and natural insemination, in *Marshall's Physiology of Reproduction*, A. S. Parkes, ed., Longman's Green, London, 3rd ed.

Wilson, E. O., 1975, *Sociobiology: The New Synthesis*, Harvard University Press, Cambridge, Mass.

Zarrow, M. X., and J. H. Clark, 1968, Ovulation following vaginal stimulation in a spontaneous ovulator and its implications, *J. Endocrinol.* **40**:343–352.

36

Reprinted from Anat. Rec. **58**:217–224 (1934)

ARTIFICIALLY INDUCED OVULATION IN THE CAT (FELIS DOMESTICA)[1]

WILLIAM WALTER GREULICH

Department of Anatomy, Stanford University, California

The domestic cat is generally considered to be a form in which ovulation normally occurs only after coitus. Coste (1847), quoted by Hensen (1876), von Winiwarter and Sainmont ('09), Vander Stricht ('11), and Longley ('11) have presented evidence in support of this view. Bonnet (1897) however, reported finding a tubal ovum in a cat which had been so carefully segregated that there was, in his opinion, no possibility of its having mated. More recently, Sadler is said by Evans and Swezy ('31) to have established the occurrence of spontaneous ovulation in this form. Since Sadler's observations apparently have not yet been published, the writer cannot comment upon them. The fact that Bonnet considered his single case of spontaneous ovulation worth reporting may be taken as indicating that he, too, recognized the rarity of its occurrence. The present paper outlines the results of an attempt to induce ovulation in the cat by stimulating the distal portion of the genital tract with a glass rod.

MATERIAL AND METHOD

Twelve female cats were kept isolated in a separate cage, and confined beyond the possibility of any contact with males, for periods varying from 13 days to 32 weeks. During much of the time of observation they were in fully screened, outside quarters, where they received ample sunshine and fresh

---

[1] The writer gratefully acknowledges his indebtedness to Prof. C. H. Danforth for unfailing helpfulness throughout this investigation. His sincere thanks are due, also, to Dr. A. W. Meyer for his kindness in reading and criticising the manuscript.

air. Their food consisted of pasteurized milk and fresh raw meat with occasional additions of cod-liver oil. They were observed almost daily by the writer; and, when an animal was seen to be definitely in heat, it was segregated from the rest and subjected to the following procedure.

The cat was held firmly by the skin of the nape and its back was stroked with a gloved hand. This stroking, plus gentle prodding of the perineal region with a glass rod, was usually sufficient to cause the animal to 'present' itself—that is, to assume the posture which it would normally take when being covered by the male. In this position, the hind parts were somewhat elevated and turned in such a way that the vulva was directed almost dorsally. The stroking and prodding evoked, also, characteristic treading movements of the hind legs and a lateral deflection of the tail which uncovered the vaginal orifice.

A glass rod, of appropriate diameter and tapering to a rounded point, was then introduced into the vagina for a distance of about 1.5 cm. In the case of animals which had borne young, and in which the somewhat greater diameter of the vaginal canal made the precedure feasible, it was noted that pressing the rod upward, downward or to either side produced no marked response. When, however, the rod was pressed forward, in the direction of the cervix, it at once elicited the characteristic cry of pain which attends the normal sexual act in the cat. This cry was invariably followed by snarling and by a vigorous attempt to free itself from the grasp which held it. Once free, the animal began to roll from side to side—almost frantically, in some cases. The periods of rolling were interrupted by intervals which the animal devoted to licking the vulva and to rubbing itself vigorously, especially its head and back, against any suitable, available object. This period of activity was followed by one during which the animal lay quietly on its side, often purring softly and 'opening and closing' its front paws—behavior which is usually interpreted as indicative of a feeling of contentment and well-being. If the animal had only just come in heat, or if it had been stimulated only once or twice, it would ordinarily submit willingly to a repetition of the procedure within 15 minutes to a half-hour after the beginning of the inactive phase just described. This behavior pattern was checked by mating other females and observing their reactions to normal coitus. The reactions in the latter, so far as it was possible to determine, differed in no essential detail from those elicited by the artificial stimulation dedescribed above.

The effectiveness of the stimulations, as gauged by the intensity of the females' reactions to them, was not the same in all the animals, nor in any one animal at different times during the same period of heat. A stimulation was considered to have been effective only when it was followed by behavior on the part of the female such as described above.

Each of the twelve animals which had been subjected to the artificial stimulation was subsequently anaesthetized with either urethane or ether, its abdomen opened, and the ovaries and tubes carefully removed while it was still alive. A lethal dose of the anaesthetic was then administered. (The interval between the stimulation and the removal of the ovaries and tubes varied from 25 to 144 hours.) All possible care was exercised in handling the ovaries, in order to prevent injury to them or possible, accidental rupture of ripe follicles. The material removed was fixed immediately in warm or cold Bouin's fluid and later embedded in paraffin by the usual technique. Complete serial sections were made of one ovary and tube of each of ten animals and of both ovaries and tubes of the other two. The sections were cut about 12 μ in thickness and were stained with Delafield's haematoxylin and eosin.

FINDINGS AND DISCUSSION

The presence of rupture points on the ovaries of nine of the twelve cats was noted by inspection of those organs at the time of their removal and confirmed by subsequent study of the serial sections. In seven of these nine cases, one or more tubal ova were also found. Since, in the other two ovaries, the ruptured follicles were quite definite and unmistakable, there can be no doubt that ovulation had occurred in them also, even though no tubal ova were seen. In nine of the twelve animals, therefore, ovulation followed stimulation with a glass rod.

Of the three animals which did not ovulate, two had not reacted favorably to the stimulations. In neither of them had oestrus, as indicated by their behavior, been the typical, unequivocal state that had been noted in the other cats of the series. Both of them had shown some signs of beginning heat, a week or more before they were finally stimulated. The stimulations had in each case been postponed in the hope that the intensity of the oestrus would increase. Instead of becoming more marked, however, it disappeared entirely within 2 or 3 days, only to reappear a week or two later, somewhat increased in intensity, but still not the typical oestrus that had been seen in the other animals. They were both stimulated while in this atypical heat period.

A study of the serial sections of the ovaries of these two cats disclosed the presence of follicles of a size strikingly larger than that of any of the ruptured follicles noted in the entire series. Sections were made of only one ovary of the third cat which had not ovulated, and which had been killed 26 hours after stimulation. Since that ovary contained a follicle almost as large as those seen in the ovaries of the two animals just mentioned, the conclusion seems justified that it, too, would have failed to rupture, even if more time had been allowed to elapse before it was removed. It is interesting to note that Robinson ('18) reported finding similar abnorally large follicles in the ovaries of thirteen ferrets which had failed to ovulate after what he considered to have been normal matings.

The definitely greater average size of those unruptured follicles suggests that the attainment of a certain size is not of itself sufficient to cause the vesicles to rupture. Walton and Hammond ('28), on the basis of their observations on ovulation in the rabbit, which ovulates only after mating, conclude that the rate of secretion and consequent increase in pressure of the follicular fluid is of primary importance in causing the follicles to rupture. They point out that the rate of secretion of follicular fluid increases following coitus; and they believe that the eventual rupture of the follicle is due to the inability of its walls to keep pace with the rapidly increasing volume of fluid. If one may assume that a similar condition obtains in the cat, it must be concluded that no such increase in the rate of secretion of follicular fluid occurred in the case of the three animals of our series which failed to ovulate after stimulation. The follicles, consequently, failed to rupture and the continued, slow secretion of fluid resulted in their abnormally large size.

The writer's experience with the cat leads him to the conclusion that ovulation in that form is not only dependent upon coitus, or some other equally adequate stiumlus, but that it is also contingent upon such stimulation occurring at the proper time in the follicular cycle of the female. He believes, further, that the presence of oestrus, even in such a degree as to cause the female to accept the male, or to submit willingly to artificial stimulation, does not necessarily imply the existence of the requisite ovarian conditions. Sadler, according to Evans and Swezy (l.c.), has demonstrated the independence of the follicular and oestrous cycles in the cat. The findings in three of our cases would seem to favor that view.

That the failure of ovulation to occur under these circumstances is due to the condition of the female at the time of stimulation, rather than to any inadequacy of the stimulation itself, is suggested by the fact that ovulation in a cat in oestrus does not invariably follow even coitus itself. The following is a case in point which was observed in our colony. A female cat which gave every indication of being in heat was covered by a potent, fertile male five times on two successive days. Mobile spermatozoa were seen in the vaginal smears made immediately afterward, and the female's behavior clearly indicated that the matings had been satisfactory. The animal was killed 72 hours later, yet ovulation had not occurred. Longley's (l.c.) data, too, show that, of ten cats killed at periods ranging from 23 to 50 hours after coitus, only six had ovulated, though the seventh (the first of the series) would, in the opinion of that author, also have done so within the longer time. As stated above, a similar experience was reported by Robinson (l.c.) in the ferret, which is also a form in which ovulation is normally dependent upon coitus. He instances thirteen cases in which ovulation failed to occur after normal matings. It is, therefore, definitely established that, in the cat and ferret, follicular rupture and consequent liberation of ova do not necessarily follow copulation.

Our series is not sufficiently large to enable us to state precisely how soon after stimulation ovulation occurs in the cat. Assuming that, as in the rabbit, the interval is fairly constant in different individuals, the minimal length cannot be more than 25 hours; for a tubal ovum was found in one of the animals which had been killed that number of hours after stimulation. The fact that, in another female sacrificed 28 hours after that procedure, four tubal ova were demonstrated would seem to confirm this. Though Longley (l.c.) concluded that ovulation in the cat takes place at "about the end of the second day after pairing," his data do not preclude the possibility of its occurring sooner, if, in interpreting them, one recall that ovulation is not an invariable sequel to mating in this form.

Ovulation was seen to follow artificial stimulation of the distal portion of the genital tract in nine out of twelve cats. Two interpretations of these results suggest themselves: either spontaneous ovulation occurs with considerable frequency in the cat or the artificial stimulation was responsible for it. There is nothing in the reported observations of other investigators, so far as the writer has been able to determine, which would support the first assumption. In the absence, therefore, of any conclusive evidence for the frequent occurrence of spontaneous ovulation in the cat, the results of these experiments are interpreted as indicating that ovulation in our animals was induced by the artificial stimulation to which they had been subjected.

The precise mechanism of that reaction remains to be determined. These results, however, serve to eliminate all constituents of seminal fluid as necessary factors.

SUMMARY

1. Ovulation was induced in nine out of twelve domestic cats by stimulating the distal portion of the genital tract with a glass rod. The minimal length of the interval between stimulation and ovulation was found to be no more than 25 hours.

2. Evidence is adduced to show that ovulation does not invariably follow normal coitus in the cat, even if the female is clearly in heat when the mating occurs. This is interpreted as favoring the idea that the follicular cycle and the oestrous cycle do not necessarily coincide in this form.

3. Though the exact mechanism effecting ovulation in the cat remains to be elucidated, the present investigation definitely eliminates all constituents of seminal fluid as necessary causative factors in that process.

LITERATURE CITED

BONNET, R. 1897 Beiträge zur Embryologie des Hundes. Anatomische Hefte, Bd. 9, S. 421.

COSTE, J. J. 1847 Histoire du développement des corps organisés. Paris. (Quoted by Hensen.)

EVANS, H. M., AND OLIVE SWEZY 1931 Ovogenesis and the normal follicular cycle in adult Mammalia. Memoirs of the U. of California, vol. 9, no. 3. U. of Calif. Press, Berkeley, California.

HENSEN, V. 1876 Beobachtungen über die Befruchtung und Entwicklung des Kaninchens und Meerschweinchens. Ztschrt. f. Anat. u. Entwickelungsgeschichte, Bd. 1, S. 213.

LONGLEY, W. H. 1911 The maturation of the egg and ovulation in the domestic cat. Am. J. Anat., vol. 12, p. 139.

ROBINSON, ARTHUR 1918 The formation, rupture, and closure of ovarian follicles in ferrets and ferret-polecat hybrids, and some associated phenomena. Trans. Roy. Soc. Edinburgh, vol. 52, part II, no. 13. Robt. Grant & Son, Edinburgh.

Vander Stricht, R. 1911 Vitellogenèse dans l'ovule de chatte. Arch. de Biol.,
      T. 26, p. 365.
Walton, A., and J. Hammond 1928 Observations on ovulation in the rabbit.
      Brit. J. Exp. Biol., vol. 6, no. 2, p. 190.
Von Winiwarter, H., and G. Sainmont 1909 Nouvelles recherches sur l'ovo-
      genèse et l'organogenèse de l'ovaire des mammifères (chat). Arch.
      de Biol., T. 24, p. 1.

DEMONSTRATION OF A QUANTITATIVE RELATION BETWEEN STIMULUS AND RESPONSE IN PSEUDOPREGNANCY IN THE RAT

JOSEPHINE BALL

From the Psychobiological Laboratory, Phipps Psychiatric Clinic, Johns Hopkins Hospital

Received for publication July 29, 1933

Within the past fifteen years a series of experiments has been reported which were designed to discover the stimulus mechanism of pseudopregnancy in the rat. The recent papers by Vogt (1931, 1933) and Haterius (1933) contain certain observations for which no very satisfactory explanation has been offered. Some data incidental to the work on sex behavior of the rat done in this laboratory seem to the writer to be pertinent to the problem and to offer a possible explanation for these results.

Vogt (1931) sought to isolate the hypophysis from nervous stimulation by removing the superior cervical sympathetic ganglia and adjoining interganglionic fibers. She found that after this operation the incidence of pseudopregnancy following glass rod stimulation of the cervix uteri was greatly reduced but that copulation with vasectomized bucks was as effective as ever. She suggested that the vaginal plug might be the cause of this differential reaction.

However, Haterius (1933), in repeating and extending her experiments with essentially the same results, attempted to duplicate at least the mechanical features of the plug by inserting cotton pellets into the vagina after stimulating the cervix artificially. But this did not increase the effectiveness of artificial stimulation. He concluded that sexual excitement was the cause of the greater success of the copulatory stimulus.

Vogt (1933) tested the chemical effects of plugs by injecting extracts of plugs or secretion of seminal vesicles or prostate glands, or by implanting fragments of plugs, and found all these procedures entirely without effect. In this paper she also described another way of differentiating between what she now called "glass rod pseudopregnancy" and "sterile copulation pseudopregnancy." She found that rats which had had vaginal smears taken daily for several weeks were less sensitive to glass rod stimulation than were those whose smears had been taken for only a few days, while this differential sensitivity was not exhibited in response to copulatory stimulation.

It is the purpose of this paper to show that glass rod stimulation is probably different from the sterile copulation stimulus in a quantitative way only.   Some observations made by the writer suggest that the rats used by Vogt and Haterius received a considerably stronger stimulus in their mating tests than in the artificial stimulation tests.   It is believed that their results can be explained as a difference in quantity of the same stimulus without the necessity of postulating a "psychic factor" (Haterius) or a function of the plug beyond that of mechanical cervical stimulation.

EXPERIMENTAL DATA.   In connection with a study of sex behavior in the male rat, 9 females were sterilized by removal of the cervix and 12 by ligation, section, or removal of small parts of the uterine horns.   It was expected that the first of these operations, cervisectomy, would prevent not only pregnancy but pseudopregnancy as well, thereby permitting the usual short cycles in the females that were used for testing the sex activity of the males.   Since each female could then be used more often, such a result would have made possible a reduction in the size of the colony of females necessary for testing the males.   However, cervisectomy did not prevent pseudopregnancy but only reduced its incidence.   Injuries to the uterine horns, moreover, had much the same effect.

The animals were mated for brief periods of time, usually from 6 to 15 minutes.   Each copulation and each ejaculation was recorded.   No mating was permitted except during the tests.

When it was found that the long intervals indicative of pseudopregnancy were occurring in the cervisectomized animals, but that it was not the invariable result of copulation, their records were examined with a view to determining why they reacted at one·time and not at another.   It was discovered that the animals had invariably become pseudopregnant after they had received two or more plugs in one evening, but only 20 per cent of the 10 times when they had received only one plug and never after copulation that had not been carried to the point of ejaculation, even when intromission had taken place as many as 29 times (table 1).   Thus it became apparent that in these animals the amount of stimulation was the factor which determined the frequency with which pseudopregnancy followed mating tests.

The records of the animals with uterine horn injuries but with intact cervices were now examined and found to be approximately the same, i.e., 100 per cent response after two or more plugs, 14 per cent (7 cases) response after one plug, and 12 per cent (34 cases) after copulation without ejaculation (table 1).   This response to copulation without ejaculation is probably due to the fact that intromission took place a greater number of times in these cases.   In all the cases where pseudopregnancy did not occur after no ejaculation, both in this group and the cervisectomized group, intromission had taken place from 1 to 29 times (median 13), whereas these 4 positive cases followed 2, 30, 41 and 44 intromissions, respectively. Only one of these cases, therefore, cannot be explained on the basis of a greater number of intromissions.

Since these two groups were so similar, both showing this quantitative relationship between amount of stimulus and frequency of response, it seemed desirable to test normal females in a similar manner to see if they also responded only occasionally to limited stimulation. Accordingly, 27 normal females were mated to vasectomized bucks. In this group, instead of 14 per cent and 20 per cent, there was 65 per cent response in the

TABLE 1

Percentage of long and short intervals between heat periods in cervisectomized, sterilized and normal rats after various amounts of copulatory stimulation

OPERATION	NUMBER OF RATS	LENGTH OF INTERVALS BETWEEN HEAT PERIODS							
		After no copulation				After copulation, no plug			
		4–5 days		11–19 days		Total intervals	4–5 days		11–19 days		Total intervals
		No.	%	No.	%		No.	%	No.	%	
Cervisectomy........	9	24	100	0	0	24	17	100	0	0	17
Uterine occlusion....	12	96	100	0	0	96	30	88	4	12	34
Normal females......	27	61	100	0	0	61	30	100	0	0	30

OPERATION	NUMBER OF RATS	LENGTH OF INTERVALS BETWEEN HEAT PERIODS							
		After one plug				After several plugs			
		4–5 days		11–19 days		Total intervals	4–5 days		11–19 days		Total intervals
		No.	%	No.	%		No.	%	No.	%	
Cervisectomy........	9	8	80	2	20	10	0	0	6	100	6
Uterine occlusion....	12	6	86	1	14	7	0	0	8	100	8
Normal females......	27	7	35	13	65	20	—	—	—	—	—

20 cases of a single plug. No pseudopregnancy followed any of the 30 cases of copulation without ejaculation in which intromission took place 1 to 34 times (median 9) (table 1). Since the response to one plug was so much higher than in the operated animals, and since even the operated rats had never failed to respond to two or more plugs, it seemed safe to assume that the normal females would also respond invariably to this increased stimulation, and hence it was thought unnecessary to subject the animals to the stimulus of more than one plug.

The details of these findings are summarized in table 1. In this table the column headed "After no copulation" constitutes the controls. Since all of the non-copulatory cycles were short, it is safe to consider the animals essentially normal and the long intervals the result of the experimental procedures.   In other words, the long intervals are interpreted as true pseudopregnancy, and not due to failure of ovarian activity even though no check was made by laparotomy or autopsy.

Each interval has been treated as a separate case.   In no instance have rats been tabulated as stimulable or non-stimulable, since there was no indication that they were so grouped.

Cervisectomy would not, of course, preclude the possibility of nervous stimulation.   Such an operation would merely injure the peripheral end of the nerve path, leaving the cut ends embedded in the surrounding tissue.   A decrease in response to submaximal stimulation, such as the formation of a single plug, instead of complete repression of the pseudopregnancy response should, therefore, have been expected.

It is more difficult to understand why injury to the uterine horns alone should also depress the response.   It might have been due to a pathological condition of the uterus caused by retention of fluid.   Marked hydrometra was discovered in several rats at autopsy a year after uterine ligation. However, the tests reported in this paper were made within two months after the operation, during which time the rats gave every appearance of normality, including short and regular cycles.   With regard to the possible injury to the nerve supply of the cervix in these operations, it may be stated that in no case was the uterine artery cut or ligated, although the branches supplying an excised section of uterus were always tied off.   In general, the nerves in this region follow the blood vessels.   No main pathways, therefore, should have been interrupted.

DISCUSSION.   It is apparent from the preceding data that there is a quantitative relation between the amount of stimulation and success in producing pseudopregnancy, since delayed oestrus always follows a heat period in which two or more plugs have been formed, it is less apt to follow oestrus in which only one plug has been formed, and it occurs very seldom if the copulation has not resulted in ejaculation.   This proportionate effect is found not only in normal animals but also in rats from which the cervix has been removed or the uterus injured, which operations merely reduce the effectiveness of the stimulus.

This quantitative relationship between stimulus and response suggests a possible explanation of the fact discovered by Vogt (1931, 1933) and Haterius (1933) that artificial stimulation of the cervix is less effective than the copulatory stimulus in operated rats and also of the fact that glass rod stimulation differentiates rats accustomed to vaginal smears from those that have not had smears taken (Vogt, 1933).   The argument, in outline, is that the glass rod provides a relatively weak stimulus corresponding roughly to the one-plug mating, while their mating tests probably involved in most cases at least two plugs, shown above to be a much stronger stimulus.

That the glass rod stimulus is approximately equal in effectiveness to one-plug mating may be seen by comparing the percentage of long intervals after one-plug mating in the group of normal rats reported above, with the percentages obtained by various investigators using glass rod stimulation. These one-plug matings gave 65 per cent (20 cases) of long intervals, a figure which is roughly comparable to the 70 per cent (151 cases) found by Long and Evans (1922, calculated from table 26, p. 80) and the 68.8 per cent (45 cases) reported by Meyer, Leonard and Hisaw (1929). Vogt's (1931) 91 per cent of 11 cases is unusual and, in view of the small number of cases, may have been due to chance.

That most of the mating tests used by Vogt and Haterius probably involved the formation of more than one plug, is deduced from the fact that they left the females with males for several hours or all night. From an extended study of sex behavior in rats (unpublished) the writer has found that a healthy young rat is capable of making as many as six plugs in two hours and that he seldom makes less than two unless he has been recently bred. Vasectomized males do not differ noticeably from normal animals in sexual potency (unpublished data). It is, therefore, probable that the females under consideration received more than one plug.

Since several-plug matings induce pseudopregnancy invariably it is evident that such tests constitute too strong a stimulus to discover moderate injuries to the mechanism. To test injuries less severe than total obstruction, mechanical or electrical stimulation must be employed unless copulation is limited, and it was when these less effective stimuli were used that Vogt and Haterius were able to show a reduced percentage of long intervals in their experimental animals.

The fact that Haterius was entirely unable to elicit pseudopregnancy by mechanical stimulation in his operated animals, whereas Vogt attained partial success by the same method, does not vitiate this argument because the former was markedly less successful than Vogt with normal rats also. It is evident that his stimulus was less potent or his rats less responsive.

The evident quantitative relation between the strength of stimulus and frequency of response as here set forth would seem to render Haterius' postulate of a "psychic factor" in copulation superfluous. In order to prove definitely that the greater success of the mating tests was due to another kind of (or at least more profuse) stimulus, it would be necessary first to show that the copulation permitted was equal in stimulus value to artificial stimulation in normal animals, and then to show that artificial stimulation was less effective than that same amount of copulatory activity after operation.

With regard to the differential response of rats much or little subjected to vaginal examination ("smearing") it would seem possible to explain this in much the same way, namely, that long "smearing" renders them less sensitive but that several-plug mating is too strong a stimulus to bring out such a slight change in sensitivity.   Only the more delicate test of glass rod stimulation brought it out.[1]

In conclusion, it may be stated that these data serve to point out the necessity of quantitative records of mating tests in such experiments. They also show the need of equating the various kinds of stimuli with respect to their effectiveness in eliciting pseudopregnancy and of choosing a stimulus suitable to the extent of the injury experimentally produced.

SUMMARY

Removal of the cervix uteri of the rat does not prevent pseudopregnancy.

There is a quantitative relationship between the amount of stimulation and the percentage of success in eliciting pseudopregnancy.

The data reported by Vogt (1931, 1933) and Haterius (1933) are analyzed in the light of this quantitative relationship to show that it is not necessary to assume a "psychic factor" in the copulatory stimulation of pseudopregnancy nor any function of the vaginal plug beyond its mechanical stimulation of the cervix in order to explain their results.   In the absence of further details, their findings can as well be explained as being due to differences in amounts of the same kind of stimulus

REFERENCES

HATERIUS, H. O. 1933. This Journal, **103**, 97.
LONG, J. A. AND H. M. EVANS. 1922. Memoirs Univ. Calif., no. 6.
MEYER, R. K., S. L. LEONARD AND F. L. HISAW. 1929. Proc. Soc. Exper. Biol. and Med., **27**, 340.
VOGT, M. 1931. Arch. f. exper. Path. u. Pharm., **162**, 197.
          1933. Arch. f. exper. Path. u. Pharm., **170**, 72.

---

[1] The writer has found that rats that have not been accustomed to having smears taken probably respond more readily to the relatively delicate stimulus of copulation that is not carried to the point of ejaculation than do those accustomed to smears.   In 22 such matings, which were made some time after the data in this paper were collected and in a different laboratory, 7 (31 per cent) delayed the following oestrus although the average number of intromissions was the same as that which was entirely ineffective for the group of normal rats.   While the groups may not be strictly comparable, Vogt's findings suggest that the reason for this is that the later group which responded had never been smeared until the week of the tests, whereas the earlier, unresponsive animals had been submitted to the smear-taking routine for several weeks.

The author is indebted to Dr. E. V. McCollum for the use of the rats in his breeding colony in collecting these data.

38

Reprinted from *Proc. Natl. Acad. Sci.* **53**:1392–1395 (1965)

THE EFFECTS OF INTROMISSION FREQUENCY ON SUCCESSFUL PREGNANCY IN THE FEMALE RAT

BY JAMES R. WILSON, NORMAN ADLER, AND BURNEY LE BOEUF

UNIVERSITY OF CALIFORNIA, BERKELEY

Communicated by Frank A. Beach, April 9, 1965

The copulatory pattern of several species of rodents including the rat, mouse, hamster, and deermouse (Peromyscus) includes a series of mounts during which the male briefly inserts the penis into the vagina but withdraws without ejaculating. Ejaculation occurs during the last insertion of a series, when intromission is maintained for a longer period of time.

The function of pre-ejaculatory intromissions has previously been investigated from the standpoint of the male. Beach[1-3] postulated that each intromission of the male rat contributed an increment toward the attainment of an ejaculatory threshold; a series of studies developing this model have subsequently been carried out.[4-10]

Another function of the multiple intromission pattern, however, may apply to the female. A variety of neurogenic factors are known to trigger endocrine events connected with reproductive functions.[11, 12] For example, females of some species (rabbit, cat, ferret, and mink) are induced ovulators: their ovaries do not release eggs unless there has been copulation by the male, glass rod insertion, or some other form of stimulation to the vagina and cervix.[13, 14] The neurogenic stimuli activate the hypothalamo-hypophyseal system causing the release of luteinizing hormone (LH), which in turn leads to ovulation.

In addition, there are species in which the female normally produces nonfunctional corpora lutea but upon appropriate stimulation will maintain a prolonged luteal phase, pregnancy, or pseudopregnancy.[14-16] This prolongation induces progestational changes in the uterine endometrium and delays the subsequent estrus. Stimulation in this case leads to a change in the female's endocrine status—presumably an extended secretion of prolactin (LTH). Thus, in both induced ovulation and pseudopregnancy, vaginal stimulation can play a decisive role in initiating a neuroendocrine reflex. Finally, Beach[17] has proposed the hypothesis that the female rat's secretion of progesterone, and hence the probability that the progestational changes necessary for implantation will take place after mating, is dependent upon stimulation derived from repeated insertions by the male during copulation. Based on this hypothesis, one of the present authors, Wilson, performed a pilot study. He found that 5 of 6 females receiving 4 or more intromissions prior to ejaculation became pregnant, whereas only 1 of 6 receiving less than 4 intromissions became pregnant. The purpose of the present study was to confirm and extend these findings demonstrating a correlation between the multiple penile intromissions of the male rat and the maintenance of a successful pregnancy in the female.

Materials and Methods.—Twenty male and 20 female Long-Evans rats obtained from Diablo laboratories served as *S*'s. They were 130 days old at the beginning of the experimental tests. All animals had continuous access to food and water.

Observation cages were cylindrical, with 24-inch glass sides and 30-inch diameter wooden floors covered with wood shavings. Responses were scored with an Esterline-Angus events recorder.

The animals were obtained when 60 days old and were maintained on a 12-hr light (6 AM–6 PM)–12-hr dark cycle. Each female was paired in a cage with a male until a litter was born. This constituted the fertility test which ended 9 weeks later when every female had delivered at least one litter. For one week during the fertility test (after most S's had delivered), vaginal smears were taken. These data are not reported.

After each female had proved fertile, she was housed with 4 or 5 others and tested 3 times daily (7 PM, 9 PM, and 11 PM) for behavioral estrus. Males were caged individually. In initial tests for estrus, androgenized females were used as stimulus partners, but these animals mounted only infrequently, and active males were subsequently employed.

When experimental females came into estrus, they were mated with either an experimental or a control male. The experimental males were allowed several preliminary intromissions with a stimulus female brought into behavioral estrus by means of hormone injections. When the observer considered the male ready to ejaculate, the stimulus female was replaced by an experimental female (5–30 sec delay), and the male was allowed to continue until he ejaculated, usually on the first or second intromission with the experimental female. For 9 females the male's ejaculation occurred soon enough so that the female received a total of 3 or fewer intromissions, counting the ejaculatory intromission and any prior intromissions she may have received during the tests for estrus; these females constitute the Low Intromission group (LI). Five of the 10 females in the High Intromission or control group (HI) underwent regular mating series with untreated males, i.e., no preliminary intromissions were given, and they received a mean of 7.4 intromissions. The other 5 members of the HI group were females who had been given an experimental male who was considered ready to ejaculate, but who did not do so quickly enough for our purposes; these females received a total of 4 or more intromissions prior to or with the ejaculate (mean intromissions = 9.4). One female did not come into estrus during any test session and is not included in the results. A mating pair was left undisturbed for 3 min after ejaculation, then the female was examined for the presence of a vaginal plug. Mated females were housed individually for 20 days, at which time they were palpated to diagnose pregnancy. Those judged pregnant were sacrificed and subjected to laparotomy. Counts were made of viable fetuses, implantation sites, and corpora lutea. Nonpregnant females were returned to colony cages and given daily tests for behavioral estrus. When each became receptive, another mating session was held using the same male that had been used in the experimental test. In this postexperimental test, the male experienced no preliminary intromissions with a stimulus female and delivered his full complement of intromissions to the experimental female (mean = 11, range = 8–14). Each female was again caged individually and examined for pregnancy on the 20th day.

Results.—The main results are presented in Table 1. The probability that the reduction in pregnancies in the LI group is due to chance is 0.0046, by Fisher's exact probability test.[18]

Vaginal plugs were found in all S's of the LI group but were not seen in 4 of the S's in the HI group. Despite this discrepancy, 3 of the latter became pregnant.

Considering those pregnancies which were successful, no relation was found between number of intromissions received and number of viable or resorbed fetuses, or between intromissions and corpora lutea.

HI S's were with a male for a median time of 5 min 27 sec during the mating tests; LI S's were with a male for a median time of 4 min. This difference is statistically significant ($p < 0.02$, Mann-Whitney U test).

Discussion.—In female Long-Evans rats of demonstrated fertility, successful pregnancy depends in part on the number of intromissions delivered by the male

TABLE 1

RELATION BETWEEN SUCCESSFUL PREGNANCY AND INTROMISSION FREQUENCY

Group	Number Pregnant		
	Fertility test	Experimental test	Posttest
Low intromission (LI)	9 of 9	2 of 9	7 of 7
High intromission (HI)	10 of 10	9 of 10	0 of 1

prior to ejaculation. When 3 or fewer intromissions were permitted (LI group), 7 of 9 females failed to initiate and/or maintain a normal pregnancy. The presence of a plug in all LI S's after mating indicates that insemination was highly probable, and argues against the likelihood of nonfertilization as the primary cause of the failure of pregnancy. Moreover, since all 7 of the nonpregnant, LI S's did become pregnant during posttest fertility checks with the same male, we conclude that intromission frequency is a crucial factor influencing nidation. However, other variables such as olfactory stimuli, time between successive intromissions, effects of mounts, and state of receptivity may also be relevant. Indeed, the HI females not only received more intromissions than the LI S's, but also spent more time in the presence of the male.

Ball[14] reported a quantitative relationship between an aspect of the male's sexual pattern and the production of pseudopregnancy in female rats. In one experiment, delayed estrus always occurred after 2 vaginal plugs, sometimes after one plug, and rarely after intromissions without ejaculation. Our results might therefore be due to differences in LTH-progesterone production (pseudopregnancy). Without the requisite number of intromissions, the female's pituitary LTH is not released, the secretion of progesterone is not maintained, and implantation of the fertilized ovum (dependent on a progestational uterus) does not occur.

An alternative explanation might be provided by recent work of Shelesnyak and co-workers.[15, 16] In addition to the requisite progestational uterus, implantation of the rat blastocyst depends on an estrogen surge which occurs on the third day after mating. It is possible that stimulation arising from the male's successive intromissions is related to this estrogen surge—without stimulation, no estrogen is secreted and implantation does not occur. In either case, implantation of a fertilized ovum would be prevented by lack of neurogenic stimuli.

In its regular estrus cycle, the female rat has a very short luteal phase. During pregnancy or pseudopregnancy this phase is lengthened, leading to: (1) progestational proliferation of the uterine endometrium which forms an environment suitable for the implantation and nourishment of the fertilized egg, and (2) suppression of the subsequent ovulation which would otherwise occur 4–5 days later.[19]

A tie between a behavioral event and the lengthened luteal phase seen in pregnancy or pseudopregnancy seems adaptive reproductively. Without such a tie, a second ovulation would occur before the first fertilized ovum had traveled to the uterus, become implanted, and signaled suppression of further ovulation. Although superfetation occurs in other species, it apparently does not in the rat. Also, without such a tie, LTH production would follow LH production and would lead to progesterone formation and endometrial proliferation during every cycle, thus necessitating a 13–14-day estrus cycle rather than the 4–5-day cycle seen when copulatory or equivalent stimulation is not received. It seems to us that the behavioral event which leads to uterine preparation is a function of copulatory stimulation, probably intromission frequency summated over a short period of time. Other experiments are being undertaken to investigate the details of this mechanism.

Summary.—Successful pregnancy occurred in 90 per cent of female rats who received 4 or more penile intromissions with the male ejaculate. Only 22 per cent of the females receiving the ejaculate with 3 or fewer intromissions became pregnant.   These results were independent of differential fertilization.   The hypothesis is advanced that, in addition to successful fertilization, a certain amount of neurogenic stimulation is necessary for uterine implantation of the fertilized egg. Possible neuroendocrine mechanisms are discussed.

The hypothesis tested in this study was suggested to us by Dr. Frank A. Beach, and the experiment was supported by USPHS grant MH-04000 to F. A. Beach.   The expert technical advice of Dr. Giorgio Bignami was deeply appreciated.

[1] Beach, F. A., *Psychosomat. Med.*, **14**, 261 (1952).

[2] Beach, F. A., in *Nebraska Symposium on Motivation: 1956* (Lincoln: University of Nebraska Press, 1956), p. 1.

[3] Beach, F. A., in *Biological and Biochemical Bases of Behavior* (Madison: University of Wisconsin Press, 1958), p. 263.

[4] Larsson, K., *Conditioning and Sexual Behavior in the Male Albino Rat* (Stockholm: Almqvist & Wiksell, 1956).

[5] Larsson, K., *J. Comp. Physiol. Psychol.*, **51**, 325 (1958).

[6] Larsson, K., *Animal Behav.*, **7**, 23 (1959).

[7] Larsson, K., *Behaviour, Leiden*, **16**, 66 (1960).

[8] Larsson, K., *Scand. J. Psychol.*, **2**, 149 (1961).

[9] Bermant, G., *J. Comp. Physiol. Psychol.*, **57**, 398 (1964).

[10] Carlsson, S. G., and K. Larsson, *Scand J. Psychol.*, **3**, 189 (1962).

[11] Everett, J. W., in *Sex and Internal Secretions* (Baltimore: Williams & Wilkins, 1961), p. 497.

[12] Young, W. C., in *Sex and Internal Secretions* (Baltimore: Williams & Wilkins, 1961), p. 1173.

[13] Long, J. A., and H. M. Evans, *Mem. Univ. California*, **6**, 1 (1922).

[14] Ball, J., *Am. J. Physiol.*, **107**, 698 (1934).

[15] Shelesnyak, M. C., P. F. Kraicer, and G. H. Zeilmaker, *Acta Endocrinol.*, **42**, 225 (1963).

[16] Shelesnyak, M. C., and P. F. Kraicer, in *Delayed Implantation* (Chicago: University of Chicago Press, 1963), p. 265.

[17] Beach, F. A., *Sex and Behavior* (New York: John Wiley and Sons, in press).

[18] Siegel, S., *Nonparametric Statistics for the Behavioral Sciences* (New York: McGraw-Hill, 1956).

[19] Corner, G. W., *The Hormones in Human Reproduction* (New York: Atheneum, 1963), p. 69.

39

Reprinted from *J. Reprod. Fert.* **13**:121–125 (1967)

THE EFFECTS OF THE MATING PATTERN OF THE MOUSE ON THE FORMATION OF CORPORA LUTEA

R. B. LAND AND THOMAS E. McGILL*

Institute of Animal Genetics, West Mains Road, Edinburgh 9

(*Received* 31st *March* 1966, *revised* 29th *June* 1966)

Summary. Female mice were placed with males and subjected to three different treatments. One group experienced a low number of pre-ejaculatory thrusts, the ejaculatory reflex and the formation of a copulatory plug. A second group experienced a variable number of pre-ejaculatory thrusts, the ejaculatory reflex and the formation of the copulatory plug which was immediately removed. The third group experienced a variable number of pre-ejaculatory thrusts only. The results indicated that a large number of pre-ejaculatory thrusts is neither necessary, nor sufficient, for the induction of luteal activity in the female mouse. The ejaculatory reflex and the formation of the copulatory plug are sufficient to induce luteal activity regardless of the number of pre-ejaculatory thrusts. The results are compared with those of similar work on the rat.

INTRODUCTION

Quantitative aspects of the pattern of sexual behaviour of the male mouse have been shown to differ according to genotype (McGill, 1962; McGill & Blight, 1963; McGill, 1965). Since the laboratory populations studied were originally derived from wild populations, it is perhaps reasonable to assume that genetic variation in these traits exists in wild populations and, further, that these traits have been subjected to the pressures of natural selection. If this is correct, it would appear that natural selection has resulted in the evolution of a comparatively long and complex pattern of behaviour which, superficially at least, would be expected to reduce rather than increase fitness. This communication is concerned with the examination of a possible mechanism for the correlation between this aspect of behaviour and fitness, but before discussing the problem further it is necessary to describe the pattern of behaviour both qualitatively and quantitatively.

The pattern of sexual behaviour in the male mouse (*Mus musculus*) consists of a series of mounts and mounts-with-intromission (called 'intromissions'). This series normally ends in ejaculation (which can be identified by the male clutching the female with all four limbs and falling to his side) and the formation of the copulatory plug which remains in the female's vagina for about 18 hr.

The work already cited indicates that for the particular strains studied

* Permanent address: Department of Psychology, Williams College, Williamstown, Massachusetts, U.S.A.

ejaculation took place after an average of 425 thrusts, with a mean mating time of 1500 sec. It has also been shown (McGill, 1963) that if a male has ejaculated in the previous 28 hr, both these parameters are reduced to about 20% of the above values.

One possible reason for the evolution of this pattern of behaviour is found in the physiology of the females. It is well known that ovulation occurs spontaneously during the oestrous cycle of both the rat (*Rattus norvegicus*) and the mouse (Asdell, 1964). However, in the absence of copulation, or similar stimulation, functional corpora lutea are not formed and the oestrous cycle does not have a true luteal phase. It is obvious, therefore, that the male mouse must somehow provide appropriate stimulation to the female during a normal mating in order to induce the luteal phase. Three sources of stimulation from the male may be either sufficient or necessary for the induction of the luteal phase. These are: (1) a minimum number of pre-ejaculatory thrusts, (2) the ejaculatory reflex and the formation of the copulatory plug and (3) the subsequent presence of the plug in the vagina. The present experiment was designed to study the relative importance of these three factors.

MATERIAL AND METHODS

Subjects were 10- to 20-week-old male and female mice of the outbred Q-strain which is maintained at the Institute of Animal Genetics, Edinburgh. They had been kept on a reversed light–dark cycle (08.00 hours–20.00 hours) for 6 weeks before the beginning of the experiment. Observations were made in the early afternoon under normal room illumination.

Sexually experienced males were placed individually in plastic cyclinders 10 in. in diameter and 20 in. in height. Two females were then introduced into each cylinder and about 15 min allowed for the initiation of sexual behaviour. Some females, even though in oestrus would reject, or be rejected by a particular male, and consequently, if mating did not occur, the females were offered to a second male, and two new females introduced into the cylinder of the first male.

Matings were carefully observed and controlled to prepare the following groups of females: (a) females which received a relatively low number of thrusts (6 to 55), the ejaculatory reflex and the formation of the copulatory plug; (b) females which received a variable number of thrusts (24 to 135), the ejaculatory reflex and the formation of the copulatory plug which was immediately removed (this is not easily done and one can never be sure that all the plug has been extracted); and (c) females which received a variable number of thrusts (15 to 156) only, i.e. these females did not experience the male's ejaculatory reflex or receive a copulatory plug.

Following the experimental treatment the females were caged singly for 48 hr. At the end of this period a sexually experienced 'indicator-male' was placed in the female's cage. The female was subsequently examined each afternoon for the presence of a copulatory plug and/or for the birth of a litter.

The birth of a litter was recorded in order to ensure that the number of days from treatment to mating was indeed the number of days taken to return to true oestrus, which enabled us to discount such problems as behavioural oestrus during pregnancy (Nalbandov, 1964). Females which took more than 8 days to return to oestrus were considered to have become pseudopregnant.

Four females which did not give birth to a litter around 19 to 21 days after mating were discarded.

RESULTS

Since we are interested in both the effects of the treatments and the effects of the number of thrusts within a treatment, the results are presented as a histogram (Text-fig. 1). Each female is represented by a number which indicates

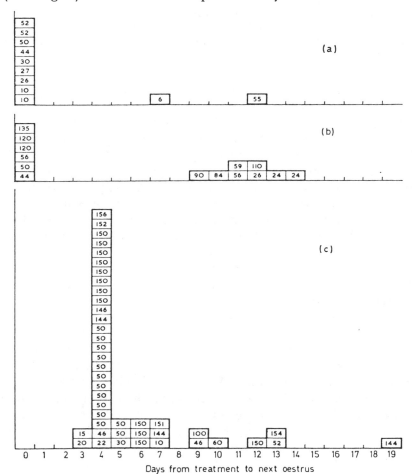

TEXT-FIG. 1. The number of days before return to fertile oestrus after experimental treatment. Each mouse is represented by a number which indicates the number of thrusts it received. (a) Normal mating; (b) normal mating, plug removed; (c) pre-ejaculatory thrusts only.

the number of thrusts it received. The abscissa shows the number of days before return to fertile oestrus. All females impregnated by the treatment are therefore indicated on Day 0. Days 1 and 2 were the aforementioned isolation days which insured that the Day 0 oestrus had passed before the introduction of the indicator males.

The results presented in Text-fig. 1 indicate that the number of thrusts without ejaculation [group (c)] has little effect on the number of days between treatment and subsequent oestrus. Most of the females returned to fertile oestrus 4 days after treatment, as would be expected if the oestrous cycle was unaffected by the treatment. By contrast, none of the females which had received a plug [groups (a and b)] returned to oestrus after the normal interval of time. All were either pregnant, or, with the possible exception of one female in group (a) that returned to oestrus after 7 days, pseudopregnant.

DISCUSSION

These results permit the following conclusions regarding the induction of luteal activity in female mice.

(1) A large number of pre-ejaculatory thrusts seems neither necessary nor sufficient, for the induction of luteal activity. It should be noted that while the number of thrusts received by the females in this study was only about 10 to 25% of that preceding ejaculation in the fully-rested (13 days) male, the treatments are similar to the number of thrusts preceding ejaculation in the fertile copulation of a male who has recently ejaculated.

(2) The observation that females of groups (a) and (b) became either pregnant or pseudopregnant regardless of the number of pre-ejaculatory thrusts they received indicates that the ejaculatory reflex, together with the formation of the plug, is sufficient for the establishment of luteal activity and that the removal of the plug only affects the proportion which become pseudopregnant rather than pregnant. The present study cannot separate the effects of the ejaculatory reflex from the effects of plug formation, but these two factors could be differentiated by the use of males which have been surgically deprived of their coagulating glands.

In the rat, intromissions are very brief, and most frequently consist of a single thrust, (Bermant, 1965). Wilson, Adler & Le Boeuf (1965) have studied the effects of number of intromissions before ejaculation on successful pregnancy in the female rat. They found that nine out of ten female rats who received four or more intromissions before ejaculation became pregnant; whereas only two out of nine females who received three or fewer intromissions became pregnant. Since an average of eleven intromissions occurs before the first ejaculation in a rested male rat (Beach & Jordan, 1956), one may conclude that at least one-third of the normal number of intromissions is necessary for the induction of a luteal phase in the oestrous cycle of the rat.

It would appear, therefore, that there is a fundamental difference between the physico-endocrinological relationships of the rat and the mouse. This indication is supported by the fact that although there is abundant evidence for the establishment of pseudopregnancy following cervical stimulation in the rat, Stone & Emmens (1964) and Finn (1965) reported that electrical and mechanical stimulation of the cervix of the mouse did not result in pseudopregnancy.

The results of this study indicate that a large number of pre-ejaculatory thrusts are not, in themselves, sufficient to initiate the neuro-hormonal reflexes which result in a luteal phase in the oestrous cycle of the mouse. By contrast, however, the ejaculatory pattern of the male, together with the formation of a copulatory plug, appears to be adequate for the induction of the formation of functional corpora lutea, regardless of the number of pre-ejaculatory thrusts. Consequently, the relationship between fitness and the complex mating pattern of the mouse does not appear to be dependent on the induction of the formation of functional corpora lutea.

ACKNOWLEDGMENTS

This research was supported in part by an Agricultural Research Council Postgraduate Studentship and further financial support from the Ford Foundation to R. B. Land, and in part by Research Grant GM-07495 from the U.S. Public Health Service to T. E. McGill. The data were collected while T. E. McGill was National Academy of Science–National Research Council, Senior Postdoctoral Fellow in Physiological Psychology at the University of Edinburgh.

REFERENCES

ASDELL, S. A. (1964) *Patterns of mammalian reproduction*, 2nd edn, p. 213. Cornell University Press, Ithaca, New York.

BEACH, F. A. & JORDAN, L. (1956) Sexual exhaustion and recovery in the male rat. *Q. Jl exp. Psychol.* **8**, 121.

BERMANT, G. (1965) Rat sexual behavior: Photographic analysis of the intromission response. *Psychonomic Sci.* **2**, 65.

FINN, C. A. (1965) Oestrogen and the decidual cell reaction of implantation in mice. *J. Endocr.* **32**, 223.

McGILL, T. E. (1962) Sexual behavior in three inbred strains of mice. *Behaviour*, **19**, 341.

McGILL, T. E. (1963) Sexual behavior of the mouse after long-term and short-term post-ejaculatory recovery periods. *J. genet. Psychol.* **103**, 53.

McGILL, T. E. (1965) *Studies of the sexual behavior of male laboratory mice: Effects of genotype, recovery of sex drive, and theory.* Sex and Behavior, p. 76. Ed. F. A. Beach. Wiley, New York.

McGILL, T. E. & BLIGHT, W. C. (1963) The sexual behaviour of hybrid male mice compared with the sexual behaviour of males of the inbred parent strains. *Anim. Behav.* **11**, 480

NALBANDOV, A. V. (1964) *Reproductive physiology*, 2nd edn, p. 269. Freeman, San Francisco.

STONE, G. M. & EMMENS, C. W. (1964) The action of dimethylstilboestrol on early pregnancy and decidua formation in the mouse. *J. Endocr.* **29**, 137.

WILSON, J. R., ADLER, N. & LE BOEUF, B. (1965) The effects of intromission frequency on successful pregnancy in the rat. *Proc. natn. Acad. Sci. U.S.A.* **53**, 1392.

40

Reprinted by permission from J. Agric. Sci. **56**:31–37 (1961)

Multiple sire mating in swine; evidence of natural selection for mating efficiency

By LAVON J. SUMPTION

Animal Husbandry Department, University of Nebraska, Lincoln

(*Received* 18 *May* 1960)

Darwin's observation (1875) that the animal breeder 'does not allow the most vigorous males to struggle for the females...does not rigidly destroy all inferior animals but protects them during each varying season', is still essentially correct. The animal breeder has not given singular attention to mating efficiency (i.e. sexual selection) except for selection against the most extreme individuals with anatomical abnormalities, atypical mating behaviour and low prenatal and postnatal survival. Multiple sire mating, the simultaneous exposure of a group of males to a population of females, was used in developing new breeds of swine (Winters, 1956) and cattle (Lasater, 1958; Rhoad, 1955) as a means of inducing natural selection for traits associated with reproductive fitness (Sumption, Rempel & Winters, 1960). It is the purpose of this paper to present evidence from the first studies designed to ascertain the nature and extent of natural selection for mating efficiency in swine.

The impact of natural selection in altering quantitative traits of farm animal populations has not been adequately measured. It is recognized that exposure to varying environmental conditions including nutrition, disease, and climate may influence phenotypic variability, and undoubtedly affects the specific genotypes actually selected, that selection pressure applied by the animal breeder is always superimposed upon some background of natural selection (Lush, 1954; McMeekan, 1941; Winters, 1938). Therefore, it would be pertinent to intensify the effects of natural selection for those traits where the objectives of the breeder and the results obtained by natural selection are compatible, and attempt to offset those effects where objectives are not compatible. It is suggested that the animal breeder has not made sufficient use of 'directed' natural selection in cases where it may be beneficial, as contrasted with the way it has been used effectively by the plant breeder (Suneson, 1956).

OUTLINE OF EXPERIMENTS

Exp. I a. 'Random' mating—this was a preliminary experiment to establish procedures and assess problems concerning multiple paternity and mating behaviour that might warrant further study Ten Duroc females were exposed daily and simultaneously during one oestrous cycle to sires of five breeds: Duroc, Hampshire, Minnesota no. 1, Yorkshire, and Line X (a population of recent cross-bred origin). Frequency of mating was recorded for each sire during 2 hr. mating periods twice daily. The breeds of sires used permitted recognition of paternity due to differences in hair colour and/or body form of the progeny, as indicated in Table 1

Exp. I b. 'Random' mating—sixteen dams of three breeds were exposed daily and simultaneously to twelve sires, four each of the Duroc, Hampshire and Yorkshire breeds. None of the sires had previous postpuberal bi-sexual mating experience General observations of mating behaviour in both sexes were made during one 3 hr. period daily for one oestrous cycle. Interference with normal be

Table 1. *Colour patterns of breeds and crosses used in Exps. I a, II and III*

Breed of sire	Colour of sire	Colour of progeny when sires are mated to Duroc females
Duroc	Red	Red
Berkshire	Black; white nose, feet and tail	Red; small black spots
Hampshire	Black, white belt around shoulder	Solid black, few with typical Hampshire marking
Minnesota no. 1	Red	Red (distinguished from Durocs by longer nose and ears and reduced pigmentation of extremities)
Yorkshire	White	White hair, occasional suggestion of red hair pigment and small red or black body spots
Line X	Variable, with tendency to black, black and white or black and red, and occasional red and white patterns	Black, or red with black spots

haviour during the time of observation was avoided. The number of matings per sire that resulted in vaginal ejaculation were tabulated. Brief descriptive notes were made of obvious deviations in male and female mating behaviour. The question of primary concern was whether or not there was sufficient variation in number of matings per male and female so that natural selection for effective mating might be of significant consequence. Paternity could not be clearly established in all cases because most of the dams were either Hampshire or Yorkshire, thus rendering colour markers partially ineffective; furthermore, the same female frequently mated with more than one male of the same breed. These data were not used in considering fertilization and the frequency of multiple paternity.

Exp. II. Controlled multiple mating—to assess the magnitude of selective fertilization, twelve Duroc dams were mated in rapid succession to sires of five breeds during the second day of oestrus only. Sires of known fertility were used. The maximum interval between the first and last mating was 65 min. ($\bar{x} = 39$ min.). Each sire was allowed only one mating; no attempt was made to control the period of ejaculation, but periods as long as 10 min. were unusual. The order of mating with respect to the breed of sire was rotated to allow equal opportunity for the sires to mate in each position relative to the other sires.

Exp. III. Double mating—twenty Duroc dams were mated during the first and second days of oestrus, on day 1 to a Duroc sire and a sire of a second breed; on day 2 each dam was remated with a Duroc sire and also to a sire of a third breed. The primary objective in this case was to produce purebred and cross-bred progeny in the same litter for

a cross-breeding study. Observations relating to selective fertilization were also made.

All of the progeny born in Exps. II and III were classified and photographed. Obviously, differential mortality of certain genotypes in early prenatal development were not detectable by the method of classification used; to that extent the observations are subject to error. The Duroc and Minnesota no. 1 × Duroc pigs were reclassified twice (because of similarity in colour) at later ages to assess the accuracy of the initial determinations. Changes in identification were made in less than 5% of the cases.

Table 2. *Distribution of progeny by sires under 'random' mating (Exp. I a)**

Litters	Sires†				P‡
	D	M	X	Y	
	No. of progeny				
110	0	5	1	0	0·01
122	—§	—	8	—	—
124	—	—	7	—	—
132	2	1	2	—	> 0·80
133	3	4	2	—	> 0·70
217	2	2	—	2	1·00
219	—	4	—	5	> 0·50
419	0	0	6	0	< 0·001
508	1	6	—	0	0·010
527	0	—	9	—	0·003
Totals	8	22	35	7	

* Arranged by breed without regard to order of mating.
† D = Duroc, M = Minnesota no. 1, X = Line X, Y = Yorkshire.
‡ P = probabilities based on χ^2 test of deviations from uniform distribution.
§ No mating.

Table 3. *Distribution of progeny by sires in litters produced by controlled multiple mating (Exp. II)*

Litters	Sires*					P†
	B	D	H	M	Y	
	No. of progeny					
10	0	6	2	—‡	3	0·070
20	0	7	2	4	0	0·005
30	0	1	8	1	0	< 0·001
40	1	1	9	2	1	0·002
50	0	7	0	1	2	0·002
60	0	0	8	0	0	< 0·001
70	0	1	10	0	0	< 0·001
80	9	0	3	0	0	< 0·001
90	0	1	6	1	0	0·004
100	0	9	2	1	1	< 0·001
110	0	2	8	1	0	< 0·001
120	0	1	3	0	3	> 0·100
Totals	10	36	61	11	10	

* B = Berkshire, D = Duroc, H = Hampshire, M = Minnesota no. 1, Y = Yorkshire.
† P = probabilities based on χ^2 test of deviations from a uniform distribution.
‡ No mating.

RESULTS

Selective fertilization. Data from Exp. I*a* raised a number of questions for further study (Table 2). The possibility of selective fertilization was suggested by the unequal representation of sires in four of the ten litters observed. Selection of another kind was also indicated in that the females were not uniformly mated by all sires.

The magnitude of selective fertilization was more effectively revealed in Exp. II (Table 3). Duroc-sired progeny were represented in slightly greater numbers than the combined frequency of the progeny sired by the Berkshire, Minnesota no. 1 and Yorkshire boars. Hampshire-sired progeny were nearly 1·7 times as frequent as those sired by Durocs. It is interesting to note that Hampshire sires were represented in eleven of the twelve litters observed, which is in sharp contrast to the Berkshire boars which produced progeny in only two litters.

Another means of examining selective fertilization is based on the average number of progeny produced per mating by each sire, under conditions of multiple mating. Data from uncontrolled 'random' mating in Exp. I and controlled multiple mating in Exps. II and III are presented (Table 4). In the two latter studies it is possible to make within-breed comparisons of four different breeds. It is interesting to note that in three breeds, the sires retained the same rank order of effectiveness. Differences between individual sires are large enough to be considered important despite the fact that observations are so few in number as to make a statistical test impractical.

When the various breeds are compared in each of the three experiments, marked differences in progeny per mating can be seen. Generally, sperm of the Berkshire, Minnesota no. 1 and Yorkshire sires were less effective 'competitors' than those of the Duroc and Hampshire sires. Observed variability within and between breeds and between experiments involving the same sires, indicates the need for cautious interpretation based on these relatively small numbers.

Mating behaviour. To evaluate the potentiality of natural selection for favourable mating behaviour, it is pertinent to determine if there is any evidence for a differential frequency of mating among certain males and females. Data from Exps. I*a* and I*b* reveal that under the conditions of 'random' or unrestricted mating there was considerable variation in the frequency of mating for both sexes (Tables 5, 6).

Variations in male activity were not as large in Exp. I*a* as in I*b*; however, there are several observations worth noting from I*a*. A clear-cut social order among the four sires was established during the first day of observation with the following descending order of dominance: Minnesota no. 1, Duroc, Line X and Yorkshire. The Minnesota no. 1 boar was the dominant sire, but this was not indicated in the relative frequency of successful mating. This sire had somewhat longer legs than the females and as a result, intromission was accomplished only with considerable difficulty. He exhibited a progressively more frustrated response to oestrus females during the course of observations. If intromission did not occur shortly after mounting, the violent thrusting action frequently prevented intromission and ejaculation.

Competition for oestrous females consisted of continual harassment of the females being mounted and mated by other males. This sexual competition

Table 4. *The relative effectiveness of sires of six breeds*

Breed of sire	Sire no.	'Random' mating (Exp. I*a*)		Multiple mating (Exp. II)		'Double' mating (Exp. III)	
		No. of matings	Progeny per mating	No. of matings	Progeny per mating	No. of matings	Progeny per mating
Berkshire	1	—	—	7	0·14	1	0·00
Berkshire	2	—	—	5	1·80	2	3·00
Duroc	1	10	0·80	—	—	—	—
Duroc	2	—	—	2	0·00	9	4·50
Duroc	3	—	—	7	4·71	6	7·50
Duroc	4	—	—	3	1·00	9	5·72
Hampshire	1	—	—	2	2·50	2	4·00
Hampshire	2	—	—	10	5·50	6	2·00
Line X	1	16	2·19	—	—	—	—
Minnesota no. 1	1	8	2·75	2	0·50	7	1·14
Minnesota no. 1	2	—	—	9	1·11	3	2·00
Yorkshire	1	7	1·00	—	—	—	—
Yorkshire	2	—	—	7	0·86	—	—
Yorkshire	3	—	—	4	1·00	2	6·50

Table 5. *Number of matings per sire during a 22-day mating period*

Breed	Exp. I*a*		Exp. I*b*	
	Sire no.	Total matings	Sire no.	Total matings
Duroc	1	10	1	8
			2	0
			3	2
			4	0*
Hampshire	—	—	1	3
			2	9
			3	8
			4	0
Line X	1	11	—	—
Minnesota no. 1	1	8	—	—
Yorkshire	1	7	1	9
			2	4
			3	4
			4	3

* Became lame on day 2; could not mate subsequently.

Table 6. *Frequency distribution of matings per female under 'random' mating (Exps. I a and I b)*

No. of matings ...	1	2	3	4	5	6	7
No. of females, Exp. I*a*	1	1	1	3	1	2	1
No. of females, Exp. I*b*	4	4	2	2	2	0	2

was not always in complete accord with social order. Low-order males would interfere with mating by a socially dominant male despite their frequent subsequent inability to dominate the female if their harassing succeeded in terminating the mating of a higher order male.

Results of Exp. I*b* do not provide clear evidence for breed differences in mating activity among males, but suggest significant variations in the number of matings per sire within breeds (i.e. the same general range of variation occurred within each breed). Table 5 reveals only a small portion of the actual variations in mating activity. In general, the boars that mated most frequently also displayed the greatest amount of lower-grade mating behaviour such as pursuit, nuzzling and trial mounts. The notable exception was Duroc sire no. 2. This boar mounted females almost continually during each observation period. Because he was incapable of producing an erection, no effective matings transpired. Duroc no. 2 actually interfered with mating attempts by other boars because of his ability to dominate the oestrous females by persistent mounting. On the other hand, Duroc sire no. 3, a litter-mate of no. 2, anatomically capable of mating as revealed by a trial mount, also failed to contribute progeny. He had

to be removed after the first 2 days of study because of severe lameness.

Among the Hampshire sires, no. 1 and no. 2 were litter-mates. Boar no. 2 began mating on the first day of observation; no. 1 did not mate until day 8 during which nine of the sixteen dams conceived. Hampshire sire no. 4 exhibited no appreciable sexual interest under the conditions of this study. Ordinarily this sire would lie down within minutes after entering the mating pen; when nuzzled by oestrous females, his only reaction was a minor aggressive response.

Variation in rapidity of learning also appeared among the Yorkshires. Sire no. 1 began mating on day 1, sire no. 2 on day 7, sire no. 3 on day 14 and sire no. 4 on day 13. None of the Yorkshires used in this study were litter-mates. In another breeding experiment not reported here, two litter-mate sires were both exposed to the same oestrous females at the same time each day under favourable mating conditions. Sire A_1 began mating on day 1, but sire A_2 did not make any attempt to mate until day 17. After the initial successful mating, A_2 was as active sexually as A_1.

Very little direct aggression was observed in Exp. I*b*. Most of the competition among sires consisted of (*a*) differences in reaction time prior to mounting after an oestrous female was identified, and (*b*) the drive expressed by the sire in mounting, together with his ability to withstand attempts by other sires to appropriate the female when she was harassed. By continued observation of the reaction among the sires that (*a*) began to mate immediately as compared to (*b*) the sires that remained sexually inexperienced for 2–13 days, it appeared that the competition encountered tended to retard the rapidity of learning an ordered mating response among the 'slower' sires.

The nature of female mating activity was similar in Exps. I*a* and I*b* (Table 6). Variation in frequency of matings per female was somewhat greater than expected and not closely associated with the number of females in oestrus on any given day, but rather apparently due to their specific behavioural response to the mating situation. Certain females were mated only once during oestrus; subsequent approaches were either nominally or totally refused. Four females in Exp. I*b* required prolonged pursuit prior to mating, especially during the first day of receptivity. The frequency of mating was reasonably high for three of these females (5, 4 and 4, respectively) but the fourth permitted only one effective mating despite persistent pursuit throughout observation periods on 2 subsequent days. The females that required prolonged pursuit emitted frequent shrill squeals and tended to draw the attention of the total group of sires away from other oestrous females.

DISCUSSION

It should be unnecessary to debate at any length about the possible contemporary value of natural selection in the evolution of improved agricultural species. If some reservation exists among animal breeders, it is probably because they do not have direct evidence available that is comparable to the classic study of Harlan & Martini (1938) or the thought-provoking results of Suneson (1956), both based on long-term experiments with plant populations of high initial heterozygosity. The problem is largely one of identifying and measuring the magnitude of its influence.

Possible effects of natural selection for mating efficiency are evident from the results of the experiments reported here. Differential reproduction could occur under the conditions of multiple sire mating as a result of (a) selective fertilization; (b) differences in mating behaviour of sires, including social dominance, libido, reaction time and speed of learning; (c) mating behaviour variations among females, especially those related to receptivity; (d) anatomical differences among selected parents and (e) several combinations of the above.

Evidence of selective fertilization is advanced from three of the experiments. Variations in the number of progeny sired by different boars when provided a similar opportunity for fertilization are significant (Tables 3, 4). Determining the physiological basis for these effects was not a part of this particular series of studies. Selective fertilization has been reported previously in swine (Klabukov, 1955; Libizov, 1956) and cattle (Rowson, 1956), also without resolution of the mechanisms involved.

Differences in competitive ability and/or concentration of viable sperm are plausible hypotheses for differential fertilization based on Beatty's work (1957) with rabbits. Production of equal numbers of progeny per sire was dependent upon use of relatively unequal semen volumes; however, another report from the same laboratory (Edwards, 1955), and research elsewhere with poultry (Bonnier & Trulsson, 1939), indicates that more than a similarity in total sperm number is involved. Substantial correlations between sperm quality and number of progeny per sire have been observed in poultry (Allen & Champion, 1955). An excess of progeny by heterospermic males caused King (1929) to postulate an affinity between genetically unlike gametes. This could explain the excess of Hampshire-sired cross-bred progeny in the present study (Table 3) but would not account for the deficiency of Berkshire, Yorkshire and Minnesota no. 1 sired progeny as compared to the pure-bred Durocs. Gametic affinities were not observed in several other recent experiments (Allen & Champion, 1955; Edwards, 1955). Genetic-physiological factors conditioning the selective fertilization that is operative in these studies may be similar to those found in plants (e.g. self-sterility) where mechanisms are somewhat better known (Hayes, Immer & Smith, 1955).

In these swine experiments the progeny of one sire tended to predominate in any one litter, although sperm from as many as four other sires were present in the female tract at the time of fertilization. Factors conditioning this effect have not been determined. One possible explanation may be related to the fact that boars typically ejaculate a large volume of dilute semen (about 200 ml.). It is not uncommon to observe loss of semen from the female tract during and following normal single matings. Under the conditions of these experiments, frequent mating in close succession could have led to major sperm loss via some sort of 'flushing' action effected by subsequent matings. The precaution of a continual rotation in the order of mating of the series of sires was designed to nullify any consistent effect of one sire. In fact, the order of exposure had no effect on selective fertilization in this experiment (Sumption & Adams, 1960) as contrasted with the dramatic results observed in some poultry studies (Bonnier & Trulsson, 1939). Several existing questions regarding selective fertilization in swine could best be elucidated by artificial insemination studies involving variations in concentration and volume of sperm mixtures.

Previous studies of mating behaviour in swine have been limited largely to a description of typical behaviour patterns (Quinto, 1957) and factors associated with the expression of oestrus (McKenzie, 1926), except for the extensive studies of Burger (1952). Limited evidence of genetic differences in mating behaviour has been advanced (Green & Winters, 1945; Hodgson, 1935) based on observed variations in breeding behaviour which were associated with differences in endocrine secretion. The extreme difficulties encountered by Hodgson (1935) in conducting a continual brother–sister mating system, provides a clue to the intensity of psychological problems involved in the expression of swine behaviour. Sexually mature boars that would breed non-litter-mates readily, refused to mate with full sisters despite attempts to break down recognition factors by the use of nostrums or isolated rearing of the sexes from weaning until postpubera exposure.

The present preliminary data indicate considerable variation in frequency of mating and time of onset of typical mating response among sexually mature males as well as differences in receptivity among the females. At the low extreme, three sires failed to mate successfully, one because of low libido, a second because of lameness, and the third because of an anatomical defect in the genitalia. The delay in onset of mating exhibited by certain boars automatically prevented their potential contribution to those litters conceived previously during the mating period. It might be questioned whether these differences in 'rate of learning' among sires lacking postpuberal bi-sexual contact are heritable or somehow related to chance differences in pre-exposure unisexual experiential factors. The influence of heredity on selective mating is effectively illustrated in critical studies of *Drosophila* populations by Merrell (1953).

Variations in sexual receptivity among the females were revealed by differences in mating frequency in two experiments (Table 6). It was interesting to note that the least receptive females were mated by the sires with the highest mating frequencies. It is not possible to more than speculate about the adaptive significance of this finding at this time. If the above trend is generally true and is not reversed by the most receptive females, the reproductive fitness of the population should be enhanced. Natural selection should favour the persistent sire that would seek and mate successfully regardless of the level of receptivity of the female.

An adequate case has been made for the evolutionary significance of sexual selection (Darwin, 1875; Jakway, 1959; Levine, 1958). It appears that reproductive fitness could be altered by some scheme of 'directed' natural selection for the combined effects of favourable mating behaviour and selective fertilization. This conclusion is drawn from the evidence in this study and extensive research in other species (Guhl, 1953; Hultnäs, 1959; Levine, 1958). 'Directed' natural selection is meant to imply that certain criteria of effective mating would be used to reduce the size of the sire population to a select group that would be exposed to the selected dams.

In the group-mating situation, mating frequencies were sufficiently non-random to allow for differential reproduction, particularly in the light of selective fertilization. The success of combined selection for these several phenomena will depend upon the magnitude of their separate heritabilities and the nature of their genetic correlation. The difficulty of the simultaneous modification of the mating behaviour–fertility complex is clear from work in which sexual activity and sperm quality were not correlated (Jakway & Young, 1958; McKenzie & Berliner, 1937). Furthermore, the rank order of the same males for separate components of behaviour influencing effective mating can be quite diverse (Guhl 1953). Dramatic supporting evidence was provided by Duroc Sire no. 2 in Exp. I*b*, a socially dominant boar that was incapable of intromission and ejaculation. To test the value of

multiple sire mating adequately, a comprehensive experiment is needed that would combine observations of mating behaviour and selective fertilization. Further analyses of mating behaviour needed to place the present observations in proper perspective with the previous work are being conducted under conditions of group (multiple sire) mating, individual and paired sire exposure to single oestrous females and with specific stimulus alterations (Sumption & Jakway, 1959).

Mating efficiency is an important economic factor in the evolution of genetically improved populations of swine. The fertility complex is affected by mild rates of inbreeding more adversely than growth rate (Dickerson *et al.* 1952). Sex drive is among the factors influenced. To what extent the conditions of selection possible under multiple sire mating might aid in offsetting the effects of inbreeding on fertility are subjects of future research.

At this point one may only speculate regarding the relationship between libido and selective fertilization on the one hand and the expression of fertility via ovulation rate and embryonic survival on the other. It is not possible to determine the benefit of natural selection for mating efficiency through multiple sire mating, until more is known of the relationship between factors associated with reproductive fitness and other quantitative traits with which the animal breeder is concerned. Nevertheless, it is clear from the second 'random' mating study that three sires exhibiting phenotypic superiority for growth were of no genetic value to the population because they were excluded from reproduction.

It is difficult to arrive at a reasonable working definition of natural selection because many factors are usually involved in differential reproduction. Until the limits can be defined more critically, the statement that 'natural selection is without purpose' (Lerner, 1958) will lack clarity as it applies to animal breeding. The term 'directed' natural selection was used in this study as it applies to multiple sire mating, to draw attention to the fact that selection for certain quantitative traits other than reproductive performance was practised before the selected parents were brought together as a freely inter-breeding population. On the other hand, the relative numbers of progeny produced by each sire was governed by selective forces not under man's direct control, operating at the levels of gametogenesis, mating behaviour, fertilization and survival (prenatal and postnatal). If the efforts of the animal breeder were to be classified as another of the series of natural biotic phenomena, would the term 'artificial' selection convey any effective meaning? It is suggested that directed selection is a more meaningful term than the autonym of natural selection.

The fact that mating appeared to be dramatically non-random in these small populations, whether examined at the level of mating or gametic combination prompted the qualification of the term 'random' mating.

SUMMARY

Evidence of natural selection for certain aspects of mating efficiency in swine are advanced based on preliminary studies with thirty-one sires, fifty-eight dams and their progeny. Selective fertilization was conclusively demonstrated. Variations in male and female mating behaviour were sufficiently large to indicate considerable non-randomness of mating frequency under the conditions of multiple sire mating (i.e. group exposure of dams to selected sires). The combined effects of the separate phenomena of selective fertilization and mating behaviour are discussed in relation to their evolutionary significance in animal breeding.

Published in co-operation with U.S.D.A., A.R.S., A.H.R.D., Regional Swine Breeding Laboratory; Ames, Iowa. Exps. I*a*, II and III of this study were conducted at the North Platte Experiment Station, University of Nebraska; the full co-operation of Supt. J. C. Adams, Robert Haney, herdsman, and Carl Swanson, assistant herdsman, is gratefully acknowledged. Appreciation is expressed for the counsel of Dr Loyal C. Payne, Associate Professor of Veterinary Science, and the assistance of Eldon Svec, student in Animal Husbandry.

REFERENCES

ALLEN, C. J. & CHAMPION, L. R. (1955). *Poult. Sci.* **34**, 1332.

BEATTY, R. A. (1957). *J. Genet.* **55**, 325.

BONNIER, G. & TRULSSON, S. (1939). *Hereditas*, **25**, 65.

BURGER, J. F. (1952). *Onderstepoort J. Vet. Res.* **25**, Suppl. no. 2.

DARWIN, C. (1875). *The Variation of Animals and Plants Under Domestication. I.* 2nd ed. Murray: London.

DICKERSON, G. E., BLUNN, C. T., CHAPMAN, A. B., KOTTMAN, R. M., KRIDER, J. L., WARWICK, E. J. & WHATLEY, J. A. JR. (1952). *Bull. Mo. Agric. Exp. Sta.* no. 551.

EDWARDS, R. G. (1955). *Nature, Lond.*, **175**, 215.

GREEN, W. W. & WINTERS, L. M. (1945). *J. Anim. Sci.* **4**, 55.

GUHL, A. M. (1953). *Bull. Kansas Agric. Exp. Sta.* no. 73.

HARLAN, H. V. & MARTINI, M. L. (1938). *J. Agric. Res.* **57**, 189.

HAYES, H. K., IMMER, F. R. & SMITH, D. C. (1955). *Methods of Plant Breeding.* New York, N.Y.: McGraw Hill Book Co., Inc.

HODGSON, R. E. (1935). *J. Hered.* **26**, 209.

HULTNAS, C. A. (1959). *Acta agric. Scand.*, Suppl. no. 6.

JAKWAY, J. S. & YOUNG, W. C. (1958). *Fertil. and Steril.* **9**, 533.

JAKWAY, J. S. (1959). *Anim. Behav.* **7**, 150.

KING, H. D. (1929). *Arch. EntwMech. Org.* **116**, 202.

KLABUKOV, P. G. (1955). *Anim. Breed Abstr.* **24**, 182.

LASATER, T. (1958). Personal communication. The Lasater Ranch, Matheson, Colorado.

LERNER, I. M. (1958). *The Genetic Basis of Selection.* New York, N.Y.: John Wiley and Sons.

LEVINE, L. (1958). *Amer. Nat.* **92**, 21.

LIBIZOV, M. P. (1956). *Anim. Breed. Abstr.* **24**, 277.

LUSH, J. L. (1954). *Proc. 9th Int. Congr. Genet.*, 589.

McMEEKAN, C. P. (1941). *J. Agric. Sci.* **31**, 17.

McKENZIE, F. F. (1926). *Bull. Mo. Agric. Exp. Sta.* no. 86.

McKENZIE, F. F. and BERLINER, V. (1937). *Bull. Mo. Agric. Exp. Sta.* no. 265.

MERRELL, D. J. (1953). *Evolution*, **7**, 287.

QUINTO, M. G. (1957). *Philipp. Agric.* **41**, 319.

RHOAD, A. O. (1955). *Breeding Beef Cattle for Unfavorable Environments.* Austin: University Texas Press.

ROWSON, L. W. A. (1956). *Vet. Rec.* **68**, 484.

SUMPTION, L. J. & ADAMS, J. C. (1960). Unpublished manuscript. University Nebraska.

SUMPTION, L. J. & JAKWAY, J. S. (1959). Unpublished. University Nebraska.

SUMPTION, L. J., REMPEL, W. E. & WINTERS, L. M. (1960). *J. Hered.* **50**, 293.

SUNESON, C. A. (1956). *Agron. J.* **48**, 188.

WINTERS, L. M. (1956). *Minn. Fm Home Sci.* **13**, (3), 3.

WINTERS, L. M. (1938). *Proc. Amer. Soc. Anim. Prod.*, 278.

41

SPERM COMPETITION AND ITS EVOLUTIONARY CONSEQUENCES IN THE INSECTS

By G. A. PARKER

Department of Zoology, University of Liverpool, Liverpool L69 3BX

(*Received* 18 *May* 1970)

I. INTRODUCTION

The possible advantages to a species of internal rather than external fertilization have frequently been stressed, though one important point appears persistently to have escaped comment. In terms of sexual rather than natural selection, copulation with internal fertilization may have arisen by a selective advantage conferred upon males able to eject sperm nearer to the ova than did their fellow males. Such a selective advantage would still operate in situations where virtually all ova are certain of fertilization with a minimum of time delay. This view would not preclude a natural selection pressure (favouring increased chances of fertilization in minimum time) as an important determinant of internal fertilization, but the sexual selection pressure of competition between males could have been much more important in the evolution of the process. With the onset of very close range (virtually internal) fertilization, different selective pressures would operate on the female so as to favour the evolution of special sperm storage structures and other modifications of the female reproductive system. Thus a series of behavioural, followed by anatomical and physiological steps, may be envisaged as intermediaries between a simple shedding of gametes into the sea and the present internal fertilization found in many marine species. However, the fact that several marine groups still swarm and spawn implies that external fertilization is, for them, still the most advantageous system.

Whatever the selective pressures which brought about internal fertilization, it is usually held that insemination is a prerequisite for terrestrial life and that only those species preadapted in this way could make the transition to the land. The method employed during this transition, in insects as well as in other groups, appears to be by use of the spermatophore, i.e. any kind of container which encloses the semen during its transfer from the male to the female. The presence of a spermatophore is usually associated with the transition to terrestrial life (Khalifa, 1949), but as Alexander (1964) points out, spermatophores are also prominent among many kinds of aquatic animals—both freshwater and marine. Copulation appears to have evolved considerably later than the spermatophore if one considers the most primitive arthropod groups (Hinton, 1964). For instance, males of *Trombicula splendens* Ewing (Acarina) deposit their spermatophores on the ground and the females inseminate themselves on finding the spermatophore irrespective of whether or not the male is present (Lipovsky, Byers & Kardos, 1957). Not all insects show true copulation. In the Lepismatidae, the male deposits a spermatophore on the ground and spins signal threads nearby which serve to direct the female (Stürm, 1956). In Machilids, the male deposits sperm droplets directly on a thread extending from the tip of his abdomen to the substrate (Stürm, 1952). The male twists his body around the female and guides her genitalia to the droplets, a procedure very close to true copulation. In Alexander's view (1964) deposition of spermatophores without relation to the presence of females evolved secondarily to direct interaction between males and females. Copulation itself is clearly not an essential feature of terrestrial life and, as Davey (1960) points out, internal fertilization seems to be the actual prerequisite for the transition. In the Pterygota, copulation is the rule though several advanced forms of many groups have adopted the transfer of free sperm (Hinton, 1964). Ghilarov (1958, 1961) also concludes that spermatophore formation is a primitive character and that forms transferring free sperm developed from spermatophore-producing ancestors. Alexander believes that nearly all arthropod copulatory acts have evolved from indirect spermatophore-transferring acts. As Hinton suggests, the complex copulatory processes of the Odonate may well have evolved independently of those in the rest of the pterygotes; once again it is much more reasonable to postulate an origin from an indirect spermatophore transfer pattern (Brinck, 1962) rather than from intromittent copulation (e.g. Fraser, 1939; Moore, 1960b).

Thus it seems likely that the reproductive behaviour of insects followed a path from external fertilization to internal fertilization with spermatophores, and later to copulation with the spermatophore deposited inside the reproductive tract of the female. A further stage has been reached in some groups with the introduction of free sperm transfer. Natural selection might have had a role subordinate to intrasexual selection during most of the earlier stages of this pathway.

The case for sexual selection as a highly significant force directing adaptation in insect reproduction does not end here. Whereas in primitive forms sexual selection would have favoured males able to deposit spermatophores nearest to, or in the place most likely to be found by, receptive females, the copulatory act of the pterygotes opens up a whole series of possibilities for intrasexual competition. The excellent and extensive review of sexual selection in the insects by Richards (1927) is still one of the most important works on the subject. Intrasexual competition arises from the large disparity in absolute reproductive potential between the two sexes and is characterized by the contribution of males to the next generation being much more variable than that of females (Bateman, 1948). A character of intrasexual advantage to a male can be assessed in terms of its effect on the male's fertilization rate, assuming the character to have equal survivorship in terms of natural selection. Male insects can achieve a sexually selective advantage by being more attractive to females (e.g. in colouration or courtship), or by aggressive behaviour and the establishment of territories or other mechanisms which serve to increase the chances of meeting a receptive female. Such adaptations are essentially precopulatory; much less attention has been given to adaptations which appear to serve as means of avoiding loss of progeny sustained by second inseminations. When a male mates with a virgin female, all the viable progeny she produces will have been fertilized by that male, providing parthenogenesis is absent and a second mating does not occur. Should a second male mate before the sperm from the first have been exhausted, conditions of sperm competition may prevail in which the subsequent progeny are of mixed parentage. Such a situation results in two opposing evolutionary forces. On the one hand, evolution will favour males which, when mating with already-mated females, displace as much as possible of the previously stored sperm or place their sperm in such a position in the female that the contribution of their own sperm in subsequent fertilization is maximized. On the other hand, a high selective advantage will be gained by any male able to avoid or reduce subsequent competition from the sperm of another male. Adaptations arising from these selective pressures might be either syn- or postcopulatory. Two predictions are therefore possible: (*a*) that where a second insemination occurs, evolution would favour the last male's fertilizing a high proportion of the subsequent offspring, and (*b*) that where the probability of further insemination is high, mechanisms will have evolved by which successful males reduce the occurrence or success of subsequent inseminations. The two adaptations are diametrically opposed and the outcome of a multiple mating can be interpreted as the extent to which adaptation in one direction outdoes adaptation in the other.

It is the aim of the present paper to examine the evidence for syn- and post-copulatory adaptations which appear to have arisen, as predicted above, through the intrasexual selective pressure of sperm competition. The literature concerning the outcome of multiple matings (in terms of the paternal derivations of the offspring) is reviewed first. Reasons are given why the insects may be considered as preadapted to sustain a high level of sperm competition.

Sperm competition may be defined as the competition within a single female between the sperm from two or more males for the fertilization of the ova. There is every reason to suppose that selection acts on individual sperm—those which physiologically 'outdo' the sperm from other ejaculates in competition for the fertilization of a given ovum would confer a selective advantage upon the male which produced them. However, the present review is concerned mainly with mechanisms by which the male himself can increase the chances that it is his sperm which fertilize most eggs.

II. METHODS USED TO DETERMINE THE DERIVATIONS OF OFFSPRING AFTER MULTIPLE MATINGS

In the present paper, the terms 'mating' and 'copulation' are regarded as synonyms. 'Insemination' is used to describe effective transfer of sperm by the male into the female reproductive tract; mating does not necessarily imply insemination. 'Fertilization' relates only to the process of entry of the spermatozoon into the ovum.

Investigations to determine how sperm from different matings are used in the fertilization of the ova have so far concerned two methods of sperm labelling. Earlier attempts used crosses between different strains which produced well-known heritable characteristics. The usual procedure consists of mating females with a recessive character to both normal and recessive males so that progeny possessing the normal characteristics can be attributed to sperm from the normal male, and *vice versa*. More recently, males whose sperm induce 'dominant lethality' during development of the ova which they fertilize have been used in competition with normal males. Males which induce sterility in this way may be obtained either by sublethal doses of chemosterilants or irradiation, or by hybridization. Although sperm from such males are often equally or almost equally competitive at fertilization as are sperm from normal males, their chromosomes are altered so as to induce lethal abnormalities during development at a rate proportional to the dose applied.

Because ejaculates from mutant or sterilized males are not always equally competitive with normal ejaculates (e.g. Ômura, 1939; Parker, 1970 c), as a control to a multiple mating experiment, it is necessary to determine the relative competitiveness of both labelled and normal sperm by reversing the order of mating of the two types of male (L, N). On the assumption that the same number of offspring are produced in each experiment, i.e. after series in which the orders of mating are NL and LN, and if L sperm are as effective as N, then the total offspring from each should be equal.

With the sterile-male method of labelling sperm used in compound matings involving several different males, a repeat series of experiments must be conducted with the labelled mating in each position in the mating order. With several different males of known hereditary constitution, this procedure is unnecessary if the relative competitiveness of the different sperm types has been fully investigated. The sterile male method is, however, the only method yet devised for species which have attracted little genetic study. The most convenient technique is to choose the lowest chemical or radiation dose which induces (virtually) total (100%) mortality before egg hatching. Thus in a double mating, all eggs failing to hatch (above the normal % infertility) may be attributed to the labelled sperm. Males that recover from up to 16 kiloroentgens of gamma irradiation often show apparently normal mating behaviour and competitiveness in procuring mates (Henneberry & McGovern, 1963; Potts, 1958).

[*Editor's Note:* Material has been omitted at this point.]

VII. CONCLUDING DISCUSSION

The prediction that insects, as a result of extreme sperm longevity within the female and an extremely high efficiency of sperm utilization at fertilization, are preadapted to sustaining a very high level of sperm competition is confirmed by the results of progeny analyses after multiple matings with males containing 'labelled' sperm. Furthermore, females often sustain more than one mating. Males usually outnumber females at the time of mating and can usually mate successfully several times. There is substantial evidence that the two conflicting sexually selected adaptations (to achieve precedence over previous sperm, and to prevent subsequent successful insemination of the female by other males) have arisen as a result of this situation by favouring males which achieve the highest fertilization rate. These two lines of adaptation, like many others, must be in a sort of 'evolutionary balance'. Suppose adaptation in the direction of prevention of subsequent insemination became totally effective after matings of all males in the population, there would then be no advantage to any male in attempting to mate with an already-mated female. Were second matings to cease altogether, there would be a definite selective disadvantage in the adaptation because all such adaptations have their 'cost' in terms of time or energy waste. Hence an observable amount of failure of the mechanism must always occur in order that it be maintained, the result being a sort of balance which yields the highest fertilization rate (assuming that the character has no secondary disadvantages in terms of natural selection).

Fights between males for the possession of females are very common in insects and (when looked for) a low but measurable rate of 'take-over' has been found. Throughout the present review only individual selection on a given genotype has usually been considered, as the situation where males are competing predominantly against siblings of identical genotype at the time of mating is probably rare except in certain parasites. Hamilton (1967) states that intramale aggression is here a clear indication that outbreeding occurs. The view that this behaviour and territoriality have evolved pre-dominantly through a form of group selection (Wynne-Edwards, 1962) is not con-sidered to be tenable for situations involving outbreeding, in which the competitors are not often closely related.

One of the most fruitful directions for future research may lie in the field of quanti-tative analyses of adaptive values, which have been attempted much too rarely in the past. It is relatively easy to speculate over the possible functions of a given adaptation, and very often a selective advantage may arise from several of the functions postulated. What is of interest is the relative importance of each function: i.e. the relative selective advantages attributable to each set of selective pressures arising in respect of each function. To this end, it is often necessary only to assess the correct order of magnitude of selective advantage. An investigation of this nature into insect mating plugs might be especially valuable.

The female cannot be regarded as an inert environment in and around which this form of adaptation evolves. Supposing, for example, a mating plug reduces a female's reproductive rate, this natural selection disadvantage will affect both male and female. Provided that the plug confers a sexual selective advantage on the male which out-weighs its natural selective disadvantage, it should evolve or be maintained. Resultant modifications within the female to prevent or reduce the disadvantageous effects of the plug might be expected; these adaptations may conflict with the line of adaptation in the male sex. The female, however, is not isolated from intramale selection because a female mating with a male possessing a character of sexual selective advantage will gain if the character is present in her male offspring.

VIII. SUMMARY

1. Intrasexual selection may have played a large part in insects in the evolution of copulation with internal fertilization from indirect spermatophore-transferring acts.

2. 'Sperm competition' may be defined as the competition within a single female between the sperm from two or more males over the fertilization of the ova.

3. There is considerable evidence from sperm-marking experiments that sperm competition is very common in insects as a result of multiple matings. Insects so far examined show that sperm from all inseminations can be used (to a varying extent) in the fertilization of subsequent offspring, but mating does not always result in successful insemination. In most cases (so far examined), the last male to mate tends to predominate in fertilizing the offspring.

4. Insects are preadapted to sustaining a very high level of sperm competition, compared with several other animal groups. The main preadaptations may be sum-marized as follows: (*a*) Females often mate several times within the duration of effec-tiveness of a given ejaculate. There may be several reasons for this. Though most usually females are unreceptive for some time after mating, some species appear 'promiscuous'. Unreceptive females are also sometimes raped. Male persistence is sometimes prolonged, so that full female receptivity is advantageous. It is advan-tageous for a female to become receptive again when fertility first begins to decline;

this is not when all the first ejaculate is used up. (b) Female insects typically possess specialized sperm storage organs in which sperm can be maintained in a viable condition for a very long time, often until the death of the female. (c) Extremely efficient utilization of stored sperm at the time of fertilization appears to be an insect characteristic. In *Drosophila* the number of stored sperm virtually equals the possible number of fertilizations. Overlapping of effective ejaculates is therefore high. Second inseminations almost invariably reduce the fertility which would have been experienced by the first male to mate.

5. It is argued that this preadaptation to a very high level of sperm competition has led to intense intrasexual selective pressures on the male. In response to these pressures, two main lines of sexually selected adaptation are predicted: (a) towards mechanisms by which a male inseminates a female in such a way as to achieve precedence over previously stored sperm and (b) accentuated by the above adaptation, mechanisms will evolve by which a male which mates with a given female will reduce the occurrence or success of subsequent inseminations with that female. These two forms of adaptation are diametrically opposed; a high selective advantage would be gained by a male which superseded previous sperm and prevented any subsequent successful inseminations.

6. Several adaptations in male insects can be interpreted in the light of the above predictions, though many of these may also have other adaptive values through natural selection. The adaptation which evolves is not necessarily that which yields the maximum possible egg gain to a given male (i.e. total sperm precedence), but that which results in the highest fertilization rate.

7. Sperm precedence is achieved in *Drosophila* by sperm displacement, where sperm from a second male predominate over previously stored sperm by directly displacing them from the sperm stores. Sperm displacement may occur in many insects.

8. Several behavioural and physiological adaptations of male insects may help to reduce the effectiveness, or occurrence of second inseminations of the same female by other males. These include: (a) Mating plugs (sphragis, spermatophragmata)—male accessory gland secretions, usually transferred after insemination, which coagulate and form plugs within the female genital tract. Plugs may often serve to 'guard' the female until unreceptivity is initiated. In many insects, agents in the seminal fluid or male accessory gland secretions induce unreceptivity in the female. (b) Prolonged copulation—sometimes copulation takes much longer than seems necessary merely to transfer the sperm. This may have the same functions as that of a mating plug, but renders the male unfree to search for further females. (c) Passive phases (amplexus, tandem behaviour)—stages of the male's reproductive behaviour during which he remains mounted on or otherwise attached to the female but without true genital contact between the two sexes. Postcopulatory passive phases sometimes serve to guard the female during oviposition where high densities of searching males are prevalent. The postcopulatory passive phase of *Scatophaga* has an extremely high intrasexual selective advantage which exceeds any apparent natural selective advantage by two orders of magnitude. (d) Non-contact guarding phases—reproductive behaviour

phases during which the male remains close but not in contact with the female, guarding her from other males. Postcopulatory non-contact guarding phases appear to have the same selective advantage as postcopulatory passive phases.

9. Mechanisms to avoid 'take-over' during copulation, passive, and non-contact guarding phases also serve to reduce sperm competition. These include increased efficiency of grasping apparatus, specialized rejection reactions which serve to dispel or 'trick' the recognition mechanism of the attacker, and emigrations from the site of highest probability of 'take-over'. There is quantitative evidence in *Scatophaga* that the emigration threshold of copulating males is determined by this form of intrasexual selective pressure.

10. Precopulatory passive phases may serve mainly to keep the sexes together until the female becomes receptive, but share several features in common with postcopulatory passive phases. Territoriality of male insects may have arisen primarily through sexual selection as a mechanism by which a male guards an area into which a female is most likely to enter.

I should like to express my gratitude in a general way to all those people, especially from the Department of Zoology, University of Bristol, who have by discussion over the past 5 years indirectly contributed a great deal to the present review. I am also indebted to Miss D. S. Paterson, Miss G. Robinson and Miss S. Carter for their help in preparing the manuscript.

[*Editor's Note:* Only the references cited in the preceding excerpts have been included here.]

REFERENCES

Alber, M., Jordan, R., Ruttner, H., (1955). Von Der Paarung der Honigbiene. *Z. Bienenforsch.* **3**, 1–28.

Alexander, R. D. (1964). The evolution of mating behaviour in arthropods. *Insect Reproduction* (Ed. K. C. Highnam), pp. 78–94. *Symp. R. ent. Soc. Lond.* No. 2.

Brinck, P. (1962). Die Entwicklung der Spermaübertragung der Odonaten. *Int. Congr. Ent. XI* (Vienna) I. 715-18.

Davey, K. G. (1960). The evolution of spermatophores in insects. *Proc. R. ent. Soc. Lond.* A, **35**, 107–13.

Fraser, F. C. (1939). The evolution of the copulatory process in the order Odonata. *Proc. R. ent. Soc. Lond.* B **14**, 125–9.

Gary, N. E. (1963). Observations of mating behaviour in the honeybee. *J. apicult. Res.* **2**, 3-13.

Ghilarov, M. S. (1958). Evolution of Insemination in terrestrail arthtopods. *Zool. Zh.* **37**, 707–35.

Ghilarov, M. S. (1961). Evolution des modes d'insemination chez les insects au cours de leur phylogénèse. *Scientia, Bologna* **96**, 386–91.

Hamilton, W. D. (1967). Extraordinary sex ratios. *Science, N.Y.* **156**, 477–88.

Henneberry, T. J., & McGovern, W. L. (1963). Effects of gamma radiation on mating competitiveness and behaviour of *Drosophila melanogaster* males. *J. econ. Ent.* **56**, 739–41.

Hinton, H. E. (1964) Sperm transfer in insects and the evolution of haemocoelic insemination. In *Insect Reproduction* (Ed. K. C. Highnam), pp. 95–107. *Symp. R. ent Soc. Long.* No. 2.

Khalifa, A. (1949). Spermatophore production in the Trichoptera and some other insects. *Trans. R. ent. Soc. Lond.* **100**, 449–71.

Lipovsky, L. J., Byers, G. W., & Kardos, E. H. (1957). Spermatophores—the mode of insemination of chiggers (Acarina : Trombiculidae). *J. Parasit.* **43**, 256–62.

Moore, N. W. (1960b). (Discussion of the evolution of mating behaviour in Odonata). In *Dragonflies* (Ed. P. Corbet, C. Longfield and N. W. Moore), pp. 158–161. London:Collins.

Omura, S. (1939). Selective fertilisation in *Bombyx mori*. *Jap. J. Genet.* **15**, 29–35. (In Japanese with English resume.)

Parker, G. A. (1968). *The reproductive behaviour and the nature of sexual selection in* Scatophage stercoraria *L.* Ph. D. thesis, University of Bristol.

Parker, G. A. (1970c). Sperm competition and its evolutionary effect on copula duration in the fly *Scatophage stercoraria*. *J. Insect Physiol.* **16**, 1301–28.

Potts, W. H. (1958). Sterilisation of tsetse-flies (*Glossina*) by gamma irradiation. *Ann. trop. Med. Parasit.* **52**, 484–99.

Richards, O. W. (1927). Sexual selection and allied problems in the insects. *Biol. Rev.* **2**, 298–364.

Stürm, H. (1952) Die paarung bei *Machilis* (Felsenspringer). *Naturwissenschaften* **39**, 308.

Stürm, H. (1956). Die paarung bei Siberfischen *Lepisma saccharina*. *Z. Tierpsychol.* **13**, 1–12.

Wynne-Edwards, V. C. (1962). *Animal Dispersion in relation to Social Behaviour*, pp. 1–653. Edinburgh and London:Oliver and Boyd.

AUTHOR CITATION INDEX

SUBJECT INDEX

About the Editor

DONALD ALLEN DEWSBURY is a professor of psychology at the University of Florida. His research interests relate to comparative animal behavior with emphasis on the study of adaptive significance, reproductive behavior, the use of the comparative method, and the behavior of rodents.

Dr. Dewsbury received his A. B. degree from Bucknell University in 1961 and his Ph. D. from the University of Michigan in 1965. He was an N.S.F. post-doctoral fellow in the laboratory of Dr. Frank A. Beach at the University of California, Berkeley in 1965-1966. His published scientific papers number over 120. Dr. Dewbury's books include *Comparative Psychology: A Modern Survey* (co-edited with D. A. Rethlingshafer, McGraw-Hill, 1973), *Comparative Animal Behavior* (McGraw-Hill, 1978), and *Sex and Behavior: Status and Prospectus* (co-edited with T. E. McGill and B. D. Sachs, Plenum, 1978). He is a member of a dozen scientific organizations and has served as Treasurer and President of the Animal Behavior Society.